MEDIA PRODUCTION, DELIVERY AND INTERACTION FOR PLATFORM INDEPENDENT SYSTEMS

MEDIA PRODUCTION, DELIVERY AND INTERACTION FOR PLATFORM INDEPENDENT SYSTEMS

FORMAT-AGNOSTIC MEDIA

Editors

Oliver Schreer
Fraunhofer Heinrich Hertz Institute, Technical University Berlin, Germany

Jean-François Macq
Alcatel-Lucent Bell Labs, Belgium

Omar Aziz Niamut
The Netherlands Organisation for Applied Scientific Research (TNO), The Netherlands

Javier Ruiz-Hidalgo
Universitat Politècnica de Catalunya, Spain

Ben Shirley
University of Salford, MediaCityUK, United Kingdom

Georg Thallinger
DIGITAL – Institute for Information and Communication Technologies, JOANNEUM RESEARCH, Austria

Graham Thomas
BBC Research & Development, United Kingdom

WILEY

This edition first published 2014
© 2014 John Wiley & Sons, Ltd

Registered office
John Wiley & Sons Ltd, The Atrium, Southern Gate, Chichester, West Sussex, PO19 8SQ, United Kingdom

For details of our global editorial offices, for customer services and for information about how to apply for permission to reuse the copyright material in this book please see our website at www.wiley.com.

Library of Congress Cataloging-in-Publication Data

Media production, delivery, and interaction for platform independent systems : format-agnostic media / by Oliver Schreer, Jean-François Macq, Omar Aziz Niamut, Javier Ruiz-Hidalgo, Ben Shirley, Georg Thallinger, Graham Thomas.
     pages cm
  Includes bibliographical references and index.
  ISBN 978-1-118-60533-2 (cloth)
  1. Video recording.   2. Audio-visual materials.   3. Video recordings–Production and direction.
I. Schreer, Oliver, editor of compilation.
  TR850.M395 2014
  777–dc23

                                                                              2013027963

A catalogue record for this book is available from the British Library.

ISBN: 978-1-118-60533-2

Set in 10/12pt Times by Aptara Inc., New Delhi, India
Printed and bound in Malaysia by Vivar Printing Sdn Bhd

1   2014

Contents

7 Scalable Delivery of Navigable and Ultra-High Resolution Video **260**

Jean-François Macq, Patrice Rondão Alface, Ray van Brandenburg,
Omar Aziz Niamut, Martin Prins and Nico Verzijp

List of Editors and Contributors

Editors

Dr. Oliver Schreer
Scientific Project Manager, Fraunhofer Heinrich Hertz Institut and Associate Professor Computer Vision & Remote Sensing, Technische Universität Berlin, Berlin, Germany

Dr. Jean-François Macq
Senior Research Engineer, Alcatel-Lucent Bell Labs, Antwerp, Belgium

Dr. Omar Aziz Niamut
Senior Research Scientist, The Netherlands Organisation for Applied Scientific Research (TNO), Delft, The Netherlands

Dr. Javier Ruiz-Hidalgo
Associate Professor, Universitat Politècnica de Catalunya (UPC), Barcelona, Spain

Ben Shirley
Senior Lecturer at University of Salford, Salford, United Kingdom

Georg Thallinger
Head of Audiovisual Media Group, DIGITAL – Institute for Information and Communication Technologies, JOANNEUM RESEARCH, Graz, Austria

Professor Graham Thomas
Section Lead, Immersive and Interactive Content, BBC Research & Development, London, United Kingdom

Contributors

Werner Bailer
Key Researcher at Audiovisual Media Group, DIGITAL – Institute for Information and Communication Technologies, JOANNEUM RESEARCH, Graz, Austria

Dr. Johann-Markus Batke
Senior Scientist, Research and Innovation, Audio and Acoustics Laboratory, Deutsche Thomson OHG, Hannover, Germany

Malte Borsum
Research Engineer, Image Processing Laboratory, Deutsche Thomson OHG, Hannover, Germany

Ray van Brandenburg
Research Scientist, The Netherlands Organisation for Applied Scientific Research (TNO), Delft, The Netherlands

Dr. Arvid Engström
Researcher, Mobile Life Centre, Interactive Institute, Kista, Sweden

Ingo Feldmann
Scientific Project Manager at 'Immersive Media & 3D Video' Group, Fraunhofer Heinrich-Hertz Institut, Berlin, Germany

Rene Kaiser
Key Researcher at Intelligent Information Systems Group, DIGITAL – Institute for Information and Communication Technologies, JOANNEUM RESEARCH, Graz, Austria

Axel Kochale
Senior Development Engineer, Image Processing Laboratory, Deutsche Thomson OHG, Hannover, Germany

Marco Masetti
Networked Media Team Leader, Research & Innovation, Softeco Sismat Srl, Genoa, Italy

Frank Melchior
Lead Technologist, BBC Research & Development, Salford, United Kingdom

Dr. Rob Oldfield
Audio Research Consultant, Acoustics Research Centre, University of Salford, United Kingdom

Martin Prins
Research Scientist, The Netherlands Organisation for Applied Scientific Research (TNO), Delft, The Netherlands

Dr. Patrice Rondão Alface
Senior Research Engineer, Alcatel-Lucent Bell Labs, Antwerp, Belgium

Richard Salmon
Lead Technologist, BBC Research & Development, London, United Kingdom

Dr. Johannes Steurer
Principal Engineer Research & Development, ARRI, Arnold & Richter Cine Technik GmbH & Co. Betriebs KG, München, Germany

Marcus Thaler
Researcher at Audiovisual Media Group, DIGITAL – Institute for Information and Communication Technologies, JOANNEUM RESEARCH, Graz, Austria

Nico Verzijp
Senior Research Engineer, Alcatel-Lucent Bell Labs, Antwerp, Belgium

Wolfgang Weiss
Researcher at Intelligent Information Systems Group, DIGITAL – Institute for Information and Communication Technologies, JOANNEUM RESEARCH, Graz, Austria

Dr. Goranka Zorić
Researcher, Mobile Life Centre, Interactive Institute, Kista, Sweden

List of Abbreviations

2D	Two-dimensional
3D	Three-dimensional
3GPP	3rd Generation Partnership Project
4D	Four-dimensional
4K	Horizontal resolution on the order of 4000 pixels, e.g. 3840×2160 pixels (4K UHD)
7K	Horizontal resolution on the order of 7000 pixels, e.g. 6984×1920 pixels
AAML	Advanced Audio Markup Language
ACES	Academy Color Encoding System
ADR	Automatic Dialogue Replacement
ADSL	Asymmetric Digital Subscriber Line
AIFF	Audio Interchange File Format
API	Application Programming Interface
AMPAS	Academy of Motion Picture Arts and Sciences
APIDIS	Autonomous Production of Images based on Distributed and Intelligent Sensing
ARMA	Auto Regressive Moving-Average model
ARN	Audio Rendering Node
ASDF	Audio Scene Description Format
ATM	Asynchronous Transfer Mode
AudioBIFS	Audio Binary Format for Scene Description
AV	Audio-visual
AVC	Advanced Video Coding
BBC	British Broadcasting Corporation
BWF	Broadcast Wave Format
CCD	Charge Coupled Device
CCFL	Cold Cathode Fluorescent Lamp
CCIR	Comité Consultatif International des Radiocommunications
CCN	Content-Centric Networking
CCU	Camera Control Unit
CDF	Content Distribution Function
CDFWT	Cohen-Daubechies-Feauveau Wavelet Transform
CDN	Content Delivery Network
CG	Computer Graphics
CGI	Computer Generated Imagery

CIF	Common Intermediate Format
CMOS	Complimentary Metal-Oxide Semiconductor
COPSS	Content Oriented Publish/Subscribe System
CPU	Central Processing Unit
CRT	Cathode Ray Tube
CUDA	Compute Unified Device Architecture
DASH	Dynamic Adaptive Streaming over HTTP
dB	Decibel
DBMS	Data Base Management System
DCI	Digital Cinema Initiative
DLNA	Digital Living Network Alliance
DLP	Digital Light Processing
DMD	Digital Micromirror Device
DMIPS	Dhrystone Million Instructions Per Second
DOCSIS	Data Over Cable Service Interface Specification
DONA	Data-Oriented Network Architecture
DPX	Digital Picture Exchange
DSL	Digital Subscriber Line
DSLAM	Digital Subscriber Line Access Multiplexer
DSLR	Digital Single-Lens Reex
DSP	Digital Signal Processor
DTAK	Dynamic Time Alignment Kernel
DTW	Dynamic Time Warping
DVB	Digital Video Broadcasting
DVD	Digital Versatile Disc
EBU	European Broadcasting Union
EBUCore	Basic metadata set defined by the EBU
EMD	Earth Mover's Distance
ENG	Electronic News Gathering
EOFOV	Edges Of Field Of View
EOTF	Electro-Optical Transfer Function
EPG	Electronic Program Guide
ESPN	Entertainment and Sports Programming Network
ESS	Extended Spatial Scalability
EXR	High Dynamic Range Image Format
FascinatE	Format-Agnostic SCript-based INterAcTive Experience
FCC	Fast Channel Change
FMO	Flexible Macro-block Ordering
FRN	Flexible Rendering Node
FSM	Finite State Machines
FTTH	Fibre-to-the-Home
FullHD	HD resolution of 1920×1080 pixels
GB	Gigabyte
GOP	Group Of Pictures
GPU	Graphical Processing Unit
GUI	Graphical User Interface

HAS	HTTP Adaptive Streaming
HBB	Hybrid Broadcast Broadband
HBBTV	Hybrid Broadcast Broadband TV
HD	High-Definition
HDMI	High-Definition Multimedia Interface
HDR	High Dynamic Range
HDTV	High-Definition Television
HEVC	High Efficiency Video Coding
HI	Hearing Impaired
HLFE	High-Level Feature Extraction
HMM	Hidden Markov Model
HOA	Higher Order Ambisonics
HOG	Histograms of Oriented Gradients
HQ	High Quality
HRTF	Head Related Transfer Function
HTML5	HyperText Markup Language 5
HTTP	HyperText Transfer Protocol
IBC	International Broadcasting Convention, annual industrial fair, Amsterdam, The Netherlands
IBR	Image-Based Rendering
ICP	Iterative Closest Point
ID	Identity
IEEE	Institute of Electrical and Electronics Engineers
IETF	Internet Engineering Task Force
IGMP	Internet Group Management Protocol
IMAX	Image Maximum (motion picture film format)
I/O	Input/Output
IP	Internet Protocol
IPTV	Internet Protocol Television
IROI	Interactive Region-Of-Interest
ISO	International Standards Organisation
IT	Information Technology
ITU	International Telecommunications Union
iTV	Interactive TV
JND	Just Noticeable Difference
JPEG	Joint Photographic Experts Group
JPIP	JPEG2000 over Internet Protocol
JSIV	JPEG2000-based Scalable Interactive Video
JVT	Joint Video Team
kB	kilo Bytes
KLT	Tracking approach proposed by Kanade, Lucas, Tomasi
KLV	Key, Length, Value; a binary encoding format used in SMPTE standards
kNN	k-Nearest Neighbour
LBP	Local Binary Patterns
LCD	Liquid Crystal Display
LCS	Longest Common Subsequence

LDR	Low-Dynamic Range
LED	Light Emitting Diode
LF	Light Field
LFE	Low Frequency Effects
LIDAR	Light Detection And Ranging
LSR	Layered Scene Representation
MAD	Mean Absolute Difference
MAP	Mean Average Precision
MDA	Multi-Dimensional Audio
MLD	Multicast Listener Discovery
MOCA	Multimedia over Coax
MP4	MPEG-4 Part 14
MPD	Media Presentation Description
MPEG	Moving Picture Experts Group
MPLS	Multiprotocol Label Switching
MVC	Multiview Video Coding
MXF	Material eXchange Format
NAB	National Association of Broadcasters, synonym for the annually held industrial convention in Las Vegas, USA
NAT	Network Address Translation
NDN	Named Data Networking
NHK	Nippon Hoso Kyokai (Japan Broadcasting Corporation)
NTSC	National Television System Committee (analogue television standard used on most of American continent)
NTT	Nippon Telegraph and Telephone Corporation (Japanese Telecom)
NVIDIA	an American global technology company based in Santa Clara, California
OB	Outside Broadcast
OLED	Organic Light-Emitting Diode
OmniCam	Omni-directional camera by Fraunhofer HHI
OpenCV	Open source Computer Vision libary
OpenEXR	a high dynamic range (HDR) image file format
OPSI	Optimized Phantom Source Imaging
OSR	On-Site Rendering
OTT	Over-The-Top
OVP	Online Video Platform
OWL	Web Ontology Language
P2P	Peer to Peer
PC	Personal Computer
PCI	Peripheral Component Interconnect (standard computer interface)
PDP	Plasma Display Panel
PiP	Picture-in-Picture
PSE	Production Scripting Engine
PSIRP	Publish-Subscribe Internet Routing Paradigm
PSNR	Peak Signal-to-Noise Ratio
PTS	Presentation Time Stamps
PTZ	Pan-Tilt-Zoom

pub/sub	Publish/subscribe
PVR	Personal Video Recorder
QoE	Quality of Experience
QoS	Quality of Service
RADAR	Radio Detection and Ranging
RAID	Redundant Array of Independent Disks
RANSAC	Random Sample Consensus
RF	Random Forest
RGB	Red-Green-Blue colour space
RGBE	RGB with a one byte shared exponent
RO	Replay Operator
ROI	Region-of-Interest
RSS	Rich Site Summary
RTP	Real-time Transport Protocol
RUBENS	Rethinking the Use of Broadband access for Experience-optimized Networks and Services
SAOC	Spatial Audio Object Coding
SD	Standard Definition
SHD	Super High-Definition
sid	Spatial Identifier
SIFT	Scale-Invariant Feature Transform
SLA	Service-Level Agreement
SMIL	Synchronised Multimedia Integration Language
SMPTE	Society of Motion Picture and Television Engineers
SN	Scripting Node
SNR	Signal to Noise Ratio
SpatDIF	Spatial sound Description Interchange Format
SQL	Structured Query Language
STB	Set-Top Box
SVC	Scalable Video Coding
SVM	Support Vector Machine
SXGA	Super eXtended Graphics Adapter referring to resolution of 1280×1024 pixels
SXGA+	SXGA at resolution of 1400×1050 pixels
TCP	Transmission Control Protocol
TDOA	Time Difference Of Arrival
TIFF	Tagged Image File Format
TOF	Time Of Flight
TRECVID	TREC (Text Retrieval Conference) Video Track
TV	Television
UCN	User Control Node
UDP	User Datagram Protocol
UHD	Ultra High Definition
UHDTV	Ultra High Definition TV
UI	User Interface
UPnP	Universal Plug and Play
USB	Universal Serial Bus

VBAP	Vector Based Amplitude Panning
VBR	Video Based Rendering
VDSL	Very High Speed Digital Subscriber Line
VFX	Visual Effects
VM	Vision Mixer
VOD	Video On Demand
VRML	Virtual Reality Modelling Language
VRN	Video Rendering Node
VTR	Video Tape Recorder
VVO	Virtual View Operator
WF	Wave Field
WFS	Wave Field Synthesis
XML	Extensible Markup Language
XPath	XML Path Language
xTV	Explorative TV
YUV	Luminance and chrominance color space

Notations

General

- Scalar value x; y in italic lower case. Coordinate values are scalars.
- 2D homogeneous vector **m** as lower case standard bold mathematical font.
- 3D homogeneous vector M as italic upper case standard mathematical font.
- Matrix **M** as upper case boldface font.
- |·| denotes the norm of a vector, length of a sequence or number of bins of a histogram.
- Vector of arbitrary dimension $\vec{x}$ as lower case standard bold math font with arrow.
- $X = (\vec{x}_1, \ldots, \vec{x}_n)$ is an ordered sequence of n feature vectors.
- $\chi = \{X_1, \ldots, X_k\}$ denotes a set of k feature vectors of sequences.

Specfic Symbols

Chapter 3 Video Acquisition

$\mathbf{m} = (x, y)^T$	Euclidean 2D point
$M = (x, y, z)^T$	Euclidean 3D point
$\mathbf{m} = (u, v, 1)^T$	Homogeneous 2D point
$M = (x, y, z, 1)^T$	Homogeneous 3D point
A	Intrinsic matrix
R	Rotation matrix
I	Identity matrix
t	Translation vector
f	Focal length
κ	Radial distortion coefficient
k_u, k_v	Horizontal/vertical scale factor
u_0, v_0	Horizontal/vertical offset
α_u	Focal length in multiples of the pixel width
C	Optical center
$\mathbf{P} = \mathbf{A}[\mathbf{R}|\mathbf{t}]$	Camera projection matrix
$I_{1,2}$	Image plane of camera 1 = left and 2 = right camera
B	Baseline, interaxial distance between two cameras
$\mathbf{m}_{1,2}$	Corresponding 2D points
$[\mathbf{t}]\times$	Skew-symmetric matrix of vector **t**
H	Projective transformation/homography

$\mathbf{m}_1, \mathbf{m}_2, \ldots, \mathbf{m}_N$	Corresponding 2D points
δ	Disparity
π	Projective plane
H_π	Homography related to a plane π
λ	Projective parameter
w_h	Sensor width
N_p	Horizontal pixel resolution
Δq	Pixel width
$H_{R,G,B}$	Histogram of the R, G, B colour component

Chapter 5 Semi-Automatic Content Annotation

$\kappa(\vec{x}, \vec{y})$	Kernel function applied to a pair of feature vectors $\vec{x}$, $\vec{y}$.
$\kappa_f(\vec{x}, \vec{y})$	Appropriate kernel function for feature f applied to a pair of feature vectors $\vec{x}$, $\vec{y}$.
$\mathbf{H}$	Histogram
T_c	Runtime complexity of component c
$\mathbf{U}$	Support vector of a model c
$O(\cdot)$	describes the upper bound of the runtime complexity of an algorithm ("big O notation").
τ	Time point
δ	Time offset

Chapter 7 Scalable Delivery of Navigable and Ultra-High Resolution Video

r_i	Bitrate assigned to tile i
s_i	Aggregated saliency score of tile i
α	Multiplicative factor that converts saliency values into rate values
λ	Impact factor of saliency on rate
$\mathbf{BW}$	Bandwidth budget
$\mathbf{M}$	Number of columns of a regular grid of tiles
$\mathbf{N}$	Number of rows of a regular grid of tiles
$\mathbf{Z}$	Overlapping factor of tiling scheme

1

Introduction

Oliver Schreer[1], Jean-François Macq[2], Omar Aziz Niamut[3], Javier Ruiz-Hidalgo[4], Ben Shirley[5], Georg Thallinger[6] and Graham Thomas[7]

[1] *Fraunhofer Heinrich Hertz Institute, Berlin, Germany*
[2] *Alcatel-Lucent Bell Labs, Antwerp, Belgium*
[3] *TNO, Delft, The Netherlands*
[4] *Universitat Politècnica de Catalunya (UPC), Barcelona, Spain*
[5] *University of Salford, Manchester, United Kingdom*
[6] *Joanneum Research, Graz, Austria*
[7] *BBC Research & Development, London, United Kingdom*

The consumption of audio-visual media has changed rapidly in the past decade. Content is now viewed on a variety of screens ranging from cinema to mobile devices. Even on mobile devices, today's user expects to be able to watch a personal view of a live event, for example, with a level of interactivity similar to that of typical web applications. On the other hand, current video and media production technology has not kept up with these significant changes. If we consider the complete media processing chain, the production of media, the delivery of audio-visual information via different kinds of distribution channels and the display and interaction at the end user's terminal, many challenges have to be addressed. The major challenges are the following.

Due to reuse of video content for different distribution channels, there is a *need for conversion and post-production* of the content in order to cope with different screen sizes. It is widely accepted that a movie production for cinema is recorded in a significantly different way to that intended for smaller screens. However, production budgets are limited; hence complex and costly re-purposing must be avoided. A good example is the production of 3D movies, where the aim is to develop camera technologies that allow 2D and 3D capture at the same time. Approaches to multiformat production that require parallel shooting or significant manual re-editing are no longer financially viable.

Media Production, Delivery and Interaction for Platform Independent Systems: Format-Agnostic Media, First Edition. Edited by Oliver Schreer, Jean-François Macq, Omar Aziz Niamut, Javier Ruiz-Hidalgo, Ben Shirley, Georg Thallinger and Graham Thomas.
© 2014 John Wiley & Sons, Ltd. Published 2014 by John Wiley & Sons, Ltd.

The convergence of broadcast and Internet requires future media production approaches *to embrace the changes brought by web-based media*. The habits of media consumption have changed drastically, partially due to the availability of user interaction with users freely navigating around web pages and interactively exploring maps and views of the street for example. Hence, future media production and delivery must support interactivity.

Although the overall bandwidth available for media delivery is continuing to increase, future media services will still face limitations, particularly if the end user at home or on-the-go is considered. Hence, new distribution formats are required to allow for *the provision of audio-visual media beyond current HDTV formats*, to support interactivity by the end user and to support intelligent proxies in the network that are capable of performing processing, which cannot be offered by low capacity devices. First developments towards resolution beyond HD are already appearing commercially, such as 4K camera and display technologies.

In addition, the user wants to decide when, where and on which device to watch audio-visual media as nowadays a variety of devices are available (including mobiles, TV at home and immersive large projection systems in cinemas). All of these devices must be supported by media delivery and rendering. Therefore, *a large variety of audio-visual formats* must be provided for the full spectrum of terminals and devices taking their special capabilities and limitations into account.

Even in live events, a lot of human operators such as directors or cameramen are involved in content creation and capturing the event from different viewpoints. Due to the increasing number of productions, *automated viewpoint selection* may be able to make a significant contribution to limiting production costs.

A new concept appearing on the horizon that could provide answers to these issues and challenges is referred to as *format-agnostic media production*. The basic idea is to define a new approach to media production that supports the necessary flexibility across the whole production, delivery and rendering chain. A key aspect of this approach is to acquire a representation of the whole audio-visual scene at a much higher fidelity than traditional production systems, and to shift closer to the user-end the decision of how the content is experienced. This idea allows end users to experience new forms of immersive and interactive media by giving them access to audio-visual content with the highest fidelity and flexibility possible. This book discusses current challenges, trends and developments along the whole chain of technologies supporting the format-agnostic approach. This approach could lead to a gradual evolution of today's media production, delivery and consumption patterns towards fully interactive and immersive media.

In Chapter 2 "State-of-the-art and Challenges in Media Production, Broadcast and Delivery", we give an overview on the current situation in audio-visual acquisition, coding and delivery and the evolution of terminal devices at the end-user side in current media production and delivery. Based on the review of the state-of-the-art and a summary of current and upcoming challenges, the format-agnostic concept is explained. This concept offers the capability to deal successfully with the new requirements of current and future media production.

The acquisition and processing of audio-visual media following a format-agnostic approach is discussed in two separate chapters, Chapter 3 and Chapter 4. In Chapter 3 "Video Acquisition", the three major video format parameters, spatial resolution, temporal resolution and colour depth (i.e., the dynamic range) are investigated with respect to the benefits they offer for future immersive media production. Due to the large variety of future video formats moving towards higher resolution, frame rate and dynamic range, the need for a format-agnostic concept is particularly helpful in supporting media production and rendering independent of the specific format. The composition and merging of visual information from different sensors will lead to more appealing and higher quality images. In Chapter 4 "Platform-Independent Audio", the current challenges faced in audio broadcast using a channel-based approach and sound scene reproduction techniques such as wave field synthesis are reviewed. The problem of having many competing audio formats is addressed at both the production and reproduction (user) ends. The concept of object-based audio representation is introduced and several example implementations are presented in order to demonstrate how this can be realised.

In Chapter 5 "Semi-automatic Content Annotation", both manual and automatic content annotation technologies that support format-agnostic media production are discussed. The specific requirements on those tools, in particular under real-time constraints of live scenarios are investigated. Relevant video processing approaches such as detection and tracking of persons as well as action detection are presented. Finally, user interfaces in media production are discussed, which help the production team to perform semi-automatic content annotation.

One of the advanced concepts of media production currently under discussion and development is presented in Chapter 6 "Virtual Director". This concept builds on various audio-visual processing techniques that allow for automatic shot framing and selection to be used at the production side or by the end user. Approaches are discussed for addressing the semantic gap between data from low-level content analysis and higher-level concepts – a process called *Semantic Lifting*, finally leading to content and view selection that fulfils the desires of the user.

Chapter 7 "Scalable Delivery of Navigable and Ultra-High Resolution Video" deals with the main challenges in delivering a format-agnostic representation of media. As the final decision on how content will be presented is moved closer to the end user, two factors have a significant impact on delivery: higher data rate at the production side and higher levels of interactivity at the end-user side. The chapter focuses on coding and delivery techniques, which support spatial navigation based on the capture of higher resolution content at the production side. Methods for content representation and coding optimisation are discussed in detail. Finally, architectures for adaptive delivery are presented, showing how ultra-high resolution video can be efficiently distributed to interactive end users.

Chapter 8 "Interactive Rendering" starts with a list of challenges for end user devices resulting from increased interaction with the content supported by the format-agnostic media production and delivery concept. Gesture-based interaction is one of the recent trends in interactive access to media, and this is discussed in detail. A number of technologies already on the market and currently under development are presented. This chapter concludes with user studies of gesture interfaces showing that technology development must coincide with continuous evaluation in order to meet user requirements.

Finally, Chapter 9 "Application Scenarios and Deployment Domains" discusses the format-agnostic concept from an application point of view. Based on the technologies described in the previous chapters, various application scenarios are derived. An analysis is presented of the impact of the format-agnostic concept and related new technologies in the production, network, device and end user domains. Based on this future outlook, this chapter concludes the book.

This book offers a comprehensive overview of current trends, developments and future directions in media production, delivery and rendering. The format-agnostic concept can be considered as a paradigm shift in media production, moving the focus from image to scene representation and from professionally-produced programmes to interactive live composition driven by the end user. Therefore, this will influence how media is produced, delivered and presented leading to more efficient, economic and user-friendly ways for media to be produced, delivered and consumed. Offering new services, better accessibility to content and putting the user in control are the main aims.

The idea for this book was born in the European FP7 research project FascinatE (Grant agreement no.: FP7 248138, http://www.fascinate-project.eu), which was proposing and investigating the format-agnostic concept for the first time. Beside the editors and the co-authors, which contributed to this book, there are several other colleagues to be mentioned. Without their expertise, their ideas and the fruitful discussion over more than 5 years, this book would not have been possible. Therefore we gratefully thank the following colleagues from several institutions and companies in Europe: R. Schäfer, P. Kauff, Ch. Weissig, A. Finn, N. Atzpadin and W. Waizenegger (Fraunhofer Heinrich Hertz Institute, Berlin Germany); G. Kienast, F. Lee, M. Thaler and W. Weiss (Joanneum Research, Graz, Austria); U. Riemann (Deutsche Thomson OHG, Hannover, Germany); A. Gibb and H. Fraser (BBC R&D, London, United Kingdom); I. Van de Voorde, E. Six, P. Justen, F. Vandeputte, S. Custers and V. Namboodiri (Alcatel-Lucent Bell Labs, Antwerp, Belgium); J.R. Casas, F. Marqués and X. Suau (University Politecnica Catalunya, Barcelona, Spain); O. Juhlin, L. Barkhuus and E. Önnevall (Interactive Institute, Stockholm, Sweden); I. Drumm (University of Salford, Manchester, United Kingdom); and F. Klok, S. Limonard, T. Bachet, A. Veenhuizen and E. Thomas (TNO, Delft, The Netherlands).

The editorial team, August 2013

2

State-of-the-Art and Challenges in Media Production, Broadcast and Delivery

Graham Thomas[1], Arvid Engström[2], Jean-François Macq[3], Omar Aziz Niamut[4], Ben Shirley[5] and Richard Salmon[1]

[1] *BBC Research & Development, London, UK*
[2] *Interactive Institute, Stockholm, Sweden*
[3] *Alcatel-Lucent Bell Labs, Antwerp, Belgium*
[4] *TNO, Delft, The Netherlands*
[5] *University of Salford, Manchester, UK*

2.1 Introduction

To place the current technological state of media production and delivery in perspective, this chapter starts by looking at some of the key milestones in the development of the world of broadcasting, taking the BBC as an example. The BBC started its first radio broadcasts in 1922 over 25 years after Marconi first demonstrated the transmission of pulsed outdoor radio transmissions in 1895. It was the first broadcaster in the world to provide a regular 405-line 'high definition' television service, starting in November 1936. The BBC launched a 625-line colour service in June 1967, although the 405-line monochrome TV service continued until January 1985. Teletext services started to appear in the 1970s, with the BBC's Ceefax service launching in September 1974. The BBC started digital widescreen (16:9) broadcasting terrestrially and by satellite in 1988, including digital text services based on MHEG (ISO, 1997), and turned off the last analogue 625-line transmissions (and thus also Ceefax) in October 2012. The UK was by no means the first country to make the switch to fully-digital TV, with the Netherlands completing the transition as early as 2006, although some other countries do not plan to switch until 2020 or beyond. Experiments in high definition television (HDTV)

Media Production, Delivery and Interaction for Platform Independent Systems: Format-Agnostic Media, First Edition. Edited by Oliver Schreer, Jean-François Macq, Omar Aziz Niamut, Javier Ruiz-Hidalgo, Ben Shirley, Georg Thallinger and Graham Thomas.
© 2014 John Wiley & Sons, Ltd. Published 2014 by John Wiley & Sons, Ltd.

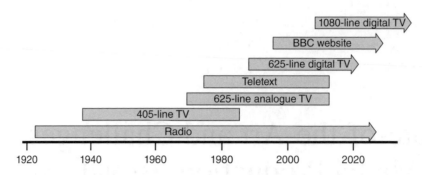

Figure 2.1   Timeline of broadcast-related technologies.

were underway in the early 1980s (Wood, 2007), although the BBC's first digital HDTV (now meaning 1,080 lines rather than 405!) did not start as a full service until December 2007. This was about the same time that the iPlayer online catch-up service was launched. At first, it was just available through a web browser, but was rapidly developed to support a wider range of devices at different bitrates and resolutions, and is currently available for media hubs, game consoles, smartphones, portable media players, smart TVs and tablets.

From the dates of these milestones, illustrated in Figure 2.1, it can be seen that that TV standards have tended to develop slowly and last a long time (49 years for 405-line, 46 years for 625-line), but that recent online services tend to adapt much more rapidly to changes in device technology and improvements in delivery methods. The way in which viewers interacted with TV changed little for many years, other than an increase in the number of channels and the arrival of the wireless remote control in 1956. The additional interactive possibilities offered by today's TVs and 'connected' devices give both the content creator and the audience many possibilities to take TV beyond its heritage of linear programmes designed to fit a 4:3 or 16:9 screen in the corner of the living room.

While delivery and display methods have evolved rapidly since the arrival of internet delivery and associated viewing devices, the way in which content is captured and edited has not changed fundamentally for many years. Cameras have become smaller, lighter and higher definition, and video editing now takes place using files on a PC (personal computer) rather than on tape or film. However, the premise of creating a single version of a programme (possibly with a few minor variants) in a given format (e.g., 16:9 1080-line HDTV) that could be format-converted for viewing on smaller screens or lower-resolution devices, has been with us for many years. The programme is edited to 'tell a story', attempting to appeal to a range of viewers on a broad selection of devices.

This chapter describes some of the fundamental technical aspects of current TV production. It also explains some of the limitations that the current approaches place on producing content that can be used in more flexible ways.

Section 2.2 starts by giving an overview of the ways in which technical limitations inherent in image capture have become a part of the 'grammar' of storytelling. Current video acquisition technology is then discussed, including a summary of current standards and some fundamentals of video such as gamma correction and the use of colour difference signals. An overview of the types of camera currently available is presented, and the issues of shooting for multiple aspect ratios are briefly discussed.

Section 2.3 summarises current approaches to audio capture, including a summary of current operational practices for both non-live production (typically single camera) and live production (for applications like an outside sports broadcast or a concert).

Section 2.4 explains how live TV is produced, including the roles of staff in a typical live sports production gallery, and the importance of their role in telling the story of an event.

Section 2.5 presents a summary of current delivery methods, including managed networks such as terrestrial, satellite and cable broadcasting, and unmanaged networks using content delivery networks or peer-to-peer distribution. It also reviews common codecs and transport protocols.

Section 2.6 describes the radical changes over the past decade in display technology, as the world has seen a shift from the almost universal use of the Cathode Ray Tube (CRT) to a variety of different flat panel display technologies. It describes the effect this has had on television, on the look of the pictures and the resulting effect on production techniques. The arrival of flat-panel displays has been the main driver for the introduction of HDTV, and consumer display technology developments now appear to drive broadcasting forward in a way that was unheard of just a decade ago.

Section 2.7 reviews current audio reproduction technology, starting with stereo and surround sound, and describing the ways in which further speakers can be added to reproduce 3D audio. Holophonic systems including wavefield synthesis and Ambisonics are then described; these seek to recreate the original sound waves. Binaural systems, which attempt to reproduce the experience of 3D sound over headphones by modelling the acoustic response of the listener's head, are then summarised.

Section 2.8 discusses how material stored in audio-visual archives is handled in today's production pipeline, and explains the principal of maintaining material in its original representation (rather than up-converting to the latest video standard as a part of the archiving process). The importance of maintaining as much data about the content format as possible is illustrated through several examples.

Section 2.9 summarises the problems highlighted in previous sections that can be tackled by a *format-agnostic* approach: keeping more of the original content (rather than 'flattening' it into a particular delivery format during production). It highlights the importance of metadata to allow repurposing, and outlines a new concept, *format-agnostic media*, that can allow content to be produced in a more flexible way, better suiting the interests of the viewer and the capabilities of their device.

2.2 Video Fundamentals and Acquisition Technology

2.2.1 How Real Should Video Look?

Before reviewing the technical standards and kinds of cameras used for today's TV production, it is worth briefly considering the fundamental question of what a production team is trying to achieve when producing video.

It is tempting to assume that the role of a TV system should be to faithfully reproduce the image as captured, with the aim of giving the viewer an experience as close as possible to actually watching the scene first-hand. For some kinds of programmes, especially live programmes such as sport and news, this is indeed generally the aim of the production team. Inherent limitations to the technology used will inevitably limit how 'real' the scene can look, and it is generally necessary to make trade-offs between system parameters such as integration time, iris, gain and gamma settings to try to squeeze the most important features of the scene into the range of brightness, depth and motion that the system can convey.

However, in many kinds of production, a realistic representation of the scene is not the aim. Instead, the production team may use characteristics of the imaging system to help 'tell the story' that is being conveyed. This is particularly the case for content traditionally captured on film, where some characteristics that could be viewed as inherent shortcomings of cameras are used as tools. The following paragraphs give some specific examples.

Motion portrayal: Film cameras operate at 24 frames per second (or 25 if shooting for 50 Hz TV), and this rate continues to be used for digital cinematography. Each image is displayed twice (or sometimes three times) when showing in a cinema, as 24 Hz is too low a frequency to prevent flicker from being seen. This results in the rapidly-moving objects that the eye is tracking appearing to judder or have a double image, making them difficult to see clearly. Cinematographers sometimes use this effect to guide the viewer's attention to the object they want them to be looking at; for example, in a panning shot following a walking person, it is difficult to study details in the background, while the person will suffer from relatively few motion artefacts. In 2012, Peter Jackson shot *The Hobbit* at 48 Hz, triggering a significant amount of debate on the pros and cons of increasing the realism of a film production. At the time of writing it is too early to tell whether such 'high frame rate' films will catch on. High frame rates are discussed further in Chapter 3, Section 3.4.

Depth-of-field: The larger the aperture of a camera, the smaller is the depth-of-field in the image (the range of depths over which objects are in focus). Although a large depth-of-field allows more of the scene to be captured in sharp focus, this is not always what is wanted. A shallow depth-of-field helps to draw the viewer's eye towards the most important object or person in the scene, and can provide a very effective story-telling tool. Large-format sensors (with sizes similar to that of 35mm film) tend to lead to a shallower depth-of-field, and are often preferred for cinema-style productions.

Transfer function: Most cameras and displays are incapable of capturing the full range of brightness that can be found in many scenes, particularly those with areas lit by the sun or having specular highlights, where a dynamic range may be as high as 20 stops (where one stop is a factor of two in brightness). It is therefore necessary to decide how the dynamic range of a scene should be reduced in order to fit it to what the camera and display system can cope with. The way in which this is done can have a significant effect on the 'look' of an image, for example either accentuating or hiding details in bright or dark areas. There is a further discussion of high dynamic range imaging in Chapter 3, Section 3.5.

Sharpness: From a signal processing point-of-view, a digital camera and display system can be viewed as a sampling system, which should be able to support spatial frequencies up to a limit of half the sampling frequency (Nyquist, 1928). Film, on the other hand, is inherently a non-sampled medium and its frequency response is determined by the structure of the film (grain size; relative depths of the R, G and B layers). A part of the 'film look' includes the gentle way in which the response drops off with increasing frequency, and the poorer detail response of red (as it is in the lowest layer of the film); this latter effect is responsible for film softening the details in faces, which can be seen as a desirable property as it can help reduce the visibility of wrinkles! Video cameras, on the other hand, are often set up to provide a flatter response, and may even include 'detail enhancement' that provides a boost to frequencies just below the cut-off point. This makes images look subjectively sharper, but can introduce a noticeable 'ring' next to sharp edges, for example, creating a black halo around a bright edge due to the overshoot from the sharp-cut filter. Different kinds of production prefer different 'looks', not always seeking to maximise the flatness of the frequency response as would be expected in other contexts, for example, audio.

Further details of what constitutes the 'film look' and how these may be reproduced by electronic cameras can be found in Roberts (2002).

2.2.2 Fundamentals of Video

Sampling and Aspect Ratio

Digital video signals are represented according to the relevant standard (ITU-R Recommendation BT.601 for standard definition images, BT.709 for high definition).

All TV production now uses widescreen (16:9) images (for 4:3 images replace '16:9' with '4:3' in the explanation below). A widescreen 16:9 image in its standard definition digital form, as specified by ITU-R Recommendation BT.601 and shown in Figure 2.2, actually has an aspect ratio wider than 16:9, due to the presence of 'non-active' pixels. The non-active pixels are there principally because the edge pixels could be corrupted by ringing in the analogue-to-digital conversion process, and because it is difficult to guarantee that exactly the right part of the signal is captured by the digitiser. These make the aspect ratio of the whole image slightly wider than 16:9. The choice of 13.5 MHz for the sampling frequency, which determines the pixel aspect ratio, was the result of an international standardisation process that considered many factors including finding a value that gave a whole number of samples per line for existing analogue TV systems. An interesting account of the process is given by Wood (2005). The figure of 702 active pixels comes from the 'active' portion of the 625-line 'PAL' (European TV format) analogue signal being 52 µs, and $52 \times 13.5 = 702$. Using the figures of 702×576 corresponding to a 16:9 image, we can calculate the aspect ratio of a pixel (W/H) as follows:

$$702 \times W = 16/9 \times 576 \times H \tag{2.1}$$

therefore

$$W/H = (16 \times 576)/(9 \times 702) = 1.458689 \tag{2.2}$$

The equivalent pixel aspect ratio for 4:3 standard-definition 625-line TV is 1.094:1.

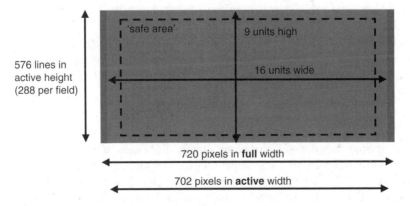

Figure 2.2   A standard-definition 16:9 '625 line' 50 Hz image.

For 'NTSC' (US TV format) images at 59.94 Hz, there are only 480 active lines, and 704 pixels are 'active', although there are still 720 pixels per full line. The pixel aspect ratio is therefore 1.21212. Note also that when two fields are 'paired up' to store as a frame, the first field occupies the lower set of lines, rather than the upper set as in 'PAL'.

Images should be composed under the assumption that the whole image is unlikely to be displayed, with all important information lying in the so-called 'safe area'. Historically this is because it is difficult to set up the image on a CRT to exactly fill the display – even with careful setting-up, poor regulation of the high voltage supply tends to make the displayed picture size change as the average image brightness varies. Rather than shrinking the image so it is guaranteed to be all visible, and thus end up with a black border around the image (as was common with CRT monitors used with PCs), CRT TVs usually slightly over-fill the display ('over-scanning'), so the outer part may not be visible on all displays. Thus, important picture content (such as captions) should not be placed so close to the edge that it may not be visible on some displays. Ironically, many modern flat screen displays also over-scan, because broadcasters sometimes do not fully fill the active part of the image as they do not expect the outermost edges to be displayed! There are actually two 'safe areas': one for 'action', and the other for graphics, with the graphics safe area being smaller. Details can be found in the European Broadcasting Union (EBU) Recommendation R95 (EBU, 2008).

Some rationalisation has happened in the standardisation of HDTV. The image format is 16:9, contains 1920 × 1080 pixels and has no 'non-active' pixels. The pixels are square. However, interlace will still be common for the short-to-medium term, except for material originated on film or designed to mimic film, where instead of 50 Hz interlaced, the standard supports 24 Hz (and 23.98Hz) progressive. At least, all interlaced formats (50 Hz and 59.94Hz) have the top field first. Unfortunately, some HDTV displays still over-scan, hence the continued need for the concept of a safe area. For a discussion of over-scan in the context of flat-panel displays, see EBU (2007).

New standards have recently been developed to support systems beyond HDTV resolution, known as *Ultra-High Definition Television*, or UHDTV:

- UHDTV1 has 3840 × 2160 pixels – twice the resolution of HDTV in each direction. This is commonly referred to as '4k', although should not be confused with the 4k cinema standard established by the Digital Cinema Initiatives consortium for 4k film projection, which has a standard resolution of 4096 × 2160.
- UHDTV2 has 7680 × 4320 pixels, commonly referred to as the 8K system.

Both levels have an aspect ratio of 16:9, progressive scanning, and currently have 24, 50, 60 or 120 frames per second, although other frame rates are under investigation (see Chapter 3, Section 3.4). The parameters are specified in ITU-R Recommendation BT.2020.

Chrominance Sampling

Although computer-based image processing usually deals with RGB (Red-Green-Blue) signals, the majority of broadcast video signal routing, recording and coding operates on luminance and colour-difference signals. Colour difference signals were first used in TV systems when adding colour to existing monochrome systems, for example, the PAL system carried

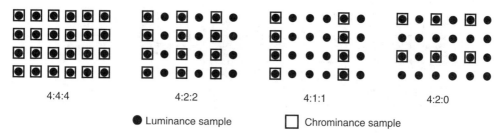

Figure 2.3   Chrominance sampling patterns.

two colour difference signals known as *U* and *V* on a subcarrier at the upper end of the part of the signal spectrum carrying the luminance information. However, this approach has continued with digital systems designed to convey colour images from the outset, because the human visual system is less sensitive to resolution loss in the chrominance channels, so the best use of the available bandwidth can be made by having more resolution in luminance than in chrominance. The chromaticities of the colour primaries and the equations for forming the colour difference signals are slightly different for standard definition, high definition and ultra-high definition, and can be found in the corresponding ITU-R Recommendations referred to above (601, 709 and 2020 respectively).

There are a variety of ways of subsampling the chrominance channels, which have potentially misleading names. Some of the more common formats, shown in Figure 2.3, are as follows:

- 4:2:2 format subsamples the colour difference signals horizontally by a factor of 2. This is the format used in digital video signals according to ITU Rec. 601 (for standard definition) or 709 (for high definition) and is the format most commonly encountered in TV production.
- Some lower-quality cameras (e.g. DVC Pro) produce '4:1:1', with colour difference signals subsampled horizontally by a factor of 4.
- Some compression systems (e.g. broadcast MPEG-2 and H.264) convey '4:2:0', where the colour difference signals are subsampled both horizontally and vertically by a factor of 2.
- Some high-end production systems handle '4:4:4' YUV or RGB, which sample both the luminance and chrominance (or the signals in their original RGB format) at the same resolution. These tend to be used primarily for electronic film production, or high-end post-production. The use of full-bandwidth chrominance is particularly valuable in processing such as chroma-keying, where any reduction in the bandwidth of the chrominance channels can affect the luminance channel and thus become more visible.

Black and White Levels

In the world of computer graphics, 8 bit signal levels range from 0 (black) to 255 (white). Digital broadcast-standard video defines black level as 16 and white level as 235, allowing headroom for undershoots and overshoots due to filtering. Although such overshoots may not be visible on a display, high frequency harmonics would be introduced, if they were clipped in the signal path, potentially causing problems in subsequent processing.

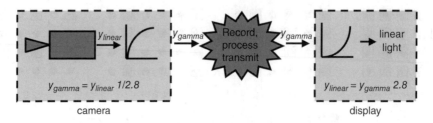

Figure 2.4   Gamma correction in the camera and the nonlinear response of a display.

Gamma

Video signals are generally *gamma-corrected*, that is, they are not linearly-related to actual light levels, but follow a power law, as shown in Figure 2.4. The display brightness is generally modelled as being proportional to the input signal raised to the power of a constant, referred to as *gamma*. Video signals are gamma-corrected by nonlinear processing in the camera, so that they will appear correctly on the display. Historically this came about because of the inherent nonlinear response of CRT displays.

Perhaps surprisingly, performing image processing such as compositing on gamma-corrected signals (rather than linear ones) does not generally create problems. Since almost all cameras (except some designed for tasks such as machine vision) will generate gamma-corrected signals, and displays (CRT or otherwise) have a nonlinear response, most video signals encountered will be gamma-corrected.

The value of gamma assumed for a display varies (historically taken as 2.8 in Europe, 2.2 in the US), although it is now accepted that the physics of the CRT is in fact the same on both sides of the Atlantic, and the differences are due to the measurement techniques historically used. More recent work has pointed to a value of 2.35 or 2.4, although in practice the actual value varies according to the setting of black level (brightness), and gain (contrast), resulting in offsets from a pure exponential law. The characteristic may therefore be expressed more generally as an Electro-Optical Transfer Function (EOTF). Cameras typically do not implement an exact power law, as this would give infinite gain at black: the slope at black is often limited to around 4. In practice, the operator of the camera control unit (often referred to as the 'racks operator') will adjust the gamma value to produce a subjectively pleasing result when viewing the image on a 'grade 1' professional monitor in a dim environment. The best results are generally obtained with an overall 'system' gamma greater than 1 (i.e., the gamma value assumed at the production end is less than that of the display), and the dimmer the viewing environment, the greater the overall system gamma should be (see Appendix A of EBU Guidelines (2007) for a discussion of this). This all shows that what was written in the various international standards documents was in practice irrelevant – the reference was the 'grade 1' CRT monitor used in the making of the TV programme.

Figure 2.5 shows a close-up of a camera control unit (CCU), and four such units installed at an operator's console in an outside-broadcast truck. The job of the operator is to adjust the controls (including iris, black level, gamma, white balance, detail) to produce correctly-exposed and subjectively-pleasing images, with matching brightness and colour across all the cameras. Note that a 'grade 1' CRT monitor (rather than an LCD monitor) is available to make quality judgements, as the performance of LCD monitors is not yet good enough to be used

Figure 2.5 A close-up of a camera control unit (left), and four camera control units as used by a 'racks operator' in an outside broadcast truck (right).

as reference monitors. This photo was taken during the setting up of a stereoscopic broadcast, which is why the iris controls on pairs of control units are ganged together; this allows the operator to adjust a pair of cameras at the same time.

Although gamma correction was initially required due to the inherently nonlinear response of CRT displays, and has thus found its way into the standards for digital video, it actually performs a useful function independent of the need to compensate for display response. Applying more gain to the signal at lower brightness at the camera means that the noise introduced by subsequent processing and transmission stages (including quantisation noise from digital sampling) is proportionately less in dark areas. The response of the eye to light is logarithmic, so the overall effect of gamma correction at the camera and display, combined with the response of the eye, is to make noise appear more uniformly-distributed across all brightness values. With a fully-linear transmission and display system, noise from processes such as quantisation would be more visible in dark areas.

Failure of Constant Luminance

An interesting interplay between gamma correction and the conversion from RGB to luminance and colour difference signals is the so-called *failure of constant luminance*. Most TV systems convert to gamma-corrected RGB in the camera, in order to minimise the effects of noise and quantisation applied to the signals in subsequent processing and recording. The gamma-corrected RGB signals are subsequently converted to luminance and colour-difference signals. If these signals retained their full bandwidth before being converted back to RGB and subjected to gamma in the display, then an image with the correct brightness would be obtained, as the original RGB would be recreated. However, when a bandwidth limitation is applied to the colour difference signals, then the original RGB values will not be reproduced in areas of the picture where the colour difference filter had an effect, such as at sharp colour transitions. Crucially, the correct luminance value may not be produced either. To avoid this problem, the calculation of the luminance signal should be computed from the linear RGB signals, and then gamma-corrected. However, this would require the display device to apply gamma *before* converting to RGB, followed by gamma correction to be applied again to the reconstructed RGB before being fed into a display device with an inherent gamma response. This adds significant complication to the receiver and therefore was not used in either standard definition or HDTV systems. However, the UHDTV systems defined in ITU-R Recommendation BT.2020 have the option for using a constant luminance approach. Further details of constant luminance may be found in Roberts (1990).

2.2.3   Camera Technology

Types of Sensor

There are two kinds of sensors found in modern video cameras: CMOS (Complementary Metal-Oxide Silicon) and CCD (Charge-Coupled Device). A discussion of the details of the way in which these sensors operate is beyond the scope of this book, but can be found in references such as Taylor (1998).

CCD sensors were the main type used until recently; CMOS sensors have started to become increasingly prevalent. CCD sensors consist essentially of an array of tiny capacitors that are charged up by the effect of light landing on them. At readout time, the charges in each row of pixels are transferred to the row above, and the charges in the top row are transferred to a row known as the readout register. This is read out by shifting the charges along the row, one pixel at a time, with the charge in the end element being read out as a voltage.

CMOS can offer increased sensitivity and cheaper manufacturing, and hence is the sensor technology of choice in low-cost applications such as mobile phones. They can also more easily integrate signal processing on the sensor chip itself, and can be easier to fabricate into large sensors required for D-cinema (film-like) cameras.

However, the basic CMOS sensor has an inherent flaw: since the light-sensitive part of each pixel is the actual photodiode whose level of charge is read out, the integration period (the time over which light is being accumulated for this image) runs between the points in time when the pixel is actually read out. Since the readout process works sequentially down the image, pixels near the top of the image are sampled significantly earlier than those near the bottom. This process is known as a 'rolling shutter', and has the effect of causing moving objects to slope. This can lead to strange effects, notably a 'wobbling jelly' effect that makes the whole image become distorted by camera shake, and strange effects on rapidly-rotating objects such as propellers. Some newer CMOS designs include ways to achieve synchronous readout to eliminate this problem.

Interestingly, the first video cameras had a similar temporal characteristic to CMOS sensors, in that the image was scanned from top to bottom, with the top of the image being sampled first. Images captured from such 'tube' cameras therefore also exhibited a sloping effect in the presence of motion. However, the CRT displays used at the time had exactly the same characteristic: the upper parts of each image were displayed before the lower parts. These effects cancelled out. However, CRT displays caused moving objects to slope with film-originated material where the whole image is sampled at the same time.

Capture of Colour

Most professional video cameras use three separate sensors for R, G and B. The image is split into its components using a so-called splitter block, consisting of an arrangement of prisms and coloured filters, with three sensors positioned on appropriate faces of the prisms. This arrangement samples the R, G and B values at each pixel, although requires three carefully-aligned sensors and a bulky splitter block. The presence of the block also dictates the minimum distance from the back of the lens to the sensors, meaning that lenses designed for use on film cameras (where the only object that needs to fit between the lens and the sensor is a shutter and possibly a mirror for the viewfinder) will generally not fit.

Colour images may also be captured using a single sensor, with colour filters arranged in a mosaic over the pixels. This approach was pioneered by Bayer (1976) and the pattern of coloured filters generally used is referred to as a *Bayer pattern* (Figure 2.6). In addition to the much lower cost of a single-sensor arrangement, obviating the need for a splitter block reduces the bulk of the camera and allows lenses with a shorter back-focal distance to be used. The main disadvantage is that the values for the two 'missing' colours for each pixel need to be interpolated, meaning that the resulting resolution of the image is not as high as the raw pixel count on the sensor would suggest. The Bayer pattern uses more green pixels than red or blue, helping to reduce losses in the luminance signal due to this interpolation, as the

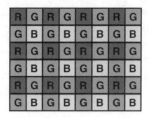

Figure 2.6 Bayer pattern typically used for colour image capture with a single sensor.

green component is responsible for the majority of the luminance signal. High-end cameras incorporate an optical anti-aliasing filter in front of the sensor to reduce artefacts caused by the sampling process.

Some other novel forms of sensor have been developed recently, for example, the *Foveon X3* sensor (Hubel, 2005), which incorporates three sensing elements and colour filters stacked vertically in each pixel.

Interlaced Capture

Interlacing is the process of scanning the odd lines of an image in one field, and the even lines in the following field, requiring both fields to make a frame with full vertical resolution. It can be viewed as a primitive form of image compression, halving the number of lines, and thus also halving the signal bandwidth needed for a given display frequency. On a CRT display it helps to hide the line structure while also providing additional vertical resolution compared to that from a progressive-scan system with the same number of lines per field, as the persistence of vision helps to make both odd and even lines appear to be present at the same time. However, in a world of digital video compression and flat panel displays that are inherently progressively-scanned, interlace is something of an anachronism. Nevertheless, it is still found in most of the HDTV systems in use today. For film-like material, shot at 24 or 25 frames per second, progressive scanning is used, and most cameras, coding systems and displays will switch to a mode that is effectively processing progressive images, even if they are delivered as two interlaced fields captured at the same time.

The most common approach to creating an interlaced output from a CCD or CMOS camera is to create the lines of each field by averaging together two adjacent lines from the progressively-captured image. This provides a degree of vertical pre-filtering necessary to avoid having excessive vertical detail in the image, which would otherwise cause 'interline twitter' effects on fine vertical detail, creating 25 Hz flicker on a 50 Hz display. The pairing of lines to be averaged is changed between odd and even fields, as illustrated in Figure 2.7. This approach is sometimes extended when producing multiformat cameras, by averaging different numbers of lines to produce a downsampled image.

Types of Camera in use Today

The kinds of cameras most likely to be encountered in today's TV and film production are briefly summarised below.

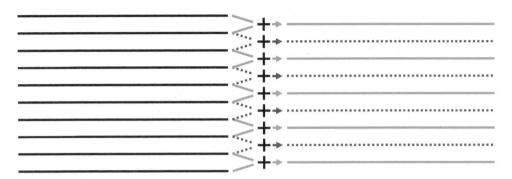

Figure 2.7   Generation of two interlaced fields (shown in solid and dotted lines) from a progressive signal by averaging pairs of lines.

'Traditional' video cameras, with three sensors and a splitter block. These are likely to offer both an interlaced and progressive capture mode. Cameras sometimes referred to as being ENG-type (Electronic News Gathering) will have all controls and recording capability built-in, and be designed to be operated by a single person. Cameras intended for use in TV studios or outside broadcasts are designed for remote recording (no built-in recorder), and control of functions such as gain and iris are handled remotely (see Figure 2.5). A standard form of optical fibre interconnect (that carries power as well) is used to connect the camera to the camera control unit (CCU), and carries signals including talkback and a 'reverse video' feed for the camera viewfinder.

Digital cinema cameras have a single larger-format sensor. They are usually designed to allow the 'raw' sensor output to be recorded with a relatively high bit depth (10–16 bits), as well as producing a live video output primarily intended for monitoring rather than as the end product. By recording the raw sensor output at full quality, processing such as colour correction and scaling of brightness can be applied in post-production, in the same way as would have happened when developing a film and digitising it via a film scanner in the past. This stage of the film production process involves a significant level of creative decision-making by a colourist, and makes use of the high dynamic range capabilities of the camera (discussed in more detail in Chapter 3, Section 3.5).

High frame rate cameras are designed to capture sequences for slow motion replay. They capture to internal storage, usually limited to sequences of around a minute or less, and then replay the captured material via a normal video output or by file transfer. Frame rates of 1,000 frames per second or more are available, at resolutions up to 4k. When used at live sports events, the workflow typically involves partitioning the internal storage into several 'chunks', with one chunk being transferred to a video server for running an action replay while another chunk can continue recording. The recording process continually over-writes material captured earlier in the chunk of memory being used, and the 'record' button will stop (rather than start) recording, leaving the last few seconds of material in the memory, ready to be played out. A more in-depth discussion of high frame rate is given in Chapter 3, Section 3.4.

DSLR (Digital single-lens reflex) cameras with a video capture mode have started to be used for some kinds of non-live production, including low-budget films. These are significantly cheaper than a professional D-Cinema camera, but struggle to match the image quality. One problem with many current-generation cameras is that the image sensor has a very high pixel count (designed for capturing still images at around 20MPixels), and it is not possible to read out every pixel value at video rates. The sensor therefore outputs a sub-set of pixels (e.g., skipping lines) resulting in significant spatial aliasing and poorer signal-to-noise ratio. DSLR cameras also tend to lack some of the other features expected in higher-end cameras, such as genlock (to synchronise the exact capture time of images when using multiple cameras) and timecode input.

It should be noted that the distinction between these types of camera is currently starting to blur, with additional functionality being added to allow multiple application areas to be targeted by a single device.

New Types of Camera

There are several other kinds of cameras being developed to overcome some of the issues with traditional cameras:

Omni-directional cameras: these cameras capture a much wider field-of-view than is possible with a conventional lens and sensor, potentially covering all viewing directions to produce a cylindrical or spherical image. Applications include producing content for immersive very-wide-angle display systems, or images that can be explored interactively by viewers by rendering a conventional image from a small portion of the omni-directional image. A detailed discussion of panoramic video acquisition can be found in Chapter 3, Section 3.2.

High dynamic range (HDR): conventional sensors cannot cover the full dynamic range of a scene due to noise at low brightness levels and saturation of the sensor in very bright areas. This limits the ability of the colourist to make adjustments to the dynamic range in post-production to the extent that was possible with film. Also, it limits the ability to produce video that can exploit the benefits of emerging HDR displays. Cameras are starting to appear that are equipped with sensors to capture images with a higher dynamic range, using techniques such as two-stage readout of each pixel, or use of 'dark green' filters on some pixels in the Bayer mask. A more in-depth discussion of HDR cameras is given in Chapter 3, Section 3.5.

Plenoptic (or light-field) cameras: these cameras aim to capture a view of the scene from a range of different closely-spaced viewpoints across the camera aperture. This allows various kinds of processing that depend on the depth of objects in the scene, such as changing the depth-of-focus by processing the captured images (allowing re-focusing in post-production), or the production of stereoscopic or multiview images, albeit with a small baseline. One approach is to place an array of microlenses over the pixels on the image sensor, such that each microlens covers multiple pixels. Each pixel then captures light entering its lens at a specific direction. An in-depth discussion of this area is beyond the scope of this book; the reader is referred to Harris (2012) for further information.

Computer Generated Imagery (CGI)

Although the focus of this section has been on acquisition of images from the real world, it is worth briefly referring to computer-generated imagery, which is increasingly used in film production and high-end TV.

Many films now include a significant proportion of CGI, either composited with live action, or to create entire scenes or even whole films. The image rendering process usually attempts to mimic some of the characteristics of real cameras (such as depth-of-field, lens flare, noise), and the key challenges are around creating and rendering realistic scenes while minimising the time needed from designers and animators and the amount of computation time needed. Further details of CGI techniques may be found in Shirley (2009).

CGI elements are sometimes included in live TV production, for example, for 'tied to pitch' sports graphics, or virtual bar graphs in election results programmes. Here, there are additional challenges around meeting the real-time constraints, while matching camera movement and lighting conditions between real and virtual elements. Further discussion on the challenges and approaches can be found in Thomas (2011).

2.2.4   Production for Formats with Differing Aspect Ratios

As explained in Section 2.2.2, TV started with an aspect ratio of 4:3, with 16:9 now being used for HDTV. Some broadcasters started to produce standard-definition programmes in 16:9 with the introduction of digital TV, sometimes being broadcast as a 16:9 or 14:9 (with the sides slightly chopped) 'letterbox' within a 4:3 raster when being broadcast as analogue TV transmissions.

It is also possible to broadcast archive 4:3 content in its native form over digital TV systems, with the aspect ratio of the source being signalled in data accompanying the signal (ETSI, 2005). This allows viewers to choose whether to display the full image with black borders on either side to fill a 16:9 display, or in some other way such extracting a 14:9 portion from the 4:3 image to reduce the size of the black bars on either side at the expense of losing a small amount of material at the top and bottom. The signalling data can also include *pan and scan* information, which indicates the location of the best part of a 16:9 image to crop out for display on a 4:3 monitor.

During the period when a significant number of viewers were still using 4:3 analogue TV, some programmes made in 16:9 were shot with a so-called *shoot and protect* approach, whereby all the important action was constrained to the central 4:3 portion (or sometimes the central 14:9), with the sides being cropped before distribution over a 4:3 channel. This constrained the freedom of the production team to make best use of the edges of the pictures, for example, forcing graphics and captions to stay within the central area, and making the video content of the edges rather boring. With the demise of analogue 4:3 broadcasting, this practice has generally ceased.

Cinema has also had a history of different aspect ratios being used (Berger, 2012), with 2.39:1 being common at present. Cinema tends to have a wider aspect ratio than 16:9 TV, largely due to the physical layout of a cinema: it is easier to make the screen wider than to add additional height because typical cinemas tend to have much more width than height. Also, a tall screen is uncomfortable to view when sitting close, as viewers have to crane their necks; it is more natural to look to the left and right. Indeed, the natural world tends to have more objects of interest across a wide horizontal angle than vertical angle, as most things tend to be near the ground plane.

The shoot-and-protect approach is commonly used with films, with camera viewfinders including markings for particular aspect ratios so that important content can be kept in the area that would be cropped for a 16:9 TV version. However, as was the case with TV, this can constrain the creative freedom of the production team to make best use of the available image space.

In addition to the shape of the viewing screen constraining the framing of content, the size can also have an effect. Rapid camera motion can make viewers feel sick when viewed on a large screen, due to the fact that the screen occupies a significant portion of their field-of-view. This strong visual sense of head motion does not match the information from the vestibular (balance) organs in the ear (which will be reporting that the head is nearly stationary). It has been suggested that this disconnect between visual and balance response may mimic the effects that can also be induced by some kinds of neurotoxins. This might explain the triggering of an automatic response to rid the body of the presumed toxins by being sick (Hughey, 2001). To avoid these problems, material designed for viewing on large screens tends to be framed wider than it would be for small screens, with less camera panning. The fact that the important action in the screen occupies a smaller fraction of the screen area does not matter so much when the image is very large.

When material shot for a large screen is viewed on a small screen (for example, when viewing a football match shot for HD on a smartphone), wide-shot framing can lead to the important action areas being too small. This problem has been recognised by a number of researchers, such as Chamaret and Le Meur (2008), who have investigated ways of automatically reframing content. An alternative approach is to try to cut out those parts of the image that contain little useful information, even if they do not lie around the edges of the image; one such approach is known as Video Scene Carving (Grundmann et al., 2010) although this will distort the distances between scene elements. The only approach in common use to image reframing that relies on metadata produced by the content creator is the *pan and scan* approach, mentioned above.

2.2.5   Stereoscopic 3D Video

The advent of new display technologies (discussed further in Section 2.6) caused a resurgence in stereoscopic film production from around 2010. A key factor in this was that the move to digital cinema projectors allowed left and right eye images to be aligned much more accurately than was easily possible with film. In particular, high frame rate projectors became available, allowing left and right images to be presented in a time-sequential manner, without the expense or alignment issues of a dual projector system. Similarly, domestic flat panel displays that were inherently capable of operation at 100 Hz or more allowed stereoscopic images to be viewed. Of course, glasses need to be worn to allow these systems to work.

In order to meet the demand for stereoscopic content, conventional cameras are usually used, deployed either side-by-side or mounted in a mirror rig. Use of a mirror rig allows the camera viewpoints to be placed as close together as desired, without the physical limitation caused by the size of the lenses or camera bodies, although this results in a much bulkier arrangement. There are also some cameras made specifically for stereoscopic 3D capture, with two cameras and lenses in one body, or even two cameras using the same lens (resulting in a more compact camera but a very small baseline).

A detailed discussion of approaches to 3D production is beyond the scope of this book, but it should be noted that the process is generally not straightforward. There are many technical challenges such as ensuring that the cameras have matching exposure, colorimetry, focus and field-of-view, and are synchronised and mounted rigidly with respect to each other. There are also many other considerations concerned with scene composition and choice of the size of the baseline. The 'best' location for a 3D camera when covering a particular event will probably be different from the 'best' location for a 2D production. For example, a wide shot from high up gives a good view of a football match in 2D, but will give little impression of depth as everything is a long way away; for a stereoscopic 3D production it is usually better to place the cameras lower down and closer to the action. A more in-depth discussion of some of the challenges in stereoscopic 3D production may be found in Jolly (2009).

2.2.6   Challenges for the Future

As explained in this section, current video acquisition is centred around capturing video under the assumption that it is being produced for a display with a given format and size: capture for TV is aimed at a 16:9 display of a size typical for that of a domestic flat panel, while film production targets wider aspect ratios and larger screens. Compromises need to be made to allow a limited amount of aspect ratio conversion (such as 'shoot and protect', Section 2.2.4). There is the provision for a limited amount of metadata to support reframing widescreen films for less-wide TV aspect ratios. However, there is generally no support for more 'extreme' reframing; for example, if a viewer wanted to zoom in to a portion of an image, or display an image on a much larger display, there is no provision for using a higher-resolution format to provide the detail needed to support this.

Video cameras currently used need to be adjusted manually during recording to provide a correctly-exposed image. If the resolution of cameras was increased to give some flexibility for the viewer to zoom into an area, or display a much wider-angle image on an immersive display, the exposure requirements would differ depending on the choices made by the viewer; at present there is no way of handling this.

Some approaches to these challenges will be explained in Chapter 3, 'Video Acquisition'.

2.3   Audio Fundamentals and Acquisition Technology

2.3.1   Introduction

As video transmission has moved from 4:3 aspect ratio standard definition analogue to 16:9 aspect ratio high definition picture and then on to 3D stereoscopic broadcast, audio for television has been through a parallel development process. The two most significant changes in audio acquisition and production have been closely allied to the shift to digital broadcast; namely the transition from stereo production to 5.1 surround sound production, and the development of an end-to-end digital, file-based, workflow. The impact of these factors has been felt to a greater or lesser degree depending on the genre of broadcast and especially whether the broadcast is a 'live' or 'non-live' production.

2.3.2   Fundamentals of Audio

Microphones used for TV broadcast now are largely the same as have been used for many years. Highly directional shotgun microphones are used in order to pick up only intended sound and reject competing sources while keeping audio equipment at a distance so as to be out of camera shot or, in the case of sports events, out of the way of the action. Small lavalier, or lapel, microphones (see discussion on specialist microphones below) are still commonly used because they are unobtrusive and can be concealed easily from camera shot. The most important development in recent times has been the use of the Soundfield® microphone (Gerzon, 1980), which although developed as early as 1978, only became significant as TV sound moved from stereo to 5.1 surround sound broadcast.

Microphone Types

Microphones can be broadly split into three types:

Dynamic Microphones

Dynamic, or moving coil, microphones work on the principle that a coil of wire moving relative to a magnet will produce a change in voltage across the coil relative to the movement of the coil with respect to the magnet. In dynamic microphones the coil is attached to a light diaphragm that moves as sound pressure levels change. This in turn moves the coil within the magnetic field and a change in voltage is produced that is related to changes in sound pressure level at the diaphragm. Dynamic microphones have relatively low output and the mass of the coil attached to the diaphragm means that they tend to be less sensitive to higher frequencies than other types. Dynamic microphones are however robust and cope well with high sound pressure levels so are useful in live music applications and for close-miking instruments.

Ribbon Microphones

Ribbon microphones work on similar principles to the dynamic microphone but instead of a coil moving in a magnetic field, the electrical conductor used is a thin metallic film ribbon. The ribbon is open to the air on two sides and will move when subject to changes in sound pressure level at either side. A useful property of ribbon microphones is that where sound pressure levels come from the end of the ribbon, sound pressure levels are equal at both open sides and no signal is generated. This gives a *figure of eight* polar pattern rejecting sound from the front, which can be useful in close-miking instruments where rejection of other sounds is desired. They are rarely used in broadcast applications because of their fragility but are still sometimes used in studio applications.

Condenser Microphones

Condenser, or capacitor, microphones, like dynamic microphones, make use of a diaphragm that moves when subject to changes in sound pressure levels. In the condenser microphone however the diaphragm forms one plate of a capacitor. A capacitor is an electrical component consisting of two plates that can store an electrical charge related to its capacitance and to the voltage across the two plates of the capacitor. Where one plate has a fixed charge a voltage change can be observed across the two plates as the distance between the plates is altered. By fixing one plate and using the other as a microphone diaphragm that moves when subject to changes in sound pressure level a change in voltage across the plates will occur relative to the sound pressure levels applied to the diaphragm. Condenser microphones tend to be more sensitive to high frequencies than dynamic microphones as the diaphragm does not have to move a coil and is lighter and more responsive. They require a charge on one plate as described which is usually provided by a 48V DC supply known as *phantom power*. Microphones with a permanent charge on one plate are available and commonly used where microphone size is critical, for instance for small lavalier style microphones that can be hidden on performers. These permanently-charged condensers are usually referred to as *electret* microphones.

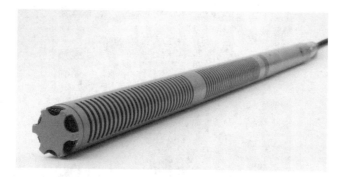

Figure 2.8   Shotgun microphone (from Wikimedia Commons, Users: PJ and Piko).

Specialist Microphones

Microphones have been developed for specialist purposes, and those which have particular relevance to TV broadcast are outlined here although these are all variations of the basic types described above.

Shotgun microphones (Figure 2.8) are highly-directional microphones designed to pick up specific sources of sound from a given direction and to reject sound from any other direction. They are commonly used wherever sound must be picked up from a distance. Shotgun microphones have a condenser microphone capsule that is seated in a casing designed to cancel out sound arriving from directions other than straight ahead.

Lavalier (lapel) microphones: These are very small, usually omni-directional, microphones that can be attached to the clothing of the person being recorded (Figure 2.9). They are used where a microphone must be unobtrusive and can easily be hidden in clothing when required.

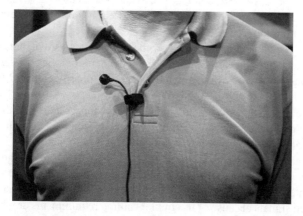

Figure 2.9   Lavalier microphone attached to clothing. Reproduced by permission of Diana Castro.

Figure 2.10   Soundfield® microphone with four capsules just visible under the protective cover. Reproduced by permission of Diana Castro.

Soundfield® Microphone: The Soundfield® microphone (Figure 2.10) was developed by Michael Gerzon and Peter Craven in 1978 to record ambisonic signals. Its usage has been adapted and it is now commonly used in recording 5.1 surround sound for TV broadcast. The Soundfield® microphone is made up of four condenser microphone capsules arranged in a tetrahedral arrangement. The capsules are sufficiently close together that for most purposes they can be considered coincident and the four output signals are manipulated in hardware or software to produce a variable polar response, a 3D (with height) ambisonic recording or, more commonly, a 5.1 surround recording.

Audio Signals

Balanced and Unbalanced Signals

Unbalanced signals have a single conductor carrying the audio signal and an earth, often consisting of a conductive sleeve, which will prevent earth loops and shield the signal from some electromagnetic interference.

Balanced signals use two conductors to carry the signal, one positive (hot) and one negative (cold) and usually also have an earthed shield. Each conductor carries the same signal but the negative one is phase inverted with the signal 180 out of phase compared to the positive as shown in Figure 2.11.

Any interference or noise from an external source will appear equally on both hot and cold wires. The receiving device responds to the difference between the two signals cancelling out any noise present on both signals and retains the transmitted signal.

Generally speaking consumer products tend to use unbalanced signals whereas professional audio installations use balanced signals in order to reduce noise on the signal caused by electromagnetic interference. There are exceptions to this and some professional installations will use unbalanced signals because of its simpler cabling requirements or when in electrically

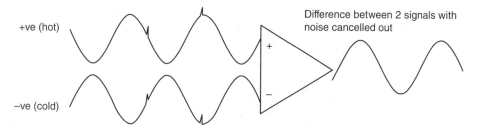

Figure 2.11   Balanced audio signal showing electromagnetic interference on the signal, which is cancelled out at the receiving device.

low-noise environments where reducing the effect of electromagnetic noise is not considered so critical.

Both analogue and digital audio signals may be balanced or unbalanced.

Analogue and Digital Audio

As already explained the movement of a microphone diaphragm caused by changes in sound pressure level directly causes changes in signal level either by alterations in capacitance or by movement of a conductor in a magnetic field. Analogue audio is therefore continuously variable and directly analogous to variations in the sound pressure level of the recorded source. The resultant electrical signal is recorded onto some analogue medium such as magnetic tape or, more often, converted into a digital audio format for storage. Digital audio, in contrast to analogue audio, consists of discrete samples of the analogue signal over time recorded as numeric values representing the signal level at each instant it is sampled. It is important to understand the process of sampling in the analogue to digital conversion process in order to have an understanding of some of the constraints of digitisation and the factors that have impact on the accuracy of the recorded digital audio.

The level of the analogue signal is first captured and stored as a single voltage using a sample and hold circuit many times per second. Common sampling frequencies are 44.1 KHz (for CD format) and 48 KHz (for film and video) although higher sample rates are also used. The sampling frequency of digital audio is important in that it defines the frequency range that can be recorded by a system; the sampling rate must be at least twice the frequency of the highest frequency to be sampled (Nyquist, 1928). The sample and hold circuit produces a 'stepped' waveform, a sequence of discrete analogue voltages that are then passed on to an analogue to digital convertor (ADC). The ADC assigns each voltage a value based on its current level, each value is stored digitally and the number of bits used to store the value determines the accuracy with which the signal value can be stored. Figure 2.12 shows the original analogue waveform, the sampled output from the sample and hold circuit and, on the y axis, the values that would be assigned to this waveform using 4 bit resolution. It is clear from this example that inaccuracies, known as *quantisation error* are present because of the limited number of available number values that can be represented. Distortion resulting from these inaccuracies is known as *quantisation distortion*. Typical bit resolutions used in digital audio are 16 bit, 24 bit and 48 bit, which introduce less quantisation error than the example

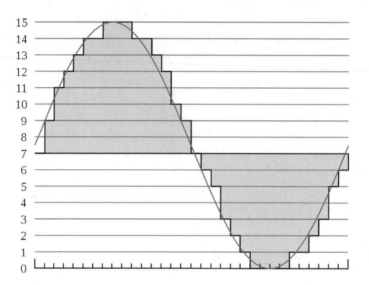

Figure 2.12 Analogue to digital conversion: the analogue wave is sampled many times per second and each sample assigned a numeric value. For illustrative purposes the sampling shown here is at a low 4 bit resolution with 16 possible values.

shown. However, quantisation error is still an issue to be considered in digital audio systems. Consider the example of a low level recording; because the levels of incoming signal are low only a much smaller number of values will be available having a similar effect to using a much lower bit resolution.

Digital audio signal formats in common use in TV broadcast are as follows:

> **AES3** (AES, 2009a), also known as AES/EBU, is a digital format used to pass audio between devices and has both professional and consumer versions. The professional version uses a balanced connection terminating in a three pin XLR connector. Consumer versions known as S/PDIF (Sony/Philips Digital Interconnect Format) are unbalanced and can be transmitted using either fibre, with TOSLINK connectors (from 'Toshiba Link', a standardised optical fibre connection system), or coaxial cable, with RCA connectors. The data formats of the professional and consumer versions are very similar and simple adapters can be used to convert one to the other. The professional version of **AES3** supports stereo and 5.1 audio with sample rates of 32 KHz, 44.1 KHz and 48 KHz directly and can be used for other rates up to and including 192 KHz.

> **ADAT** (Alesis Digital Audio Tape) Optical Interface, also known as *Lightpipe*, uses fibre optic cables with TOSLINK connectors and can carry 8 channels of 24 bit, 48 KHz audio. Originally designed to connect Alesis Digital Audio Tape machines the interface has been adopted by many other manufacturers and has outlasted the tape format it was designed for.

MADI (Multichannel Audio Digital Interface) (AES, 2008), also known as AES10, is designed to carry multichannel audio and supports both fibre optic, with STI connectors, and coaxial cable, using BNC connectors. MADI supports up to 64 channels of digital audio at 24 bit resolution and up to 96 KHz sample rate.

In addition both USB (universal serial bus) and Firewire formats can be used to transport digital audio.

2.3.3   Non-Live Production

For genres such as drama, developments in the technology required for audio content acquisition and production for television have changed acquisition considerably even as microphone techniques have remained largely consistent. Portability of high quality audio equipment mean that productions can be less 'studio-bound' and make much more use of location recording. Earlier recording equipment tended to be bulky, required access to power and was difficult to operate under sub-optimal weather conditions. This generally meant that recording on location required generators and mobile studios to power and to house equipment. Multichannel recorders required even more support and were rarely used on location. In addition to making filming on location resource intensive the fact that usually only stereo recordings were taken meant that there was limited scope for changing the recorded mix in post-production. Current digital recorders take advantage of a lack of moving parts and improved battery technology and are now light, portable and can be made very weather resistant. The move toward completely digital file based workflows with fast ingest capability combined with more portable technology has led to a situation where edit decision lists and preview screenings can also be carried out easily on location using laptop-based edit workstations.

In a single camera shoot, audio is typically recorded by a shotgun microphone held out of shot or by a lavalier microphone discreetly concealed on the actor. Often even where a wireless lavalier radio microphone is being used a boom operator will be covering the actor's dialogue (Figure 2.13) as backup in case of lavalier microphone failure, with one being sent to left channel, and one to right channel of a recorder for a single character.

Microphones are typically routed to a field mixer so that a sound recordist can monitor levels and adjust as appropriate as the recording is carried out. The field mixer output is then sent to a field recorder. Sometimes this is stereo only, but more often now a multichannel recorder is used so that each microphone can be recorded individually giving a greater degree of freedom, control and flexibility in mixing.

Depending on the setup (and the budget) the field recorder or the camera may be used to provide a common SMPTE time code to allow the audio and video to be synchronised accurately later. The field recorder will also add time metadata (and sometimes more detailed metadata such as location information) to the recorded broadcast wav (BWF) audio files. The vast majority of recording will be carried in mono with all mixing for 5.1 carried out in post-production. Left and right channels on recorders are usually used as split mono channels rather than panned left and right for stereo. Where stereo recording is carried out this will typically be done as an mid-side stereo recording (Pizzi, 1984; Dooley and Streicher, 1982) using a shotgun and a 'figure of eight' microphone so that an accurate and stable mono recording and control over stereo width are both available in post-production (see Figure 2.14).

Figure 2.13 A boom operator keeps the microphone out of camera shot while ensuring all of the actor's dialogue is recorded. Reproduced by permission of Jorge Polvorinos.

In addition to dialogue recording, the location sound itself, without dialogue, will be recorded in order to provide a 'wild track' or 'room tone' to layer under any dialogue added in automatic dialogue replacement (ADR) sessions and to cover up any minor problems, such as unwanted noise, with the recorded dialogue. This room tone may also be used as an acoustic fingerprint for noise reduction algorithms if the location is particularly noisy (Hicks and Reid, 1996). Wild track recording will typically happen as a completely separate procedure immediately after shooting and not during takes. Very occasionally wild tracks will be recorded using four channels to route to left, right, left surround and right surround (centre channel is reserved for dialogue and spot effects) but more usually wild tracks are recorded only in stereo with a 5.1 mix for broadcast created in post-processing. As an example, a character in a drama is in a large reverberant set such as a church or warehouse. For distant shots the character's voice will be made to sound distant with a high pass filter taking out low frequencies in the dialogue and with plenty of reverberation to give the impression that the viewer is hearing the

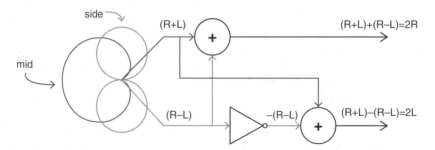

Figure 2.14 Mid-side stereo signal flow showing how a cardioid microphone and figure of eight microphone's outputs are combined to generate both a stable monoaural image and control of stereo source width.

dialogue from the camera position. For a close up the character's dialogue will have much more low frequency content, reverberation from the space will be reduced and probably moved further to the rear of the 5.1 mix. All of the character's dialogue will be produced from the same mono wireless lavalier microphone recording and processed in post-production to sound closer, or more distant. Reverberation on the character's voice will be added using artificial processing in post-production and the recorded room tone will be used to fill out the space, to give the feeling that the viewer is in the location and to disguise audio edits by giving a consistent background sound. Room tone is recorded using either a spaced pair of microphones or a coincident stereo pair of microphones although for an unusual acoustic environment an additional pair of microphones may be added at the rear to make up the left surround and right surround of the 5.1 mix.

As already mentioned, ADR is often required to add replacement, or even altered, dialogue to a scene in post-production. Care is taken in ADR sessions to use the same, or similar, microphones that were used to record the original dialogue on location so that the ADR lines can be edited into location dialogue with no noticeable change in voice characteristic. ADR is recorded in a studio environment that is likely to have very different acoustic properties than the set where the original dialogue was recorded. Artificial reverberation must be added to the dry speech recorded in the studio so that the two recordings blend well. This creates a challenge in that artificial reverberation is sometimes not a good match for the natural reverberation from a location. This is particularly true in locations with unusual acoustical properties and it makes it difficult to seamlessly blend dialogue lines so that they all *sound* as though they are being spoken at the same location.

For some television drama an additional recording of either an impulse or a swept sine wave has been made on location and this has been used to create an accurate simulation of the original location acoustics in the post-production studio using convolution reverb plugins. This is time-consuming however and only rarely done, usually only where there is a combination of an unusual and difficult-to-simulate acoustic environment and when the set is used for a series with shooting over a prolonged period of time. A recent example was the long running UK television drama series, Waterloo Road,[1] which was set in a school environment. For this series the acoustics of the long corridors and stairwells where much of the dialogue was situated was particularly difficult to match using off-the-shelf reverberation processing and the best results were obtained by using customised convolution reverb derived in this way (Figure 2.15).

Before shooting took place swept sine waves were recorded at each location using a reference monitor and stereo microphone techniques. The recordings were made at both near and far distances from the loudspeaker in each location used in the series and impulse responses for the location generated by deconvolution. The resultant impulse response (shown in Figure 2.16) was then used with a convolution reverb plugin to create accurate representations of the space artificially. These were then applied to dry dialogue added in ADR sessions to create a match for speech recorded in the natural acoustical environment.

In many cases the schedule demands for television drama are such that extensive post-production and ADR are luxuries for which no time is available. A typical example of this is the UK soap opera Coronation Street, which broadcasts three episodes a week, typically shooting 30 pages of script in a single day. In this case boom-held shotgun microphones are almost always used except for very wide establishing shots when wireless lavaliers are used. In order to avoid the need for time-consuming synchronisation in post-production or at ingest,

Figure 2.15   Recording of impulses at Hill Top School before shooting of the UK TV drama series Waterloo Road. Reproduced by permission of Tony Greenwood.

video feeds from camera and audio signals from microphones are recorded simultaneously onto a pair of P2 recorders (one as backup) thus retaining synchronisation from the point of acquisition. Where two characters are in shot and two radio microphones are used a single boom is usually in place as a safety measure, the boom being routed to the left channel and a live mix of the two radio microphones being sent to the right channel. Radio microphones are also backed up separately onto two further tracks but wherever possible the dialogue is mixed live at the time it is recorded. Because of the 'almost live' nature of this fast capture method,

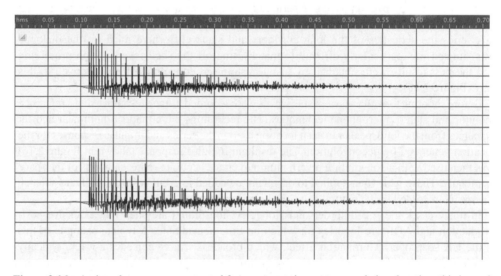

Figure 2.16   An impulse response generated from a swept sine wave recorded on location, this is used with a convolution reverb plugin to create accurate simulation of real acoustic spaces in post-production.

boom operators have to be especially attentive and rehearse boom movements to script before shooting begins. Where the script calls for a cut from a medium to a close shot the boom operator will rapidly move the boom microphone in close to the actor between the appropriate lines of dialogue. The boom operator is helped by a convention of 'fast in slow out' between close and medium or wider shots so that there is time to remove the boom from camera shot. For shoots such as this where video and audio is recorded to the same device, wild track is often recorded after filming alongside colour bars on the video so that the editor can visually identify wild track location quickly while editing.

In the studio both video and audio are recorded directly to a media server; again synchronisation is maintained from the point of record into editing stages. In some circumstances latency is present in one of the signal paths for some reason, the most common example being where a radio camera is used that will typically add a two-frame delay to the video path. Where radio cameras are used and audio is being mixed to mono 'live' as described previously, additional delays will be inserted into the audio path to ensure that the media is ingested to the server already synchronised and ready for editing. Where more tracks are recorded in a scene, synchronisation may be carried out after ingest, before editing takes place, with the sound recordist being aware that there is a small, but consistent, delay in what he is monitoring. The digital acquisition workflow is file-based and instant ingest into the server means that editing will already be underway before the day's shooting is complete.

Additional playback facilities often have to be put in place for the acting talent during shooting, for example where a character has to be singing along with some music, or where the character is in what is meant to be a noisy location and has to respond to cue sounds. In-ear wireless monitoring will be used to provide sound for the actor to react to or as loud background so that the actor naturally raises their voice as though in the noisy location. Where playback of this type is used, boom microphone and playback cues will be recorded as split channels to left and right alongside the video, again to make synchronisation quicker when editing.

2.3.4   Live Production

In non-live productions most recording is monoaural and creating a mix for broadcast is carried out in post-production. In contrast, live productions must capture and mix audio to whatever format it is being broadcast in at the point of acquisition. In order to accomplish this, outside broadcast (OB) trucks are taken to the venue for the event and act as a mobile studio for both audio and video. The details of the techniques involved vary considerably from live event to live event and two examples are discussed here, sport and music, that between them cover the range of methods currently used by production teams on live productions.

Sport

Sports broadcast is a broad term and in itself covers many sub-genres – coverage of Formula 1 racing will be very different to coverage of Premier League football or tennis at Wimbledon. The example here of football in the UK is of particular interest because the adoption of 5.1 surround sound for broadcast had a significant impact, not just on the methods used, but also on the production values and the nature of sound design for sports. Live coverage of the UK Premiere League was among the earliest programming broadcast in the UK in 5.1 surround sound so techniques developed here have informed the techniques used in other genres of sports coverage.

The key objectives for audio in football coverage are twofold; picking up sounds on the pitch clearly during the game, and utilising the 5.1 surround sound capability to give the viewer a sense of immersion and of 'being there'. These objectives are achieved by use of two separate microphone setups common to Premiere League coverage and also coverage of World Cup and other international football.

For on-pitch sounds the ball kicks and whistle blows are happening some distance from any possible microphone position so shotgun microphones are used to pick them up. Twelve shotgun microphones are positioned around the football pitch facing towards the action. If all of the microphones are live in the mix at any given time the background noise from the crowd swamps the sounds from the pitch making them inaudible. In order to prevent this from happening, microphones are mixed live so that only the microphone closest to the ball location is in the mix at any given time. This requires a skilled sound engineer to follow the action on the pitch on the mixing console in the outside broadcast truck and ensure that only the microphone nearest the ball is active in the mix. As the broadcast is live the engineer must predict where the ball is going to be next but also has to be aware of what likely camera angles will be chosen. In addition to static camera locations there are a number of handheld cameras, so-called Steadicams, around the pitch and when the producer decides to cut to the Steadicam shot there is a potential issue for the sound engineer to be aware of (see Figure 2.17). The

Figure 2.17   A Steadicam in use, the video and audio has a delay of around four frames compared to wired cameras and microphones at an event that must be taken into account while mixing audio to prevent 'double hits' of sound events. Reproduced by permission of Michael Tandy.

Steadicam camera has a microphone which can be mixed into the 5.1 broadcast audio but the video also has a built-in latency of four video frames because of video processing. As the audio is also delayed by the same amount of time there is potential for double kicks to be heard should the Steadicam microphone be active at the same time as another nearby pitch-side microphone. For this reason every time the Steadicam shot is chosen all of the nearby pitch-side microphones must be muted until the shot changes.

The crowd sound for live football coverage is key to building the atmosphere for the television viewer and is usually picked up by a single Soundfield® microphone suspended from the gantry above the crowd. The Soundfield® microphone consists of four near-coincident microphone capsules arranged as a tetrahedron. The four outputs are encoded into a B-format ambisonic (Gerzon, 1980) signal by a microphone controller on the gantry. The B-format signals from the Soundfield® microphone define the sound in three dimensions at the microphone location and these can, if desired, be decoded for periphonic (with height) reproduction. The four B-format signals (W, X, Y and Z) are sent to the OB truck as AES 3-id (AES, 2009b) signals on unbalanced BNC cables. For television broadcast the Z (height) component is ignored and the B-format signals are decoded into a 5.1 feed at the OB truck. This 5.1 crowd noise channel is mixed into the 5.1 programme audio both to give a more immersive experience for viewers and also to cover up any audible artefacts from mixing between the pitch-side microphones. The mix of crowd noise will be adjusted from time to time during the game.

Music

In principle, acquisition of live music for television follows the same principles as for live sports events. Sound events, in this case instruments, are picked up via microphones as close to the sound source as possible or via direct line signals and mixed to the format required. Ambience, audience noise and, where desirable, natural reverberation from the acoustic space where the music is played are picked up by a more distant microphone, either in the same way as for football crowd noise using a Soundfield® microphone, a surround microphone array or, where 5.1 surround sound is not required, using a spaced stereo pair of microphones. In each case the ambience is mixed in with the music. Unlike sports events, in audio acquisition for live music there is the opportunity to use microphones much closer to the sound sources being recorded. This simplifies the sound recordist's task considerably. Wherever possible and practicable, instruments will have microphones close by to reduce spillage between nearby sound sources. Where there are a very large number of sources, for example, in an orchestra, instruments may be acquired as a group, so sections of the orchestra will be recorded and treated as though they are the same sound source. For television broadcast of the BBC Proms concerts (a series of classical concerts broadcast each year on both TV and radio) two different approaches are taken depending on the target broadcast media. Only some of the Proms concerts are broadcast on television with surround sound whereas all are broadcast live on radio. For 'radio only' concerts around 40 spot microphones are in place to capture the orchestra sections and soloists. However, this setup is based around the need to capture many different ensembles over a series of concerts and not all will be in use at one time. In addition Decca Tree and Hamasaki Square microphone arrays (Hamasaki and Hiyama, 2003) are used to capture left front, centre and right front channels of the 5.1 mix, and the surround channels respectively.

2.3.5   Surround Sound

One clear driver for the rapid roll-out of 5.1 surround sound has been the expansion of *movies on demand* services, however, film production is largely beyond the scope of this book. For television, live sports broadcast has been one of the biggest drivers for 5.1 audio, largely because of its popularity and the associated budgets but also because surround sound has the potential to give the impression of being in the crowd at the game. The more immersive experience given by 5.1 audio is well-suited to the genre. In some ways this has impacted considerably on the way in which sound design for sports events is approached. Consequently there has been an impact on audio capture both in terms of methods, and of production drivers and priorities.

2.3.6   Challenges for the Future

Current techniques and methods for audio acquisition and production described here are aimed at a *channel-based* paradigm for TV broadcast. Production is carried out with reference to the reproduction system that it is intended to be played back on: a sound comes from *this loudspeaker* or from *between these two loudspeakers* with no direct reference to the sound scene being recreated. This creates some challenges to audio for TV broadcast. Developments in delivery mechanisms that are moving the industry from linear 'push' broadcast to video on demand and interactive content are likely to drive demand for more flexibility in reproduction. Programme content is now delivered to many more devices than before including mobile, tablet, computers and game consoles, each of which have different requirements in terms of reproduction systems. There are also in parallel increased levels of interaction, and increased demand for second screen audio, as well as video and text. As long as production is channel-based, delivery to increasingly diverse audio reproduction systems will require additional production time, and interaction on those devices will be limited. Chapter 4, 'Platform Independent Audio', suggests some solutions and an alternative format-agnostic approach to production which has a much more flexible approach to broadcast audio.

2.4   Live Programme Production

The characteristic qualities of live television are often considered to be immediacy, intimacy and a sense of 'being there' at an event as it happens (Dayan and Katz, 1992). Production of live television is consequently spoken about in terms of realism, of covering an event as it unfolds, or taking the viewer to the sports stadium or concert arena. Taking a closer look at the practice of production of a live event such as a sports game, however, it becomes apparent that what is taking place is far from a simple representation of 'being there'. This was discussed in Section 2.2.1 in terms of the image characteristics and how 'real' the images should appear, but the process of selecting shots also plays a key role in how the event is perceived by the viewer. Gruneau (1989), in the first detailed case studies of the practice of live television production, describes the work of representing live action through a process of image selection:

   'What is "shown" on television is always the result of a complex process of selection: what items to report, what to leave out, and what to downplay. Television sports production also involves a wide range of processes for visual and narrative representation – choices regarding

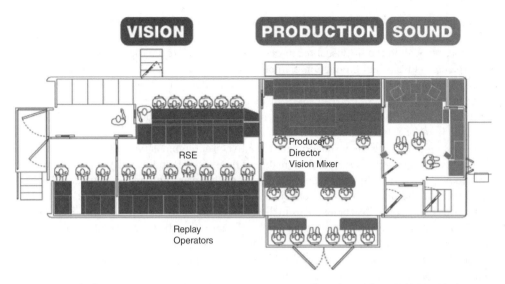

Figure 2.18 Configuration of the Outside Broadcast studio (adapted from SISLive, 2011).

the images, language, camera positioning, and story line required to translate "what happened" into a program that makes "good television"' (Gruneau, 1989, 135).

This process of selection is at the core of the vision mixer's (VM) work. In the following sections, we will describe this professional role and others that take part in producing the content for a live broadcast. When required we will use televised sports as an example of live television production. Sport is a common type of live television content, and its nature of being fast-paced yet organisationally well-structured and self-contained, highlights some of the specific characteristics of live television production.

2.4.1 The Production Area and Roles in Production

An OB truck is a mobile television control room used in most television productions of live events. The production area of the OB truck is where the actual production decisions are made and the show is created. This area is centred around the workspaces of the key people in the production: the Director, the Producer and the Technical Director or Vision Mixer (see Figure 2.18). They are aided by assistants and collaborate closely with specific production roles such as Replay Operator, Graphics Operator and Audio and Video Engineers.

Programme Production Roles: Producer, Director and Vision Mixer

The *Producer* is in charge of the planning and organisation before and during the broadcast. They are responsible for the overall concept of the programme, and their job is to ensure that the event is captured in a way that matches the viewers' interests. This work demands intimate knowledge of the activity to be covered, such as the history of the teams in a sports game and the characteristics of the venue (Owens, 2007). The Producer's responsibilities also include coordinating the crew and talent, and making sure the broadcast is on schedule. The *Director*, who reports to the Producer, is in charge of composing the overall storyline of the broadcast over time, using all the audio-visual resources produced by the team. This involves knowledge of all the crew's tasks and production elements, such as live cameras, audio, taped segments and replays, and the ability to combine them into a visually compelling whole over the course of the broadcast. The *Vision Mixer* takes instructions from the Director and translates them into mixing decisions between the visual resources at hand, in real time. Vision mixing is described in more detail below.

The task of producing the visual content for a live programme is typically shared between the director and the vision mixer, and the nature of their task separation and collaboration may vary. To generalise, the director, who is in charge of the overall broadcast, typically follows the live action and directs the shot selections, which are then executed by the vision mixer. These directions may be instrumental, as in direct orders to cut to specific cameras. Depending on the team they may also be on a higher level, as in narrating the event verbally and guiding the attention of the vision mixer, who then more freely interprets the directions into shot selections. In small productions, the two roles may be executed by a single director, who then directs the broadcast while carrying out the live mixing operations themselves.

For larger productions, each of these three key roles may be supported by additional professionals to share the workload. The *Executive Producer* manages production related matters outside the live broadcast and over time, across a series of occasions such as a season of sports broadcasts. Their responsibilities include negotiating broadcasting rights, finance and scheduling. An *Assistant Producer* or *Features Producer* may be assigned to manage a subset of the producer's responsibilities, such as supervising the insertion of pre-recorded video packages into the live broadcast. *Scripts* and *Production Assistants* support the Producer with tasks including scheduling, organising information and communication with remote team members. The Vision Mixer is sometimes aided by a *Switcher*, who executes their mixing decisions.

Replay and Technical Operators

The **Replay Operator**, or **EVS Operator** (from the brand name of the most commonly used replay production hardware) acts as an editor of replay content that is provided to the Director. The Replay Operator follows the live action on the main broadcast cameras, and can access, play back and inspect historical material using a powerful video server. Replay sequences are then edited based on that material and played back in carefully-timed slots in the live broadcast, or saved for later use. The work of replay operation is further described below. In large-scale productions, replay operators commonly work in teams, where the operators then work in parallel on different camera sources covering the action. This ensures that more replay material is readily available upon request, and makes it possible for a dedicated *Replay Subeditor* (RSE) to quickly assemble a sequence of replay shots from multiple angles.

The **Video Engineer**'s responsibility is to set the video levels, white balance and colour controls for each video source, and to maintain these values over the course of the covered event, as lighting conditions may vary over time. The Video Engineer also has to make sure these values are balanced across all cameras used in the broadcast, so that visual continuity is maintained as the Vision Mixer switches between camera feeds.

The **Audio Mixer** and **Audio Engineers** are similarly responsible for creating a coherent aural representation of the covered event. The Audio Mixer oversees the placement of microphones around the venue and creates a mix of these sources to be used by the Director. This mix is dynamically designed and adjusted over time as the action unfolds. This is carried out separately from the vision mixing, as the audio should cover ambience and key actions in the scene rather than follow cuts between broadcast cameras. The Audio Mixer takes instructions from the Director and Producer, but is typically placed in a separate but adjacent room, as their task is sensitive to outside noise.

The **Graphics Operator** operates the graphics generator machine, which holds a number of set templates for visually presenting information such as scores and names of team members, overlaid on the broadcast image.

2.4.2 The Production Workspace

This section presents a closer look at the main area where the production staff work.

The monitor wall, or gallery, is centrally placed in the production area. It is the production team's main visual resource for producing the live broadcast. The gallery includes a large number of visual resources, arranged in a matrix of video monitors (see Figure 2.19).

The main monitors in the production gallery are as follows:

> **Programme monitor**: a large centrally-placed screen displaying the camera image currently being broadcast. The selected camera image is shown immediately.

> **Preview monitor**: a screen adjacent to the programme monitor, which allows the Director and Vision Mixer to preview a video image before going to air. The

Figure 2.19 Production gallery. Reproduced by permission of Mark Perry.

preview monitor is often set up to display one of the manned cameras complementing the camera selected in the programme monitor, by default if no camera
is selected for preview. The programme and preview monitor pair are typically at
the centre of the gallery.

Camera monitors: one for each camera, typically arranged in a 'camera row'
below the Programme and Preview monitors. The main manned cameras are
located centrally. More rarely-used cameras, such as fixed cameras providing
event-specific details, are located to the sides of the monitor wall.

Replay or Video Tape Recorder (VTR) monitors: showing available pre-
recorded content such as interviews and introductions to be shown at the beginning
of the broadcast and during pauses in the event action.

Graphics monitors: a screen dedicated to graphics to be overlaid onto the broadcast image. The graphics monitor displays a preview of the graphics selected by
the graphics operator, on black background.

The layout of the gallery is generally programmable so that any camera or video source can be
routed to any monitor allowing the Director and Vision Mixer to customise the gallery. Modern
galleries can be customised to an even greater extent using large high-definition displays,
where multiple image sources are displayed on a single screen. The video control and audio
control areas are normally located in separate sections adjacent to the main production area
(Millerson, 1999). These areas include spaces for the Video and Audio Engineers, whose role
it is to oversee the technical quality of the broadcast.

2.4.3   Vision Mixing: Techniques for Directing, Selecting and Mixing Camera Feeds

A main concern in the work of mixing live television is rendering an understandable and
aesthetically appealing view of the action at all times. Visually, live television production
follows traditional film grammar, a system of rules for how to effectively tell a story in
images. Editing strives to provide multiple viewpoints on the covered action, and using these
to produce rhythm, balance between detail and overview, and a dynamic and compelling
sequence of images (Holland, 2000, Ward, 2000). For practical reasons, the positioning of
cameras in a sports arena is restricted to certain points along the side of the field and from
the balcony. Similar restrictions apply to the audio mixer's placement of microphones. The
vision mixer has to have a well-developed awareness of the location of the video and audio
resources in the venue, to be able to make quick decisions, for example, on cutting to a detailed
shot from a fixed camera. Apart from spatial concerns, there are visual grammar rules that
further prescribe how cameras should be placed and used by the Vision Mixer to build the
sequence of images that forms a live broadcast. These rules are largely derived from tried
and tested practices in editing film, but the Vision Mixer has to follow them while mixing
in real time, in order to maintain visual continuity in the broadcast. The most important
techniques include:

The '180-degree rule' This rule states that two images in sequence must be shot from the same side of an imaginary line through the covered scene. 'Crossing the line' means introducing contradictory directions in the sequence of images – action may appear to suddenly change direction between cuts, which is confusing to the viewer. This effectively means that cameras cannot be placed on opposite sides of a sports arena, for example.

Cutting 'on the action' A cut made in the middle of a significant action taking place in the covered scene disguises the edit point and helps maintain visual continuity. If it is well done, the viewer will be able to follow the flow of the action across the cut without consciously realising a cut has occurred.

Avoiding similar compositions. Two following shots should ideally diverge in both camera angle and framing. Broadcasting two similar shots in sequence may result in a disruptive 'jump cut', or in the audience losing their orientation.

Variation and shot patterns. Viewing many shots over time from similar camera angles and a similar distance from the subject is considered to be tiring for the audience. This is typically addressed by patterning of shots, the use of predefined sequences of images in order to best cover events that are known to occur (Holland, 2000). As an example, a scene often opens on a wide establishing shot showing the general setting and mood, followed by more closely framed medium and close-up shots of people and event-specific details. This way, the viewer gets both the overview and an emotional closeness to the action as the scene progresses. Similarly, predictable situations that recur throughout the production, such as a goal situation in a sports game, may have predefined patterns that support editing decisions and aid the vision mixer in producing meaningful footage of the event. These patterns may also vary between domains of covered action. Continuous action requires a steady flow of cuts that can make use of patterns that extend over time, whereas 'stop-and-go' action, such as in American football, relies more heavily on replays from different angles to complement the live feeds in pauses in the game (Owens, 2007).

To a bystander, vision mixing may appear to involve mostly moment-by-moment decisions on cuts between a handful of key camera sources. A large part of the final broadcast is indeed produced through mixing back and forth between one main overview camera and a small number of manned cameras continuously framing detailed shots of the action. This interplay between overview and detailed shots follows the above mentioned rules for visual grammar, but also relies on a number of other resources that are aligned to support the Vision Mixer's decisions. The camera team, as well as the Director and commentators, coordinate their efforts carefully and mutually focus on topics in the live action that are of concern for the on-going production. For instance, the camera operator's skilled recognition of a potential penalty situation in a sports game, observable in the way they centre on regions of interest on the field and frame their shots in a way that makes room for such potential situations, guides the Vision Mixer's attention and supports their mixing decisions (Perry et al., 2009). The cameras' placement and the pace of the live action are both resources that the Vision

Mixer uses for determining when to safely cut to a detailed shot of action. Similarly, but more rarely, the Vision Mixer may use his awareness of the placement of microphones to temporarily enhance the audio produced by the audio mixer, emphasising specific situations in the live action.

2.4.4   Audio Mixing

For live broadcast the audio production is doing more than just relaying what can be heard from the current camera position. If the audio always followed the camera shot there would be unpleasant sudden shifts in audio every time a cut was made between cameras. The audio, much like the video, is being manipulated in order to tell the story of the event and in common with film production, the sound is not simply recorded but there is a large element of *sound design* involved in the process.

A key audio element of current football coverage is that on-pitch sounds, especially ball kicks, are much louder than would be heard from the crowd itself. This sound design signature element is considered to be important in engaging viewers with the game and has been picked up and used by game developers for console football computer games. Currently all of what could be called *diegetic audio* – the on-pitch sounds related to the action – are panned centrally, either to a centre channel or to a phantom centre between left and right. Crowd sounds are panned to surround channels for 5.1 surround broadcast and used to immerse the viewer in the event, to make them feel like they are at the game.

2.4.5   Replay Operation in Live Television Production

Instant replays are a key feature in live television content. The ability to create instant replay material in the production of contemporary television relies on the use of machines that allow the replay operator to access stored video footage of action immediately after it takes place. Video and audio materials are captured to a video server, which allows recorded footage to be searched, edited, recompiled into new sequences and played back at variable speed so that they can be timed with the live footage. The use of replay operation also allows the director to cut into the live broadcast to show recorded footage from cameras that were not initially selected for broadcast, presenting additional angles on action taking place during the game and at different playback speeds. This is especially useful in providing visual explanations of fast-paced action that has just occurred, in pauses where relatively little is taking place in the live footage. The role of the instant replay operator is to assess and select sequences very rapidly as soon as they occur and to create relevant material that can be cut into the live broadcast (Engström et al., 2010). The task is not simply a technical one, to quickly produce content when requested. The operator also needs to be highly attentive to the developing game, and have an intimate knowledge of its structure, in order to select material from multiple sources and produce relevant and timely footage.

The Replay Operator contributes to the production in several ways. Most visible in the produced broadcast are the replay sequences themselves, produced immediately after a significant event takes place in the scene and inserted at an appropriate time into the live broadcast. Producing and fitting non-live replay material into a sequence of live shots is highly skilled work on the Replay Operator's part, but it is also highly collaborative work in which the replay operator, Vision Mixer, Director, commentators and camera operators all work together to solve the task of telling the story of what happened, using both live and non-live image sources. Less visibly to the TV audience, the Replay Operator also aids the production in searching through historical material, providing an understanding of past events that is then used by the Director and commentators.

In a detailed interaction analysis of replay production in a live ice hockey broadcast, previously published in its entirety in Engström et al. (2010), a number of specific techniques and ways of working emerged, which illustrate the multiple ways replay production feeds into the larger production of a live television broadcast.

Temporal coordination through media threading: at the same time as searching through logged data, the Replay Operator listens to the on-going audio commentary, using this as a resource to check the live video feeds on occasions where they talk about possible replayable topics. These occasions function as situated audio 'tags' that allow work on the live video streams to be synchronised with pre-recorded media.

Tracing historical references backwards in time: when an ambiguous situation occurs in the game, the Replay Operator traces cues in both the live broadcast and stored video content to help the production team make sense of what has occurred. In the example analysed by Engström et al. (2010), the search began with the live camera's image of the referee's arm signals as indicators of actions that had occurred previously in the encounter between two players, that is, one player illegally hooking the other player with his stick. He then continuously obtained more visual information through searching multiple stored video feeds. Here, the referee's signal, the commentators' remarks and the camera operator's selection of a player skating towards the penalty box were references that helped make sense of previous actions. It follows that the Operator acts not just on what he sees, but also reads this as references to what happened previously. In this case, the Replay Operator 'back projected', or 'retrospectively indexed' (cf. Goodwin, 1996), what he saw live onto what had gone before to identify the offender. Importantly, the result of this search informs both commentators and the Vision Mixer, who switch focus between referring to what is currently happening, and to past action in the broadcast.

Distributed and parallel search: Verbalising the Replay Operator's search out loud allowed the production team to search for an important event for replay simultaneously. This helped them cover more material in the brief time available. There is a form of 'functional separation' of the search between the Vision Mixer and the Replay Operator, bringing several more screens into use for replay production than could be used by the Replay Operator alone.

Synchronising production with game time: Replay production is oriented towards game time in that it allows the production team to fill gaps in game play. The intermittent structure of game time, and especially the pause in play, provides opportunities to focus more on editing and less on the live action. This is because it is unlikely that any new game action will emerge that is appropriate to use for replays during this time.

Narrative formats supporting replay production: The live feed of video provided by the camera operators during game intermissions is helpful for the Replay Operator, even though he is not using this material in his edited version. It is useful because the narrative format changes when the referee blows his whistle. At this point, the camera operators switch from following the game action to showing what had happened. In this case they switched from following the puck to trying to provide an account, by exposing who committed the offence earlier on. This switch in narrative formats has two consequences for the Replay Operator. First, it provides the Replay Operator with time to search and edit his material. Second, it provides him with a bridge in the narration of the game in between the actual situation and the replay of it.

A live television broadcast is an intricate mix of live camera feeds and timely inserted replay material, and the two modes of production are in constant interplay. Vision mixing relies on replay production to recapitulate and explain past events. Replay production, in turn, is possible not just because the Operator can rewind his tapes, but because the whole production apparatus can be organised to either work in synchronisation with the live action, or to revisit, present and comment on historical material. The combination of the two forms allows the director to present the TV audience with a rich, yet clear account of the covered event.

2.4.6   Challenges for the Future

Conventional live television broadcasts have both the advantages and limitations of being singular, well-defined programmes produced by highly skilled professionals. The advantages largely have to do with quality; high production value and a familiar viewing experience for the audience a high degree of quality control, reliability and by extension profitability for the broadcaster. One obvious limitation is that the viewing experience is locked into a single account of the broadcast event, narrated and presented by the production team who decides which part of the event should be viewed at any given time. The broadcast is ideally the best aggregation of all available video and audio sources, but in the process most of those sources are discarded, never to be seen by the TV audience.

The proliferation of high-bandwidth networks and interactive technologies both in the home and on mobile devices opens up a range of possibilities for new experiences of live media content, while at the same time presenting challenges for how this content should be produced. Approaches explored in interactive television (iTV) and online video broadly include giving viewers access to more of the media content and information related to that content, personalising presentation to fit individual interest, as well as supporting viewers in interacting with that content directly in interesting ways. As the amount of produced media content and interactive features continue to grow, it is not realistic to have teams of professionals produce finalised broadcasts in the conventional sense that cater to all the interests of the viewers. Automating various steps of the production workflow has been proposed as a way to overcome this problem. Where the goal is to fully replace manual production roles, a main and possibly overwhelming challenge will be to replicate the complex mechanisms through which production takes place. In the case of vision mixing and replay production, these mechanisms may be grounded in basic rules and production formats, but they also depend heavily on a complicated interplay of domain knowledge, context dependent cues, collaboration, timing, orientation to future scenarios and other factors. To represent all the decisions a skilled production team makes, if at all possible, would require artificial intelligence and image recognition far beyond today's state of the art. From a broadcaster's point of view, a more manageable challenge would instead be how to combine manual production with partial automation and end-user interactivity in order to leverage the potential of new technology, while at the same time maintaining some of the quality control of the traditional production model. This will be discussed in Chapter 5 ('Semi-Automatic Content Annotation') and Chapter 6 ('Virtual Director').

2.5 Coding and Delivery

Today's delivery networks are characterised by their increasingly heterogeneous nature. An end-to-end delivery network may consist of fibre core network elements, peer-to-peer-based content delivery networks and fully-managed digital subscriber line access networks. However, for clarity we briefly sketch the state-of-the-art in delivery networks along two main lines. We start with *managed* delivery networks, that is, networks where resource allocation and availability, the quality of service and user experience are under control of the service provider, broadcaster and/or network provider. The topics discussed in this section include broadcast networks and IP (Internet Protocol) television. Then, we continue with *best-effort* delivery networks: networks where resource allocation and availability, and the quality of service and user experience are not controlled. The topics discussed in this section include content delivery and peer-to-peer networks and mobile broadband. Additionally, we briefly consider the codecs and transport protocols that are being used over these networks.

2.5.1 Managed Delivery Networks

For managed delivery networks, we look mainly at broadcast and IPTV (Internet Protocol Television) networks. The first type includes satellite, terrestrial and cable networks. The second type relates to a form of digital TV as transmitted over the various types of digital subscriber base lines and increasingly over cable networks.

Broadcast Networks

Broadcast remains a necessity for meeting today's media distribution needs and are fundamentally unidirectional networks. The technology supports, and often also is limited to, a one-way and one-to-many data communication path. By the principle of broadcasting, where receivers can consume the content from the source without having a bidirectional connection towards this source, the media can be delivered in a cost-effective and very scalable manner to a virtually-unlimited number of receivers. After the growth of digital satellite services and the switch-off of analogue terrestrial broadcasting, today most TV services in Europe are delivered via digital broadcast networks. Digital satellite, cable and terrestrial broadcasting are the best-known technologies for digital broadcasting. These technologies have a lot in common; they share a basic functional architecture, depicted in Figure 2.20. The differences are mostly seen on a service delivery level.

Globally, *digital satellite broadcasting* is the most popular technology for TV distribution. Broadcasters can quite easily set up a satellite TV service and cover a very wide area. For satellite broadcasting no fixed operator-managed reception infrastructure is needed on the ground and satellite space is relatively cheap since much space is available on a large number of satellites. The subscriber needs a relatively large installation with a satellite dish, specific receiver and cabling. Receiving a *digital terrestrial broadcasting* service is very easy for

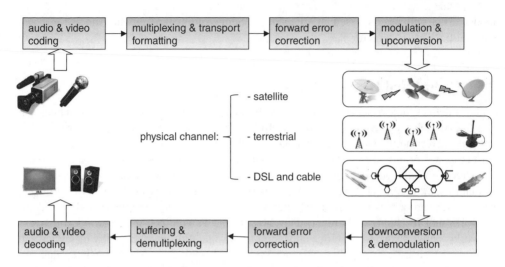

Figure 2.20   The basic functional architecture for a digital broadcasting chain.

a subscriber: only a small antenna is needed and in most cases, a passive indoor antenna suffices. Digital terrestrial receivers are commonly built into TV sets. However, frequency space for terrestrial broadcasting is very scarce and mostly licensed per country. Furthermore, for operators it is relatively expensive to set up and maintain a terrestrial antenna network. *Digital cable broadcasting* depends on the availability of a cable network that is often owned by the TV service provider or TV channel provider themselves. The subscriber needs a cable receiver, often a stand-alone set-top box (STB), which besides providing TV channels also offers on-demand and recording functionality.

Recently, broadcast technologies have been improved by using the frequency spectrum in more efficient ways: by employing more complex modulation algorithms, improved channel coding techniques, increased and more efficient error correction, and audio-visual coding techniques. This has resulted in a bandwidth gain that enables more channels and/or better picture quality in the same frequency spectrum. This is an important enabler for the delivery of HD channels over satellite, cable and terrestrial networks. These new developments can be grouped as second-generation broadcast technologies.

Most countries in Europe use the Digital Video Broadcasting (DVB) (DVB, 2010) standards for broadcasting. There are multiple variants, each detailed for a particular physical medium, such as DVB-C/C2 (DVB-C2, 2010) for cable, DVB-S/S2 (DVB-S2, 2010) for satellite and DVB-T/T2 (DVB-T2, 2010) for terrestrial network infrastructures. At the end of 2010, a total of over 520 million DVB receivers were deployed (ScreenDigest, 2010). For satellite currently DVB-S2 is gaining popularity as many service providers are switching from DVB-S; HDTV is a major driver for this. Also outside Europe DVB-S/S2 is by far the most popular standard for satellite broadcasting, with over 200 million DVB-S/S2 receivers deployed worldwide. DVB-T services are now deployed in 62 countries worldwide and five countries have DVB-T2 services deployed. The DVB-C2 standard was released in early 2010; as of 2013 the hardware is still in development and therefore there are no DVB-C2 deployments. DVB-C is the dominant cable technology in Europe. All three second generation broadcast technologies make use of two forward error correction techniques, BCH (Bose and Ray-Chaudhuri, 1960) code which is very compute-friendly and the Low-density parity-check (LDPC) code which is very useful for creating reliable broadcasting over noisy channels. As the performance of second-generation DVB technology is getting so close to the theoretical Shannon limit, it is not expected that any more disruptive improvements to spectrum efficiency will be developed that would justify the introduction of third generation DVB technologies (DVB-TM-C2, 2008).

IPTV Networks

Among the numerous IP-based video services commercially deployed over the past five years, Internet Protocol Television (IPTV) has been a particularly disruptive force in the TV industry. IPTV can be defined as a TV service delivered over a managed IP network with real-time linear programming, on-demand video, on-screen user guide and ancillary services under the control of a service provider and subscribed by an end-user.

IPTV is not the only IP video service that currently exists. Monpetit and colleagues (2010) argue that 'IPTV and other IP video services are in a co-evolutionary feedback loop, where commercial IPTV standardises and supports new technologies for marketing and delivering IP media, while complementary IP services bring Internet "anywhere" access and Web-based interactivity to IPTV'. At the same time technology advancements and architectures developed for IPTV are now commonly used for video delivery over the internet, from protecting commercial content to influencing the *Internet of Information* envisaged as the future of the internet. As depicted in Figure 2.21, IPTV services are delivered over various network segments that need to carry the video traffic at different scales of aggregation.

Core/Aggregation Networks
Since IPTV is a managed service that provides Quality of Service (QoS), it is usually transmitted in the core network over a fibre optic backbone and using Multiprotocol Label Switching (MPLS). As video traffic soars to terabytes of information per day, core networks are forced to

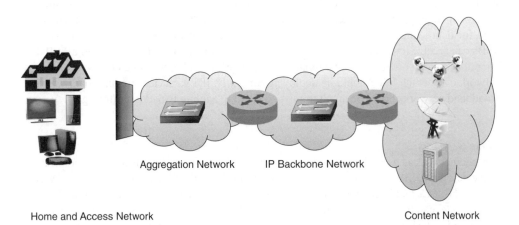

Figure 2.21   An IPTV delivery network.

become more efficient in the way they transport video. While this is still the object of research, options include better usage of existing bandwidth through network coding, and more efficient optical devices.

Access Networks

Although first deployments of Fibre-to-the-Home (FTTH) solutions have already taken place, Digital Subscriber Base Line (DSL) has been the main wired access technology over which IPTV has been deployed so far. In such networks, the DSL traffic from several homes is multiplexed in an access node called the Digital Subscriber Line Access Multiplexer (DSLAM). In the upstream direction, it aggregates multiple connections from these DSL ports and forwards this data to the aggregation network; in the downstream direction, it combines the multiple streams of service traffic, for example, voice, video, and sends it at the appropriate priority to the DSL ports. All broadcast and unicast traffic, from and to its end users, needs to be treated by the DSLAM. Several DSL technologies exist, each with different bandwidth available for the user.

IPTV originally followed the development of broadband access in telephone networks, using DSL technology. Initially broadband used Asymmetrical DSL (ADSL) but it moved rapidly to Very High Speed DSL (VDSL) or Very High Speed DSL version 2 (VDSL2) to provide the speeds necessary for a good IPTV offering. The DSL technologies co-exist with the telephone signal on copper twisted pairs. High-speed data is transmitted digitally at higher frequencies above the usual voice transmission. Nowadays, different versions of ADSL or VDSL technologies are used on the provider side. The link to the aggregation network, as well as the aggregation network itself, can be Asynchronous Transfer Mode (ATM) based or Ethernet. This choice also dictates the technology used in the aggregation network. We see a strong evolution towards Ethernet based uplink and Ethernet switches in the aggregation network because Ethernet switches are usually cheaper than ATM switches.

With demand for HD content rapidly growing, IPTV deployments now use a variety of other broadband access networks particularly FTTH. IPTV over traditional coaxial cable (coax) access networks, using a cable modem, adds capabilities for interactive TV and converged IP services for cable subscribers. IPTV can be delivered over coax frequency bands commonly used for analogue signals, replacing traditional channels, or it can be delivered over frequencies used for data using Data Over Cable Service Interface Specification, DOCSIS 3.0. Hybrid STBs can receive traditional broadcasts from terrestrial or satellite broadcasts and also IPTV. Therefore a cable operator's move to an all-IP network would represent a new type of access network over which IPTV could be deployed. Similarly, IPTV services running over wireless broadband networks could be an alternative to existing Mobile Broadcast offerings.

Home Networks

A home network is a collection of two or more terminals sharing the broadband connection to the service delivery network. A triple play user will typically have one or more TV sets with a STB, PCs and IP phones at home. Also game consoles, printers, faxes, mobile devices, for example, can be part of the home network. All these terminals are connected to a gateway that provides the broadband connection to the access node. Also intra-home communication is possible, so home networking also involves communication between devices within the home.

2.5.2   Unmanaged Delivery Networks

For unmanaged best-effort delivery networks, also referred to as over-the-top (OTT), we mainly look at content delivery and peer-to-peer networks.

Content Delivery Networks

A Content Delivery Network (CDN) is a collection of network elements arranged for more effective delivery of content to end users (Douglis and Kaashoek, 2001). On top of these network elements, CDNs provide services that improve network performance by maximising the available bandwidth. They offer fast and reliable applications and services by distributing content to cache and edge servers located close to end users (Pallis and Vakali, 2006).

A layered architecture of a CDN is shown in Figure 2.22 (Pathan and Buyya, 2007). The basic fabric, the lowest CDN layer, consists of the distributed computational resources such as file servers, streaming servers and networking hardware. It also includes system software such as operating system, content indexing and management systems. The communication and connectivity layer provides the protocols for communication, caching and delivery of content and/or services in an authenticated manner. These protocols include regular internet protocols, CDN specific internet protocols and security protocols. The CDN services layer provides for surrogate selection, request routing, caching, geographic load balancing, management of service-level agreements (SLAs) and so on. The end-user is typically a web user who connects to the CDN by specifying the location of a content provider.

Related to online video, this architecture can be completed by adding the Online Video Platform (OVP) provider as an intermediate between the end-user and the CDN services, as depicted in Figure 2.22. Examples of OVP providers are Brightcove (Brightcove, Inc., 2010) and Ooyala (Ooyala Inc., 2010). The OVP can be regarded as offering additional services on top of a CDN.

A CDN typically hosts static web content (images, on-demand video, advertisement clips, etc.) but it can offer live streaming as well. For both of these purposes, a CDN has some combination of content-delivery, request-routing and content distribution infrastructure.

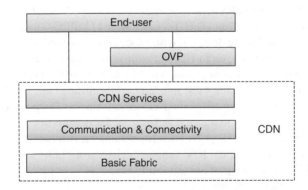

Figure 2.22   The CDN Layered Architecture.

Peer-to-peer Networks

Peer to peer (P2P) networks use nodes, called peers, for the distribution of content. Where traditional delivery techniques typically use a client–server based approach, P2P networks operate on the basis of peers being part of the delivery network: peers operate as clients for the consumption of data and act as servers to other peers at the same time, contributing to the delivery of the content to other peers. P2P networks typically implement an abstract application layer overlay network, independent of the physical topology of the underlying network. This is similar to a CDN network, and P2P techniques can be found within CDNs.

P2P delivery networks initially originated as solutions for the decentralised and distributed delivery of files via networks such as Kazaa (Liang, Kumar and Ross, 2010), Gnutella (Wikipedia, 2010) and Bittorrent (Cohen, 2003). At that time the networks were unsuitable for live video distribution due to the manner in which the content was distributed; the files are split into segments and are distributed among peers in an arbitrary order. Since then P2P technology has matured and is now also suitable for time-stringent based consumption, that is, live or on demand delivery of video. The main advantages of P2P networks in comparison to traditional delivery are the offloading of the content source (in client–server networks everyone receives the content from the server) and an increased resiliency towards failure of network nodes. Drawbacks of P2P networks include the distributed and high dynamic user behaviour (high churn) and the heterogeneous bandwidth capacity. These aspects of P2P networks make it difficult to create stringent QoS requirements. P2P networks can be characterised based on two criteria (Meshkova et al., 2008). First, resource discovery outlines how to find content and who provides the content. Then, network topology describes how the network is organised.

Resource discovery relates to the discovery of the content, but also peers that can help in the delivery of the content, or peers that might be interested in the content. A P2P network is often an application layer network, independent of the physical location of the peers. There are two basic types of network topologies: mesh-based and tree-based topologies. Figure 2.23 shows a mesh-based topology whereas Figure 2.24 depicts a tree-based topology.

2.5.3   Codecs and Transport Protocols

Regardless of the underlying delivery network, the audio-visual content needs to be compressed with suitable audio and video encoding tools, in order to be transported with suitable protocols over these networks. Most of today's video encoding is done with the standard from the Joint Video Team (JVT) of ISO/MPEG and ITU-T VCEG, H.264 / MPEG-4 part 10 (Wiegand et al., 2003). H.264 has been essentially designed with a focus on compression efficiency for a large range of bitrates and resolutions, up to HD and with some support for 4k but at constrained frame rate. For video resolutions of 4k and higher, the successor to H.264, referred to as HEVC (High Efficiency Video Coding), has been standardised (Sullivan et al., 2012).

For TV broadcasting services in IPTV, the Real-time Transport Protocol (RTP) is typically used. RTP provides functionality to transmit separate media streams belonging to one session, for instance audio and video. RTP provides a time-stamping mechanism, allowing a client to synchronise several RTP streams. RTP supports delivery of H.264 encoded video, either multiplexed in an MPEG-2 Transport Stream (MPEG2TS, 2000) or packed directly over RTP, using the H.264 Network Abstraction Layer.

For over-the-top delivery networks, HTTP Adaptive Streaming (HAS), or Dynamic Adaptive Streaming over HTTP (DASH), as explained in Pantos and May (2010) and Stockhammer

Content Source

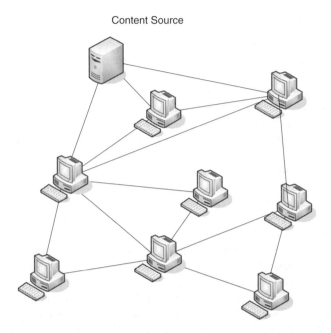

Figure 2.23 Example of a mesh-based P2P delivery network topology.

(2011) introduces segmented content streaming. With HAS, a client does not request a complete file but sequential segments. This allows for adaptive rate switching, where depending on the available bandwidth, a client can request a lower quality version of the same segment. Thus, HAS serves as a QoS mechanism for best-effort networks. HAS has been standardised in 3GPP (3GPP-AHS, 2010) and by MPEG (MPEG-DASH, 2010).

In Chapter 7, 'Scalable Delivery of Navigable and Ultra-High Resolution Video', we further consider the possibilities of HAS for format-agnostic and interactive media delivery.

Content Source

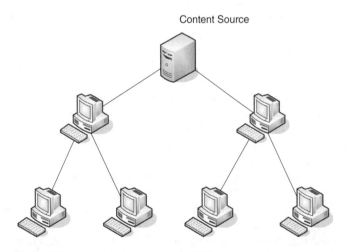

Figure 2.24 Example of a tree-based P2P delivery network topology.

2.5.4   Challenges for the Future

The format-agnostic paradigm that we advocate in this book has the ambition to change the way content is produced at the source, so that the highest level of quality and flexibility is maintained as long as possible in the delivery chain, leaving more possibility to customise the content as a function of end-device characteristics and of the end-user preferences. This approach appears to be in contrast with today's network and device diversity, whereby current approaches are actually contributing to a multiplication of formats, being deployed over several media delivery chains. The acquisition technologies for layered, panoramic or omnidirectional content enable the production of immersive content, around which the end-user may have the freedom to navigate. The delivery of format-agnostic content imposes some key challenges on the network resources, which cannot be met by today's content delivery networks and protocols. For example, because the decisions on the content formatting are pushed to the end of the delivery chain, a key problem faced in delivery is the high rate of data that has to be ingested for delivery.

In Chapter 7, 'Scalable Delivery of Navigable and Ultra-High Resolution Video', we investigate how to cope efficiently with these stringent requirements of high data rate and of fine-grained interactivity; offering better alternatives to a simple push of the whole content from one end to the other. Also, we address the central problem of high data rate for panoramic video and explain how to efficiently compress the content while still allowing a direct access to independent subsets of the content. The chapter also looks at two types of evolution in transport technologies and details to what extent they can be adapted for immersive and interactive video delivery.

2.6   Display Technology

The CRT dominated the display industry for 100 years. This dominance came to an end with the emergence of flat panel displays in the mass market during the first decade of the twenty-first century. This section provides an overview of the take-over by flat panel displays of the entire TV market, and the impact that this has had on programme production and broadcast.

In common with the rest of this book, no particular emphasis is given to stereoscopic 3D technology, so the development of 3D displays are beyond the scope of this section. However, since many kinds of 3D display are based on a flat panel display with the addition of technology such as polarisers, lenticular lenses or time-sequential shuttered glasses to deliver different images to each eye, an understanding of flat panels provides a good foundation for understanding of 3D displays.

Plasma display technology was the early leader for flat panel displays, but the LCD (Liquid Crystal Display) very rapidly moved to a position of dominance. There have been other promising technologies, but only OLED (Organic Light Emitting Diode) technology shows any real likelihood of breaking through to challenge LCD.

Through the early years of the twenty-first century it was said that the increasing sales and reduced cost of large flat panel displays would, very soon, have an impact on the way the viewers look at broadcast pictures. This revolution was rather slow in coming. Plasma Display Panels (PDPs) of 42 and 50 inch diagonal had been available for some years, and their costs fell dramatically, but their market penetration was not enough to make any serious impact.

In 2003, however, it suddenly became apparent that this slow evolution was about to be transformed by the arrival of a serious competitor, the LCD. The perception from outside the display panel industry had been that it would still be some years before large enough LCD panels could be manufactured to challenge plasma, but technological breakthroughs enabled LCD to challenge PDP, not least because it was hard to make the feature size of a PDP small enough to make a high resolution panel, which was not an issue for LCD technology.

Back projection, especially the use of the Digital Micro-mirror Device (DMD), was another type of display expected to make big inroads into the traditional TV market. Before coming on to these technologies in more detail, it is first worth reviewing the technology of the PDP.

2.6.1   Plasma Displays – Developing the Flat Panel Display Market

The heart of a plasma display is the discharge cell, as seen in Figure 2.25. Sandwiched between two sheets of glass, constrained by barriers, the cell has an anode and cathode, and a plasma discharge in the low-pressure helium/xenon gas mixture in the cell generates ultra-violet radiation, converted to red, green or blue visible light by a conventional phosphor. The PDP is therefore self-emissive, but the form of construction leads to a relatively heavy and fragile panel.

The advantage of the PDP was the ease and cheapness of manufacture, which was based on printing techniques. Conversely, the LCD required the use of photo-lithography in its production processes. However, the benefits of scale are now being felt in new LCD plants, which are making LCD production significantly cheaper.

The PDP manufacturers invested considerably in their factories. Due to high panel costs, these displays initially found a niche as public data displays in airports and railway stations, but have been found to suffer from a lack of brightness when viewed in natural light, and often exhibit image-sticking and phosphor burn-in when displaying basically static or repetitive information. Figure 2.26 shows an example of the brightness/contrast problem, which had to be cured by placing ever larger canopies above the displays, to block light coming in from the glass roof above during daylight hours.

A major problem for PDP initially was motion portrayal, with colour fringing becoming visible due to the use of pulse-width-modulation to obtain a greyscale, and low achievable brightness. While, as discussed earlier, the historic advantage of PDP over LCD was the ease of making a large panel, creating higher resolution panels was harder. This was initially reflected by the resolution of panels of around 40–42 inch diagonal being limited to 480 lines, and those around 50 inches in size having 768 lines.

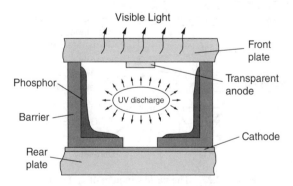

Figure 2.25   Structure of plasma display cell.

Figure 2.26   PDPs used as data displays in a major UK railway station.

Only in 2004 were full HD-resolution PDPs demonstrated in prototype form, and it was another couple of years before they entered the consumer TV market.

2.6.2   LCD – the Current Dominant Technology

The LCD consists, at a fundamental level, of a layer of liquid crystal material sandwiched between polarising sheets, shown in Figure 2.27. When a voltage is applied across the electrodes, the polarisation of light passing through the panel is altered, and hence the transmission of the cell can be varied. A colour display is achieved by adding filters, so that a triplet of

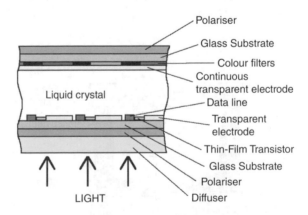

Figure 2.27   The basic structure of an LCD.

cells is used for each pixel of red, green and blue. It is, in direct view designs, a transmissive technology, requiring a back-light. The 'always-on' nature of the display prevents flicker problems, but the sample and hold presentation introduces the appearance of smearing of moving objects (and scrolling captions).

The LCD panel, starting from a natural niche market in the laptop computer, was then able to dominate desktop PC monitors, and achieved phenomenal sales growth rates. The volumes are enormous, as is the investment in new fabrication facilities (mostly in Korea, Taiwan, Japan and China). The level of investment dwarfed that in PDP manufacturing plants, and in 2002 the LCD manufacturers targeted the consumer TV as the one remaining market for their continued expansion.

It was initially observed by many that LCD was not the ideal technology for TV, but the same was true for the VHS recorder, and many early LCD TVs were far from ideal in terms of picture quality as a result of motion blur caused by slow response speeds, poor colorimetry and viewing angle, as well as high costs. As explained below, all these factors were then addressed.

Colorimetry improvements proved relatively simple to implement. Even by 2004 motion blur had greatly improved through a variety of proprietary techniques that aimed to speed the transitions between grey levels by modifying drive voltages during the transition. Improved viewing angles were achieved, and cheaper backlights, a significant part of the cost of a large display, were developed.

The key to driving down cost and increasing panel size lay in the new fabrication plants then under construction. In 2004 28-inch diagonal LCD panels were in mass production, but by 2005 new plants were coming on stream that could make batches of six 46-inch panels on a single substrate. The largest substrates are now over 3m square.

Demand for LCD panels initially outstripped supply, resulting in the manufacturers being able to sell panels with defects, which would normally cause them to be dumped. But during the course of the past decade a vast increase in manufacturing capacity saw manufacturers pushing larger display sizes as a means of utilising the massive increase in manufacturing capacity.

The greatest change in LCD technology over the past 5 years has been its ability to improve motion portrayal. With faster switching times, up-conversion of frame rates to 100 Hz and 200 Hz, 240 Hz is now not uncommon. However, what the manufacturers refer to as 'Hz' are not always what the scientist might consider as a rigorous description of the processes used. In fact '240 Hz' may be images presented 120 times per second with another 120 black fields inserted between, to reduce the 'sample-and-hold' motion blur. The consumer will also be told that the display-rate up-conversion is hugely sophisticated, making use of motion-compensation techniques, but as broadcasters know, such techniques are very hard in practice, and those used in displays are necessarily several orders of magnitude simpler than those used in the best broadcast standards converters.

2.6.3   Projection Technologies

However, the LCD/PDP battle was not the entire story. Projection, particularly rear-projection, has been of much greater interest to North American consumers than to Europeans in the past, and concurrent with the shift to flat panels in the direct view market, the projector market was also subject to just as radical a change. Gone were the fuzzy pictures from heavy CRT-based back-projectors.

The Digital Micro-mirror Device (DMD) has come a long way since the first demonstrations of Digital Light Processing (DLP) by Texas Instruments two decades ago. Licensees of the technology produced projectors at a variety of resolutions, using one chip and a colour-wheel in cheaper models, with three-chip solutions for professional applications. However, there was still a significant consumer projector market for the traditional poly-silicon LCD.

At the heart of the DMD are mirrors of just a few μm across, formed on the surface of a silicon chip. The square mirror, supported at diagonally opposite corners on a torsion hinge only a few atoms wide, tilts by means of electrostatic forces generated by writing to memory elements beneath the other two corners of the mirror. Thus the light falling on the chip is reflected either out through the projector lens, or into a 'black hole' absorber. As with a PDP, the grey-scale is generated by means of pulse-width modulation.

The best DLP-based projectors are to be found in digital cinemas. Technically, DLP has advantages over most other technologies in that it is immune from most of the ageing problems seen on other products. Thanks to the reflective nature of the technology, except for stray light falling down the slight gaps between pixels, the chip is able to handle higher levels of projector brightness that would leave many other technologies struggling to dissipate the absorbed heat. On the other hand, the sequential-colour approach used on the cheaper single-chip projectors can generate a slight feeling of grittiness on some pictures.

An alternative projection technology is JVC's DILA (Direct-drive Image Light Amplification) technology, and Sony's SXRD (Silicon X-tal Reflective Display); both forms of reflective Liquid Crystal on Silicon (LCoS) technology. This suffers a cost and complexity penalty from the need always to use three chips, one each for red, green and blue, but have been very important in the high-end domestic and digital-cinema market, including at 4k resolutions.

Predictions that LCoS technology would enable back-projection TV sets to dominate the market proved unfounded, as consumers voted with their wallets for the flat panel, so projection remains a niche.

2.6.4   Other Technologies

The Organic Light Emitting Diode (OLED), or Light Emitting Polymer (LEP), is a class of display technology that has been developing steadily. In 2003 OLED technology hit the headlines, with a million units sold as part of a Philips shaver to give an indication of battery level, and then a full colour version appeared as the display on the back of a Kodak digital stills camera. As an emissive display it is much thinner than an LCD with its back-light, and has found a ready market for secondary displays in mobile phones and for other mobile devices.

However, it has still to make a serious impact on the TV industry. Display lifetime was initially thought to be the main problem restricting development, but solutions were found, and pilot manufacturing plants built. The ability to make ultra-thin displays did not impress the domestic TV buyer, when the displays were hugely expensive and of very small size, so the Sony 11-inch TV (Figure 2.28) turned out to be a marketing dead-end, in spite of being marketed as an 'organic' panel.

The OLED display is now making its tentative way in the professional market, providing high quality monitors in sizes up to 24-inch diagonal. Larger sizes have long been promised but are yet to appear. The hurdle to be cleared appears to be reliability (and hence yield) in manufacturing as volumes increase. It might have a significant impact in the market in 5 to 10 years.

Figure 2.28   Sony XEL-1 on sale in Tokyo in December 2008.

Two decades ago, one of the big hopes for a new display technology was the cold cathode, or field emission, display. Work in this area had almost stopped, although carbon nanotube technology caused a minor re-invigoration of research effort. Various versions have come and gone, but the dominance of LCD is too big a barrier for a successful challenge from a new technology, particularly when trying to attract the enormous levels of financial investment needed to go into full-scale production.

2.6.5   Impact on Broadcasters

Over the past decade the introduction of flat screens have seen the size of TVs rise from roughly 28-inch diagonal to a now-mature market of 42, 46 and 50 inch screens, with 65 and 80 inch screens at the high end for consumer installations. From 3 per cent of households having 30-inch or larger flat-panel TVs in Europe in 2007, larger displays of higher resolution are now the norm, and have been the driver for the introduction of HDTV services.

Now the display market is again moving to higher resolutions, with '4k' (actually 3840 × 2160 pixel resolution) displays likely to become popular, and in the mobile computing/personal display realm this manifesting itself as so-called 'retina' displays.

The science behind an increase in display resolution (in terms of pixel-count) beyond the accepted norm for human visual acuity of 1 minute of arc is easy to explain. The idealised pixel is a mixed-coloured Gaussian-shaped spot, very different from the RGB triplet of the normal display 'pixel'. The eye is the reconstruction filter for the 'digitised' light output of the display, so in order for the brain to perceive an un-aliased smooth image, a higher resolution in the imperfect display is necessary than the underlying information content of the image (Drewery and Salmon, 2004). The same is true whether the display is showing textual data or TV images, and is the basis for the Apple® Retina® display. This creates a marketing problem, where the broadcast signal may be seen as lagging behind the display technology in terms of pixel count, but in practice an HDTV signal still has plenty of life in it, which is fortunate, since most broadcasters have yet to even complete their HDTV investments.

Changes in display technologies have had other impacts on the way images are created and perceived. To start with, there is the temporal smearing, which has caused broadcasters to limit the crawl speed of captions. Then there is an interesting by-product of the temporal up-conversion in the display, which can result in what the director intended as 24 Hz or 25 Hz 'jerky motion film-effect' (Roberts, 2002) to be rendered into normal smooth (TV-like) motion in the display.

There have been arguments as to whether the CRT 'gamma' characteristic (discussed in Section 2.2.2) should be retained in the new flat-panel display environment. As it happens, this exponential characteristic is pretty much ideal in spreading the visibility of noise evenly across the signal range, and coupled with the problem of making a step change in an industry where the long-tail of archive content remains an important factor, has ensured that the standard form of gamma has remained.

The way in which a 'broadcast monitor' is defined has become very important; the look of a TV picture is determined by the characteristics of the master monitor used in the TV production facility. One CRT had very similar characteristics to another, and although some aspects were standardised, some elements of the specification (the value of gamma) were simply part of the physics of the device. In moving to flat panels, in order to retain compatibility it has been necessary to characterise the display, and the EBU has produced a document outlining the broadcasters' requirements (EBU, 2010).

2.6.6   Challenges for the Future

The broadcaster now faces a rapidly-changing display environment, where the consumer TV industry is continually seeking technical advances to incorporate in their products to give them a marketing advantage. The migration from standard definition to widescreen, and then to HD was a process driven to an extent by the broadcasting industry, over a period of decades. The move to still-larger screen sizes, to 4k resolution, but also to an increasing variety of mobile/handheld devices, and potentially improved auto-stereoscopic 3D screens, driven instead by the consumer equipment manufacturers, will put pressure on broadcasters to be more flexible. They will need to provide material to an ever wider range of devices, so being able to re-purpose material for a variety of resolutions, and providing multiple views of the same scene, will become very important.

2.7   Audio Reproduction Technology

This section surveys spatial reproduction systems as they are in use today. Principally one can distinguish three approaches for the generation of a sound scene for an accurate perceptual experience.

First, and most commonly, the specific perceptually-relevant inter-aural cues can be generated artificially by the reproduction system to give the desired auditory impression, by exploiting certain psycho-acoustical features of the auditory system. This is achieved using loudspeakers or headphones and is described in more detail in Section 2.7.1.

An alternative approach is to aim for a physical reconstruction of a sound field within the listening space (see Section 2.7.2). It is assumed therein that if the sound field is reconstructed correctly the listener exposed to it will perceive the desired sound scene correctly in terms of its spatial attributes. Since both of the aforementioned approaches have limitations, and in any case, in the context of entertainment systems, only the perceptual results are relevant, most of today's research investigates hybrid solutions that are often found on a continuum between these two extremes as described in Section 2.7.4.

A third approach can also be considered, where the sound is recorded or synthesised in such a manner that the exact signals arriving at each ear can be presented directly to the listener to evoke the true impression of the sound scene. Such an approach generally uses headphones or is optimised for a single listener and is described in 2.7.3.

2.7.1   Stereophonic Sound Systems

A variety of so-called *stereophonic* reproduction systems exist today, such as two-channel stereo pairs and surround sound setups utilising five or more loudspeakers. All these loudspeaker layouts aim to produce the accurate impression of a spatial sound scene (i.e., to generate an *auditory scene* – see Section 4.2.1). Commonly for these systems, the correct sound scene is only accurately perceived in the *sweet spot* (the centre of the loudspeaker setup), outside this region the sound scene impression is distorted. The reason for this is the fundamental concept of such a system that aims to artificially reconstruct localisation cues at the listeners' ears to create an auditory scene. Since these are mainly inter-aural level differences and inter-aural time differences, a slight misalignment of the listener in the sound field or of the loudspeakers creating the sound field results in a spatial distortion of the auditory scene (the relevance of such a distortion is not discussed here).

Loudspeaker Layouts

There are many different loudspeaker layouts used for stereophonic reproduction, some of which can be seen in Figure 2.29 that shows a 9-channel surround playback system. This setup can be split into different subsets for the different layouts chosen; the two-channel stereo loudspeaker layout is the smallest of such a subset, consisting of simply the left and right loudspeaker. The surround sound system subset adds centre, left surround and right surround loudspeakers. As this specific loudspeaker layout is the one mostly used in cinemas, the screen position is also indicated. For movie theatre sound, the centre is mostly used for dialogue playback, left and right loudspeakers are used primarily for music playback and stereo effects, and the surround loudspeakers are used for room and ambient effects. The low frequency effects (LFE) channel usually addresses a subwoofer (not shown in the Figure), commonly indicated as '.1'. The placement of the 5.1 layout is described by ITU recommendation ITU-R BS.775 (ITU, 2006). Note that in practical cinema setups additional surround loudspeakers are added to cover a larger listening area.

Various extensions for the 5.1 setup have also been suggested, most of which are defined in two dimensions although there are some exceptions that define three-dimensional loudspeaker layouts. For example, Tomlinson Holman's 10.2 system, as introduced in 1996, defines an upper left and upper right channel in the front and extends the five-channel surround setup to

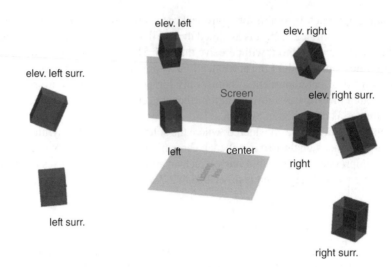

Figure 2.29 Various extensions exist based on the stereo layout with left and right loudspeaker including 5.1 and 3D layouts like 2 + 2 + 2 and 9.1.

eight channels (Holman, 2001). Furthermore two LFE channels are used. The 2 + 2 + 2 system was introduced in 2001 (Steppuhn, 2001) utilising the five-channel surround sound format with elevated loudspeakers above the left and right loudspeaker positions. The centre position in this case is omitted. The LFE channel serves as ordinary audio channel so that an elevated left surround and elevated right surround loudspeaker can be addressed (see Figure 2.29). The 9.1 layout, proposed in 2005, extends the 2 + 2 + 2 arrangement by two additional loudspeakers below the surround positions and once again includes the centre channel and LFE channel (Auro, 2013). In 2005 another system was introduced in the form of 22.2 (Hamasaki et al., 2005). It uses three layers of loudspeakers, an upper layer having nine channels, a middle layer with 10 channels, and a lower layer containing three channels and two LFE channels.

Panning Functions

A simple stereophonic playback system consisting of left and right loudspeakers can produce an auditory event centrally between two loudspeakers by playing back the same signal with both loudspeakers as described in Section 2.7.1. If it is desired that this auditory event should appear somewhere non-centrally between the loudspeakers, so-called *panning functions* are required. Depending on the desired phantom source position, the audio signal of each loudspeaker is modified by the panning tool with respect to amplitude and/or phase. The first case is known as *intensity panning* (or *intensity stereophony*), the second *time-of-arrival panning* (or *time-of-arrival stereophony*) (Blauert and Rabenstein, 2012; Gardner, 1998).

To move sources within the reproduced sound scene, panning functions can be applied to individual audio signals. Various panning functions implement intensity panning for two or more loudspeakers. Various optimisation criteria like localisation accuracy, size of the listening

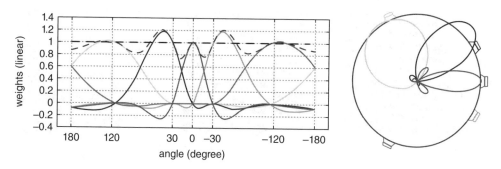

Figure 2.30 Robust panning function for a surround sound setup.

area, or colouration of sound can be taken into account. Among the existing panning functions, some have been extended for use in a three-dimensional arrangement (Batke and Keiler, 2010; Pulkki, 2001). A popular concept, which includes the case of pairwise panning for 2D as well as triplewise panning for 3D, is Vector Base Amplitude Panning (VBAP) (Pulkki, 2001). Another approach presented by Poletti (2007) optimises the panning functions with respect to the spatial arrangement of the loudspeakers. An example for a surround sound loudspeaker layout is given in Figure 2.30. The graph on the left side shows the loudspeaker weights for three of the five loudspeakers in the surround sound layout as shown on the right. Depending on the desired direction of the virtual source that is to be created, the corresponding weighted version of the source signal is assigned to the respective loudspeakers. The diagram on the right also shows these same weights as a polar plot.

The recording of natural auditory scenes representing a multitude of auditory events requires spatial microphone arrangements (Bartlett, 1991). In this case the characteristics of the microphones and their spatial placement determine the resulting panning functions.

To distribute stereo or surround content, it is common to broadcast one signal for each loudspeaker – the *channel-based* approach described in Section 2.3.6. If the loudspeaker layout is the same on the production and reproduction side the spatial impression may be recreated. This is discussed in more detail in Chapter 4, 'Platform-Independent Audio'.

Usually audio content consists of a mixture of signals representing, for example, an orchestra or a motion picture sound track. Consequently, a panning tool is a fundamental part of an audio mixing console, which in turn addresses a specific loudspeaker layout, for example, stereo or surround sound. This shows the interlinked relation of the targeted platform and the content creation process for a channel-based approach to audio.

2.7.2 Holophonic Systems

Wave Field Synthesis

Following the initial concept of the acoustic curtain as published by Snow (1953), Wave Field Synthesis (WFS) was developed by researchers at the Technical University Delft in the

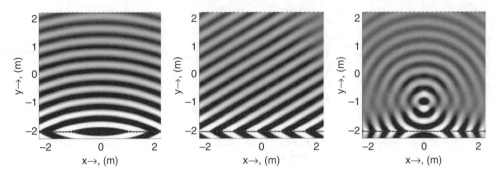

Figure 2.31   The three major source types created by typical WFS systems, (point sources, plane waves and focused sources). The wave field is created by the superposition of the array of loudspeaker sources shown at the bottom of each plot.

early 1990s (Berkhout et al., 1993). Subsequently there has been much work done in this area, an overview of which (along with the fundamental theory) can be found in Spors et al. (2008). The basic concept is to recreate a sound field in a listening room by utilising a large number of loudspeakers arranged in a horizontal array. The aim of a WFS system is to create a physically-correct sound field so as to evoke an accurate impression of the intended sound scene. Although it is possible to create more complex sources in WFS, current wave field synthesis systems are able to create principally three different types of sound sources[2] (see Figure 2.31).

- Point sources: the associated auditory event is perceived at the same position for the whole listening area.
- Plane wave: the associated auditory event is perceived coming from the same direction for the whole listening area.
- Focused source: the associated auditory event is perceived as coming from a position within the listening area.

A traditional WFS system utilises a very dense array of loudspeakers given that the loud-speaker spacing defines the upper limit up to which a sound field can be synthesised correctly. If larger loudspeaker spacings are used the reproduced sound field will exhibit unwanted contributions propagating in unintended directions known as spatial aliasing arte-facts (Spors and Rabenstein, 2006). Typical implementations therefore employ a loudspeaker spacing of approximately 10–20 cm. More recently perceptual optimisations have however been applied to reduce the number of loudspeakers required while keeping the proper-ties of WFS in terms of different source types (Melchior et al., 2011). Even though WFS typically works in horizontal layers only, extensions have been made to combine multiple layers of loudspeakers and more arbitrary layouts (Corteel et al., 2011). Since the calculation of the loudspeaker signals for a WFS system is done based on detailed prior knowledge of the loudspeaker layout, delivery of content traditionally does not use the loudspeaker signals directly but rather a description of the scene. This information consists of the necessary source audio signals and additional metadata describing the desired sound field to be recreated. The loudspeaker signals are then created on the receiver/reproduction side.

Higher Order Ambisonics

In the 1970s Michael Gerzon and others developed a series of recording and replay techniques that are commonly termed as *Ambisonics* technology (Gerzon, 1973), including the Soundfield® microphone discussed in Section 2.3.2. The mathematical foundation of this technology is the spherical Fourier transform that represents a desired sound field using a number of coefficients (Williams, 1999; Rafaely, 2005). The order of this representation specifies the spatial resolution of the sound field being described. The initial Ambisonics approach of the 1970s uses first-order signals only, more recently however research on *Higher Order Ambisonics* (HOA) representations used for the recording for example, Meyer and Elko (2002) Teutsch (2007) and reconstruction for example, Daniel (2000), Malham (2003) and Zotter (2009) of sound fields has caused a revival of this technology.

A sound source is first encoded based on metadata describing the desired spatial layout and the source signals. The resulting representation consists of the coefficients of the spherical Fourier transform (also often referred to as Fourier-Bessel-Series; see Daniel and Moreau, 2003) that is approximating the desired sound field using a certain order. The higher the order, the higher is the spatial resolution of the sound field. The coefficients are time-varying since the sound field changes over time and technically behave like audio signals that are transmitted as usual. However, on the reproduction side, the coefficients need to be decoded to generate loudspeaker signals. The decoding process depends on the loudspeaker positions and may also incorporate specific panning functions, if certain optimal loudspeaker geometries cannot be physically realised.

2.7.3   Binaural Systems

Binaural systems are a group of systems for reproducing sound over headphones. The aim of such a system is to create all the necessary perceptual cues required for the listener to perceive an auditory event at a location outside the head (Batke et al., 2011), (Breebaart et al., 2006). Headphones are often used for mobile devices, which has led to an increased interest in binaural techniques for personal spatial audio applications. Headphones provide two channels, therefore they are mostly used for playback of stereo content. Stereo produces a sound scene that spans from the left to the right loudspeaker, also referred to as a stereo image. For headphone listening, this stereo image can appear inside the head, between the ears, a phenomenon termed *in-head localisation*, and is an unnatural listening situation (Blauert, 1997). The reason for this problem is the missing acoustic transfer path information that is added by the properties of the listener's head, the so-called *head related transfer function* (HRTF). If such HRTFs relating to a pair of loudspeakers are measured carefully, a stereo signal can be processed with them. Headphone playback can then lead to a perception of virtual loudspeakers that appear as a stereo image outside the head again. Of course this approach, known as the binaural synthesis of loudspeakers, is not just limited to two loudspeakers; arbitrary numbers are also possible. If loudspeakers positioned away from the listener are used instead of headphones, this technology is also referred to as *transaural audio* (Bauck and Cooper, 1996).

2.7.4 Hybrid Systems

Due to the limitations of available reproduction techniques, there has been some research given to the hybridisation of several techniques to maximise the effectiveness of the overall system by exploiting the relative strengths of multiple systems. One such system is the *Optimised Phantom Source Imaging in Wavefield Synthesis* (OPSI) (Wittek, 2007). The OPSI method combines stereophonic and WFS playback: WFS is used to generate virtual sources below the spatial aliasing frequency, and a pair of loudspeakers creates virtual sources above this frequency. As their individual auditory event directions match, they are perceived as one common source.

Another hybrid system is the Binaural Sky that combines WFS, binaural synthesis and transaural audio (Menzel et al., 2005). Stable localisation of virtual sources is achieved for listeners even with head movement. WFS focused sources are used in this case for transaural signal reproduction. The position of the focused sources is kept constant relative to the ears of the listener for every head direction by means of a head tracking system.

2.7.5 Challenges for the Future

There are two drivers that provide key challenges for TV audio broadcast. First, 'TV' broadcast has to a large extent moved away from the living room. Increasingly TV is being viewed on mobile phones, tablets, gaming consoles and computers, each of which brings audio reproduction differing from TV broadcast. As a consequence channel-based production, based on the reproduction system's capabilities, is becoming less useful. Programming mixed for 5.1 surround may be downmixed for two channel stereo reproduction but the stereo downmix may not be ideal for headphone reproduction on a mobile device. Second, and related to the variety of devices used to view programming, TV is becoming more interactive. The growth of 'smart TV' sales and second screen usage, and the use of gaming consoles to view programming is inevitably leading to a requirement for increased interactivity. This increased interactivity is likely to impact on audio for TV broadcast and already in the UK the BBC have carried out live experiments allowing the viewer to take some control over their audio during live broadcasts from tennis championships at Wimbledon (Brun, 2011a). Further viewer choices in audio mixes have been suggested for accessibility purposes in creating sound mixes for hearing impaired viewers (Shirley and Kendrick, 2006). Increasing demand for user choice and interaction is better served by the choice of an object-based, rather than a channel-based audio paradigm. This approach is detailed further in Chapter 4, 'Platform-Independent Audio'.

2.8 Use of Archive Material

Although the focus of this book is on present and future technology for audio-visual content capture, production, distribution and consumption, it is worth briefly considering one source of content that will undoubtedly play a role in the future, even though it has already been captured: archive material. Audio-visual archives present a number of challenges, including how best to preserve the actual medium on which the content is stored, and how to transfer it to new formats as older ones become obsolete. The fundamental challenge is to preserve the original audio-visual signal (Wright, 2011), with other challenges including how to remove artefacts such as scratches, dirt and drop-out and how to index and search the content (Teruggi, 2009). However, the aspect of archive material most relevant to this book is how the content itself should be made fit-for-purpose in the context of a media landscape using standards that are significantly more advanced than when the content was originally made. The following sections consider how legacy video and audio formats are treated in the context of today's distribution formats.

2.8.1   Video Format Conversion

A principle often followed by archivists is to focus on content *preservation* (Wright, 2011), that is, to keep the content in its original form as faithfully as possible, even if the storage medium is changed. This ensures that the content can always be made available in any new standard without losses from multiple stages of format conversion, and with the latest (and presumably best) conversion technology being applied. For example, standard-definition content would not automatically be up-converted to HD resolution once standard-definition production had ceased; instead items are up-converted as required. The quality obtainable when up-converting content varies significantly depending on the approach taken. For example, the simplest approach to converting a standard-definition 576-line interlaced signal to an HD-format 1080-line signal would be to perform linear interpolation between lines of each field in order to generate new lines at the required position. This introduces aliasing and loss of vertical resolution as it makes no use of the samples from intermediate vertical locations available in the adjacent fields. A more sophisticated approach is to use a multifield filter, which may even incorporate motion adaptation or compensation (Thomas, 1988). These kinds of techniques are constantly improving, so it is best to convert the content when it is needed, using the latest technology. Up-converted content also needs more storage capacity.

There are exceptions to this principle, however, when the format is particularly hard to decode well. For example, the BBC converted the composite PAL video tapes in its archive (held on D3 format tape) to a component digital format at standard definition. This required decoding the PAL signal, which is hard to do well due to the way in which the chrominance and luminance spectra overlap. The conversion was carried out using the BBC-designed 'Transform PAL decoder' (Easterbrook, 2002). In addition to being an excellent PAL decoder in its own right, it is what is known as a 'complementary' PAL decoder, meaning that the filters used for separating luminance and chrominance are complementary, with their combined response adding up to unity. This means that if the recovered chrominance signals were to be re-modulated onto a PAL subcarrier in the correct phase, and added to the decoded luminance, then the original PAL signal would be regenerated. This leaves scope for future generations of video engineers to improve on the PAL decoding, should they be able to do so!

Occasionally, the original version of some material can be lost, and the only version available is one that has been processed or converted in an undesirable way. In such situations, knowledge of the processing that the material has undergone can be crucial in allowing the original material to be recovered. An interesting example of this happened when the BBC realised that the original recordings of some episodes of the science fiction series *Dr. Who* had been lost. However, a version was found that had been converted from its original 50 Hz version to 60 Hz for distribution in the US. The standards converter used had been designed by the BBC, and documentation existed on the interpolation method used when changing the field rate. With this knowledge, it was possible to design a filter to reverse the original standards conversion process and obtain a much higher-quality copy of the programme than would otherwise have been possible. This is one example of the importance of retaining metadata about how content has been processed: it may turn out in the future to be useful in ways that were not envisaged at the time that it was produced.

2.8.2   Audio Format Conversion

Like video, audio formats evolve over time. The intrinsic quality of the audio expected today is higher than in the past, with lower noise and higher bandwidth. Signal processing techniques exist for improving the subjective quality of audio, such as noise reduction. There have been some studies into bandwidth extension methods for 'inventing' higher frequencies based on the content at lower frequencies, such as Arora et al. (2006), although at present the approaches are limited mainly to speech (Bauer and Fingscheidt, 2009). These are analogous to the up-conversion of video from SD to HD resolution, and following the principle discussed above, if they were to be used at all they would be applied when taking content out of the archive for broadcast rather than to process content that was then stored.

Early TV and radio programmes were produced in mono; TV and radio is now mostly stereo, with 5.1 being the preferred audio format for HDTV. It may be desirable to convert audio to a higher number of channels when broadcasting in a modern format than it was originally intended for. Commercial devices exist that attempt to convert from stereo to 5.1 (Marston, 2011). These make no particular assumptions about how the content was captured. However, where there is some knowledge of the microphone configuration used, it is possible to produce a more faithful 5.1 upmix from stereo (Brun, 2011b). This again serves to highlight the importance of retaining metadata relating to content capture and processing in order to be able to make best use of it in future generations of distribution and reproduction systems.

2.8.3   Challenges for the Future

One challenge that audio-visual archives face is obtaining and storing the metadata associated with the archived content. The examples above have shown how valuable 'technical' metadata can be in allowing content to be re-purposed for future delivery formats. More conventional types of metadata (such as names of performers, locations used, and details of incidental music) are key to enabling new ways for archives to be searched, as well as to obtaining rights clearance to use material in the future. While some of this information may already be archived, it is not always stored in an easy-to-access way, and in some cases may be completely absent. It can be difficult to get programme-makers to record metadata for which they have no immediate use, as this can be seen as a non-productive use of their limited time. Automated methods of logging metadata, and standardised ways of storing it in an easily-searchable manner, are among the challenges currently being faced.

2.9 Concept of Format-Agnostic Media

2.9.1 Limitations of Current Production and Delivery Approaches

The preceding sections of this chapter have set out the current state of technology and operational practices used in today's media production industry. These have evolved over the years, often driven by particular delivery formats and reproduction systems, such as the evolution of TV systems described in Section 2.1. These systems have until recently been based on TV-type displays in the domestic living room, and larger screens in cinemas. This dichotomy of screen sizes already presents problems in optimising the reproduction of content, triggering developments such as pan and scan (described in Section 2.2.4). For audio, content is typically produced either in stereo or 5.1, although there is sometimes a requirement to distribute content in 5.1 format that was captured in stereo (Section 2.8.2).

In recent years, the range of screen sizes on which content is being consumed has greatly increased. Large-screen displays are becoming more common, while the rise of viewing on tablet devices means that a predicted 10 per cent of all TV and video consumption in US households will be on tablets by 2017 (Niemeyer, 2013). Accompanying this diversification in screen sizes is a similar broadening of audio consumption devices, with sophisticated cinema audio systems at one end and small speakers or headphones for mobile devices at the other. Even listeners that have a system capable of reproducing 5.1 surround sound are unlikely to have their set of speakers correctly positioned so as to experience the audio in the way that was intended.

The degree to which users interact with the content they are consuming is also increasing significantly, driven by the rise in the use of video games and online applications including social networking as well as general web browsing. Such interactivity has yet to reach mainstream video. Navigable 360-degree static panoramas initially became available using technology such as Quicktime VR (Chen, 1995). This used spherical images, usually created by stitching together multiple stills, and included the option of clickable 'hot spots'. Such static image panoramas are now in common use through applications such as Google Street View, where images are captured with specially-designed clusters of cameras. Some niche applications are starting to offer video panoramas, although the quality of the video is at present constrained by the bit rate needed to compress the whole panorama, and by the low-cost lightweight cameras that are typically used. Further discussion on video panoramas can be found in Chapter 3, Section 3.2.

Current production and delivery technology is ill-equipped to provide truly flexible content that can be adapted to the wide range of screen sizes found today, while also providing options for a significant level of interactivity at a visual quality matching that of conventional video and film production. Content tends to be produced for a given target format, for example, with images being framed at capture time for a particular aspect ratio, and audio being mixed during recording or editing to produce a particular format such as stereo or 5.1.

2.9.2 A New Approach: Format-Agnostic Media

The following chapters in this book present an approach to freeing up audio-visual content production and delivery to allow the material to be used in the optimum way to provide the desired user experience. We use the term *format-agnostic media* to refer to content that has been captured in a way to support such a wide range of delivery and consumption methods.

The key to this approach is to delay the final rendering of the material to as late a stage in the delivery chain as possible, with the content being retained in its original captured form during its passage through the production and delivery process. By 'rendering', we mean the process of creating a video image and set of audio channels to drive the user's audio-visual reproduction devices: this could happen in the display device itself (e.g., a PC) or during the final stages of the production process, where a version for a particular kind of device is automatically generated.

This requires that metadata describing the content elements is captured and delivered, so that the rendering and display processes know how to interpret the data they receive. In current systems, such metadata is usually implicit, or very basic in nature, and is usually specified in terms of the kind of device that the content is intended to be presented on. For example, audio tracks will be labelled 'left' and 'right' referring to the stereo speakers to which they are to be delivered, and video will have a fixed image resolution and frame rate. In the format-agnostic approach, rather than down-mixing audio to a stereo or 5.1 configuration, the audio signals would be sent in a form related to the scene they were describing rather than the envisaged reproduction system. This would allow each individual reproduction device to render the required version of the content without being constrained by assumptions about how other devices might need it.

It is interesting to note that the approach of maintaining content in its original form is already well-established in the field of content archiving, discussed in Section 2.8. This section also showed that knowledge of how the content was captured can help with re-purposing and format conversion. In that respect, the proposed format-agnostic approach for content creation and delivery can be seen as a logical extension of principles already established in audio-visual archives, where future viewing will be on devices not yet invented.

2.9.3   Metadata for Enabling Rich User Interaction

In addition to allowing optimum content formatting to suit a target device, another key problem that we would like this new paradigm to address is enabling increased user interactivity and choice when consuming the content. For some kinds of programming, such as drama, allowing viewers to choose where to look or which video feed to select could interfere with the story-telling nature of the medium, although it may also open up new ways of presenting narrative. For other kinds of content, such as a concert or sporting event, there may be many parts of the scene containing areas of interest, and the choice of where to look may depend on the preferences of individual viewers. For example, an amateur guitarist may want to closely watch the guitarist in a band, while a drummer may be more interested in the percussion section.

In order to offer this kind of interactivity, we propose to extend the audio-visual information captured so that a selection of possible areas of interest can be presented to the viewer. A key element of this is the use of panoramic video, so that the whole of the area of interest is potentially visible, in the same way as it would be for a member of a live audience in a prime seat. This video should be used to support the more traditional kinds of broadcast camera, which can obtain close-ups of key areas and take shots from a range of locations. By ensuring that the captured metadata includes the position, orientation and field-of-view of the panoramic and broadcast cameras it would be possible to allow users to select between camera views, using the panoramic image as a guide to which areas have close-ups available.

Although offering free navigation around a panoramic video scene gives significant freedom to the viewer, it could become overwhelming if they have to constantly interact to select areas of interest. Therefore we propose that the scene description should include metadata describing regions-of-interest selected at the production side. These can be thought of as 'virtual' cameras, viewing particular parts of the panoramic scene. They could be generated manually, by a 'virtual' camera operator, or make use of some (semi-)automated content analysis and tracking technology, for example, to follow the players and ball in a football match.

The metadata describing these regions of interest could also be used to produce a conventional (non-interactive) version of the programme, with shot framing or region selection targeted for a particular screen size or audience interest. For example, a region-of-interest could be framed tightly when producing content for delivery to mobile devices, and more loosely (wide shot) for delivery to larger screen devices. Shots featuring close-ups of a particular team could be produced for delivery to the TV channel run by that football club.

2.9.4 A Format-Agnostic Media Production and Delivery System

An example of a system incorporating the main components needed to implement a format-agnostic system for the production and delivery of media is shown in Figure 2.32.

The key elements of the system and chapters of this book in which they are described follows.

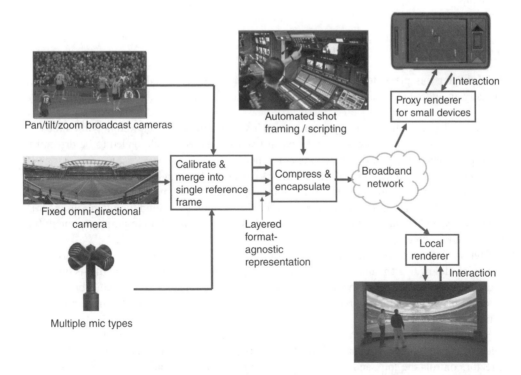

Figure 2.32 Diagram of the proposed format-agnostic system. Images reproduced by permission of The Football Association Premier League Ltd.

The scene is captured by a combination of conventional broadcast cameras and one or more fixed ultra-high-resolution panoramic cameras (see Chapter 3, 'Video Acquisition'). A selection of microphones captures the sound field (or 'ambience') plus localised sound from individual microphones covering particular areas (Chapter 4, 'Platform-Independent Audio'). Each of these sources is accompanied by metadata, including its position and orientation in a common reference frame. These are combined into a *layered format-agnostic representation*, reflecting the fact that parts of the scene (both video and audio) may be available in several separate layers (e.g., viewed by several cameras).

The content is analysed using semi-automatic content annotation tools (Chapter 5), driving a 'virtual director' (Chapter 6), which produces automated shot framing or 'scripting' decisions that offer a range of different shots to the viewer.

The layered scene representation is then compressed and encapsulated in a manner that allows end devices or intermediate network elements to dynamically access different subsets of the content (Chapter 7, 'Scalable Delivery of Navigable and Ultra High Resolution Video'), and finally viewed on an interactive display (Chapter 8, 'Interactive Rendering').

Such an approach offers the possibility of creating a flexible media production and delivery system, exploiting the features of a wide range of consumption devices. It could also add value to a conventional production and delivery system, by providing more efficient ways of producing conventional programming, particularly where several versions (tailored for particular kinds of device or viewer interest) need to be produced. Various deployment scenarios that reflect these options are discussed in Chapter 9, 'Application Scenarios and Deployment Domains'.

2.10   Conclusion

This chapter has presented some of the key elements used in today's media production industry, including video and audio acquisition, editing, and post-production, coding, delivery and display.

Section 2.1 gave an historical perspective on TV production, and showed how the industry has become used to a slow evolution between fixed standards, which tended to be driven by the capabilities of the viewers' receiving equipment.

Section 2.2 explained some of the details of current video standards, and how these have been influenced by characteristics of now-obsolete display systems, giving us features such as interlace and gamma. It gave an overview of the technology behind the range of cameras in use today, and explained how current approaches make it difficult to re-purpose content for very different screen sizes.

Section 2.3 discussed the current technology and production practices for audio, with examples of how live and non-live recordings are made and the kind of content that is typically captured. It was shown that typical productions tend to produce a single audio output to a given format (e.g., stereo), although the captured signals could potentially support a wider variety of formats.

Section 2.4 reviewed current practices in live programme production, and examined the roles of production staff in a typical sports outside broadcast. It was explained how the director controls the look and feel of the programme, and determines how the various audio and video feeds are used to 'tell a story' from the event, based on assumptions about the viewers' preferences.

The coding and delivery systems currently in common use were summarised in Section 2.5. Traditional managed networks such as terrestrial or satellite broadcasting are now being joined by IPTV systems being run on managed IP networks supplied by cable operators, allowing guaranteed throughputs for AV content. Unmanaged IP networks, with AV content being sent over the wider internet, are particularly suitable for video-on-demand services, supplanting the film rental market that was previously dominated by physical media such as DVDs, and allowing 'catch-up' services to give PVR-like (personal video recorder) functionality. Their use of content delivery networks or P2P techniques was summarised.

Display technology has been a key driver for video standards and formats (as mentioned in Section 2.2), so Section 2.6 presented a review of current display technologies and the impacts that these have had on broadcasters.

Audio reproduction technologies were summarised in Section 2.7, starting with a review of well-established systems to deliver to particular speaker layouts such as stereo and 5.1. Holophonic systems (that attempt to reproduce the sound field rather than merely feeding a given speaker layout) were then discussed, including wave field synthesis, higher-order Ambisonics, and binaural systems. Hybrid systems built from combinations of these were also discussed. It was explained that holophonic systems have a capacity to deliver immersive sound with better directionality and support for a height dimension, although these capabilities are currently only being exploited in some specialist or high-end systems.

The use of archive material was briefly discussed in Section 2.8, focusing on the challenges of re-purposing content recorded in older audio or video standards. The usefulness of metadata describing the way in which the material was recorded or processed was highlighted through several examples.

Section 2.9 proposed an approach to producing content fit for the emerging media landscape: *format-agnostic media*. An example of a production and delivery system according to this idea was presented, in which multiple sets of audio and video signals representing the scene at a range of resolutions and viewpoints are captured and retained as far through the production and delivery pipeline as possible. By delaying the final choice and formatting of audio-visual content until it reaches the viewer's device, the most appropriate choices (both in terms of content format and selection of material) can be made, tailored to that particular viewer and their reproduction equipment. The following chapters of this book look in detail at the functioning of each of the system components in this proposed format-agnostic system.

Notes

1. Waterloo Road, television series, Shed Media, Broadcast by BBC1, 2005–2012.
2. Even if these source types are not completely related to the physical sources after which they are named, the common names will be used here to provide reference to the literature.

References

3GPP TS 26.234 (2010) *Transparent End-to-end Packet Switched Streaming Service (PSS); Protocols and Codecs*, accessed 12 August 2013 at: http://www.3gpp.org/ftp/Specs/html-info/26234.htm

AES (2008) *AES10-2008: AES Recommended Practice for Digital Audio Engineering—Serial Multichannel Audio Digital Interface (MADI) (Revision of AES10-1991)*. New York: Audio Engineering Society.

AES (2009a) *AES3-2009: AES Standard for Digital Audio – Digital Input-Output Interfacing Serial Transmission Format for two-Channel Linearly-Represented Digital Audio Data – Part 1: Audio Content*. New York: Audio Engineering Society.

AES (2009b) *AES3-4-2009: AES Standard for Digital Audio – Digital Input-output Interfacing – Serial Transmission Format for Two-channel Linearly-represented Digital Audio Data – Part 4: Physical and Electrical*. New York: Audio Engineering Society.

Arora, M., Lee, J. and Park, S. (2006) 'High Quality Blind Bandwidth Extension of Audio for Portable Player Applications'. 120th Audio Engineering Society Convention, Paris, 20–23 May.

Auro (2013) Home page, accessed 12 August at: http://auro-3d.com

Bartlett, B. (1991) *Stereo Microphone Techniques*. Boston, MA: Focal Press.

Batke, J.M. and Keiler, F. (2010) 'Investigation of Robust Panning Functions for 3D Loudspeaker Setups'. 128th Convention of the Audio Engineering Society, London, 22–25 May.

Batke, J.M., Spille, J., Kropp, H., Abeling, S., Shirley, B. and Oldfield, R. (2011) 'Spatial Audio Processing for Interactive TV Services', 130th Convention of the Audio Engineering Society, London, May 13–16.

Bauck, J. and Cooper, D. (1996) 'Generalized Transaural Stereo and Applications'. *Journal of the Audio Engineering Society* 44, 683–705.

Bauer, P. and Fingscheidt, T. (2009) 'A Statistical Framework for Artificial Bandwidth Extension Exploiting Speech Waveform and Phonetic Transcription'. 17th European Signal Processing Conference (EUSIPCO 2009), Glasgow, 24–28 August.

Bayer, B. (1976) 'Color Imaging Array'. US Patent 3971065.

Berger, J.L. (2012) *Aspect Ratios*, accessed 12 August 2013 at: http://www.widescreen.org/aspect_ratios.shtml

Berkhout, A.J., de Vries, D. and Vogel, P. (1993) 'Acoustic Control by Wave Field Synthesis', *Journal Acoustic Society of America* 93(5), 2764–2778.

Blauert, J. (1997) *Spatial Hearing*. Cambridge, MA: MIT Press.

Blauert, J. and Rabenstein, R. (2012) 'Providing Surround Sound With Loudspeakers: A Synopsis of Current Methods', *Archives of Acoustics* 37(1), 5–18.

Bose, R.C. and Ray-Chaudhuri, D.K. (1960) 'On A Class of Error Correcting Binary Group Codes', *Information and Control* 3(1): 68–79.

Breebaart, J., Herre, J., Villemoes, L., Jin, C., Kjrling, K., Plogsties, J. and Koppens, J. (2006) 'Multi-Channel Goes Mobile: MPEG Surround Binaural Rendering'. 29th International Conference, Audio Engineering Society, Seoul, Korea, 2–4 September.

Brightcove, Inc. (2010) 'Home page', accessed 12 August 2013 at: http://www.brightcove.com

Brun, R. (2011a) 'NetMix: Create Your Own Sound Balance from Centre Court', BBC Internet Blog, 28 June, accessed 12 August 2013 at: http://www.bbc.co.uk/blogs/bbcinternet/2011/06/netmix_create_your_own_sound_b.html

Brun, R. (2011b) 'The Festival of Nine Lessons and Carols in surround sound', BBC Radio 3 Internet Blog, 21 December, accessed 12 August 2013 at:
http://www.bbc.co.uk/blogs/radio3/2011/12/the-festival-of-nine-lessons-and-carols-in-surround-sound.shtml

Chamaret, C. and Le Meur, O. (2008) 'Attention-Based Video Reframing: Validation using Eye-Tracking'. 19th International Conference on Pattern Recognition (ICPR), 8–11 December 2008, pp. 1–4.

Chen, S.E. (1995) 'QuickTime VR: An Image-Based Approach to Virtual Environment Navigation'. SIGGRAPH '95 22nd International ACM Conference on Computer Graphics and Interactive Techniques, Los Angeles, 6–11 August, pp. 29–38.

Cohen, B. (2003) 'Incentives Build Robustness in BitTorrent', accessed 12 August 2013 at:
http://www.ittc.ku.edu/~niehaus/classes/750-s06/documents/BT-description.pdf

Corteel, E., Rohr, L., Falourd, X., NGuyen, K. and Lissek, H. (2011) 'A Practical Formulation of 3 Dimensional Sound Reproduction using Wave Field Synthesis'. In International Conference on Spatial Audio, Berlin, Germany, 10–13 November.

Daniel, J. (2000) 'Représentation de champs acoustiques, application à la transmission et à la reproduction descènes sonores complexes dans un contexte multimédia', unpublished PhD Thesis, Université Paris 6.

Daniel, J., Nicol, R. and Moreau, S. (2003) 'Further Investigations of High Order Ambisonics and Wavefield Synthesis for Holophonic Sound Imaging'. 114th Convention of the Audio Engineering Society, Amsterdam, 22–25 March.

Dayan, D. and Katz, E. (1992) *Media Events: The Live Broadcasting of History*. Cambridge, MA: Harvard University Press.

Dooley, W.L., and Streicher, R.D. (1982) 'MS Stereo: A Powerful Technique for Working in Stereo', *Journal of the Audio Engineering Society* 30(10): 707–718.

Douglis, F. and Kaashoek, M.F. (2001) 'Scalable Internet Services', *IEEE Internet Computing* 5(4), 36–37.

Drewery, J.O. and Salmon, R.A. (2004) 'Tests of Visual Acuity to Determine the Resolution Required of a Television Transmission System', BBC R&D White Paper WHP092, January 2004, accessed 12 August 2013 at: http://www.bbc.co.uk/rd/publications/whitepaper092

DVB (2010) Home page of the Digital Video Broadcasting Project, accessed 12 August 2013 at: http://www.dvb.org/

DVB-TM-C2 (2008) DVB TM-C2 0044rev2. Second Generation Transmission Technologies for Cable Networks – Call for Technologies.

DVB-C2 (2010) 'DVB-C2 Factsheet', accessed 27 August 2013 at:
http://www.dvb.org/technology/fact_sheets/DVB-C2_Factsheet.pdf

DVB-S2 (2010) 'DVB-S2 Factsheet', accessed 27 August 2013 at:
http://www.dvb.org/technology/fact_sheets/DVB-S2_Factsheet.pdf

DVB-T2 (2010) 'DVB-T2 Factsheet', accessed 27 August 2013 at:
http://www.dvb.org/technology/fact_sheets/DVB-T2_Factsheet.pdf

Easterbrook, J.E. (2002) 'Improvements Relating to Chrominance Selective Filtering for Use in Colour Video Signal Decoding'. UK Patent Application GB2365247.

Engström, A., Juhlin, O., Perry, M. and Broth, M. (2010) 'Temporal Hybridity: Mixing Footage with Instant Replay in Real Time.' 28th SIGCHI Conference on Human Factors in Computing Systems (CHI '10), Atlanta, 10–15 April.

ETSI (2005) 'ETSI Technical Specification 101 154. Digital Video Broadcasting (DVB); Implementation Guidelines for the use of Video and Audio Coding in Broadcasting Applications based on the MPEG-2 Transport Stream', accessed 12 August 2013 at:
http://www.etsi.org/deliver/etsi_ts/101100_101199/101154/01.07.01_60/ts_101154v010701p.pdf

European Broadcasting Union (EBU) (2007) 'EBU Guidelines for Consumer Flat Panel Displays'. EBU Technical Specification 3321, 2007, accessed 12 August 2013 at: http://tech.ebu.ch/publications/tech3321

European Broadcasting Union (EBU) (2008) 'Safe Areas for 16:9 Television Production. EBU Recommendation R95-2008', accessed 12 August 2013 at: http://tech.ebu.ch/docs/r/r095.pdf

European Broadcasting Union (EBU) (2010) 'User requirements for Video Monitors in Television Production. EBU Technical Specification 3320 version 2, October 2010', accessed 12 August 2013 at: http://tech.ebu.ch/publications/tech3320

Gardner, W.G. (1998) 3-D Audio Using Loudspeakers. Dordrecht: Kluwer.

Gerzon, M.A. (1973) 'Periphony: With-Height Sound Reproduction'. Journal of the Audio Engineering Society 21(1), 2–10.

Gerzon, M.A. (1980) 'Practical Periphony: The Reproduction of Full-Sphere Sound', AES Preprint 1571.

Grundmann, M., Kwatra, V., Han, M. and Essa, I. (2010) Discontinuous Seam-Carving for Video Retargeting. IEEE Conference on Computer Vision and Pattern Recognition (CVPR), San Francisco, USA, 13–18 June.

Goodwin, C. (1996) 'Transparent Vision'. In: E. Ochs, E.A. Schegloff and S.A. Thompson, (Eds.) Interaction and Grammar. Cambridge: Cambridge University Press, 370–404.

Gruneau, R. (1989) 'Making Spectacle: A Case Study in Television Sports Production'. In L.A. Wenner (Ed.), Media, Sports and Society. Thousand Oaks, CA: Sage, 134–154.

Hamasaki, K. and Hiyama, K. (2003) 'Reproducing Spatial Impression with Multichannel Audio'. 24th AES International Conference on Multichannel Audio, Banff, Canada, 26–28 June.

Hamasaki, K., Hiyama, K. and Okumura, R. (2005) 'The 22.2 multichannel Sound System and its Application'. 118th Convention of the Audio Engineering Society. Barcelona, 28–31 May.

Harris, M. (2012) 'Focusing On Everything', IEEE Spectrum, May, accessed 12 August 2013 at:
http://spectrum.ieee.org/consumer-electronics/gadgets/lightfield-photography-revolutionizes-imaging/0

Hicks, C. and Reid, G. (1996) 'The Evolution of Broadband Noise Reduction Techniques', The 6th Regional Australian Convention of the Audio Engineering Society, Melbourne, Australia.

Holland, P. (2000) The Television Handbook, 2nd edn., London: Routledge

Holman, T. (2001) 'The Number of Loudspeaker Channels'. Audio Engineering Society Conference: 19th International Conference: Surround Sound-Techniques, Technology, and Perception, Schloss Elmau, Germany, 21–24 June.

Hubel, M. (2005) 'Foveon Technology and the Changing Landscape of Digital Camera Technology.' Keynote paper at the Thirteenth IS&T Color Imaging Conference, Scottsdale, AZ, accessed 12 August 2013 at:
http://www.foveon.com/files/CIC13_Hubel_Final.pdf

Hughey, M.J. (2001) 'Motion Sickness.' In Operational Medicine 2001 GMO Manual. California: Brookside Associates.

ISO (1997) ISO/IEC 13522-5:1997 Information Technology – Coding of Multimedia and Hypermedia Information – Part 5: Support for Base-level Interactive Applications. International Organisation for the Standardisation of International Electronics Communication, Geneva.

ITU (2006) 'ITU-R BS.775-2: Multichannel stereophonic sound system with and without accompanying picture'. International Telecommunications Union, accessed 27 August 2013 at: http://www.itu.int/rec/R-REC-BS.775/

Jolly, S., Armstrong, M. and Salmon, R. (2009) 'The Challenges of Three-Dimensional Television.' BBC R&D White Paper 173, January, accessed 12 August 2013 at: http://www.bbc.co.uk/rd/publications/whitepaper173

Liang, J., Kumar, R. and Ross, K.W. (2010) 'The Kazaa Overlay: A Measurement Study', *Computer Networks* (Special Issue).

Malham, D. (2003) 'Space in Music – Music in Space', unpublished PhD thesis, University of York.

Marston, D. (2011) 'Assessment of Stereo to Surround Upmixers for Broadcasting'. 130th Audio Engineering Society Convention, 13–16 May.

Melchior, F., Heusinger, U. and Liebetrau, J. (2011) 'Perceptual Evaluation of a Spatial Audio Algorithm Based on Wave Field Synthesis using a reduced Number Of Loudspeakers', 131st Convention of the Audio Engineering Society, New York, 20–23 October.

Menzel, D., Wittek, H., Theile, G. and Fastl, H. (2005) 'The Binaural Sky: A Virtual Headphone for Binaural Room Synthesis'. 1st Verband Deutscher Tonmeister Symposium, Hohenkammer, Germany, 31 October–2 November.

Meshkova, E., Riihijarvi, J., Petrova, M. and Mähönen, P. (2008) 'A Survey on Resource Discovery Mechanisms, Peer-to-Peer and Service Discovery Frameworks', *Computer Networks* 52(11), 2097–2128.

Meyer, J. and Elko, G. (2002) 'A Highly Scalable Spherical Microphone Array Based on an Orthonormal Decomposition of the Soundfield'. *IEEE International Conference on Acoustics, Speech, and Signal Processing*, Orlando, FL, 13–17 May.

Millerson, G. (1999) *Television Production*, 13th edn. Woburn, MA: Focal Press.

Montpetit, M.-J., Calhoun, H., Holtzman, H. and Grossman, D. (2010) 'Adding the Community to Channel Surfing: A New Approach to IPTV Channel Change'. *6th IEEE Consumer Communications and Networking Conference (CCNC)*, Las Vegas, NV, 11–13 January.

MPEG2TS (2000) ITU-T Rec. H.222.0 | ISO/IEC 13818-1:2000 Generic Coding of Moving Pictures and Associated Audio Information, Part 1: Systems, accessed 12 August 2013 at: http://www.itu.int/rec/T-REC-H.222.0-200002-S/en

MPEG-DASH (2010) ISO/IEC JTC1/SC29/WG11 N11338, Call for Proposals on HTTP Streaming of MPEG Media, April 2010, Dresden, Germany.

Niemeyer, B. (2013) 'Tablet Diffusion and Its Impact on Video Use – Forecasts and Recommendations'. Report by The Diffusion Group, March, accessed 12 August 2013 at: http://tdgresearch.com/report/tablet-diffusion-and-its-impact-on-video-use-forecasts-and-recommendations/

Nyquist, H. (1928) 'Certain Topics in Telegraph Transmission Theory.' *Transactions of the American Institute of Electrical Engineers* 42, 617–644.

Ooyala Inc. (2010) Home page, accessed 12 August 2013 at: http://www.ooyala.com/

Owens, J. (2007) *Television Sports Production*. New York: Elsevier

Pallis, G. and Vakali, A. (2006) 'Insight and Perspectives for Content Delivery Networks', *Communications of the ACM* 49(1), 101–106.

Pantos, R. and May, W. (2010) *HTTP Live Streaming IETF Internet Draft*, accessed 12 August 2013 at: http://tools.ietf.org/html/draft-pantos-http-live-streaming-05

Pathan, M. and Buyya, R. (2007) 'A Taxonomy and Survey of Content Delivery Networks', Technical Report, GRIDS-TR-2007-4, Grid Computing and Distributed Systems Laboratory, University of Melbourne, Australia, 12 February.

Perry, M., Juhlin, O., Esbjörnsson, M. and Engström, A. (2009) 'Lean Collaboration through Video Gestures: Co-ordinating the Production of Live Televised Sport'. *Proceedings of ACM CHI 2009*, Boston, MA, 4–9 April. 2279–2288.

Pizzi, S. (1984) 'Stereo Microphone Techniques for Broadcast.' 76th Convention of the Audio Engineering Society, New York, NY, 8–11 October.

Poletti, M. (2007) 'Robust Two-Dimensional Surround Sound Reproduction for Non-uniform Loudspeaker Layouts', *Journal of the Audio Engineering Society* 55(7/8), 598–610.

Pulkki, V. (2001) 'Spatial Sound Generation and Perception by Amplitude Panning Techniques', unpublished PhD thesis, Helsinki University of Technology.

Rafaely, B. (2005) 'Analysis and Design of Spherical Microphone Arrays'. *IEEE Transactions on Speech and Audio Processing* 13(1), 135–143.

Roberts, A. (1990) HDTV – A Chance to Enhance Television Colorimetry. BBC Research Department Report 1990/2, accessed 12 August 2013 at: http://www.bbc.co.uk/rd/publications/rdreport_1990_02.shtml

Roberts, A. (2002) 'The Film Look: It's Not Just Jerky Motion'. BBC R&D White Paper WHP053, December, accessed 12 August 2013 at: http://www.bbc.co.uk/rd/publications/whitepaper053.shtml

ScreenDigest (2010) 'The Global Transmission Market, A Screen Digest report for DVB', September.

Shirley, B. and Kendrick, P. (2006) 'The Clean Audio Project: Digital TV as Assistive Technology', *Technology and Disability* 18(1), 31–41.

Shirley, P., Ashikhmin, M. and Marschner, S. (2009) *Fundamentals of Computer Graphics*. Wellesley, MA: A.K. Peters.

SISLive (2011) 'OB 3 HD Production Units – Outside Broadcast', accessed 27 August 2013 at: http://www.sislive.tv/outside-broadcasts-ob3.php.

Snow, W. (1953) 'Basic Principles of Stereophonic Sound'. *Journal of the Society of Motion Picture and Television Engineers* 61, 587–589.

Spors, S., and Rabenstein, R. (2006) 'Spatial Aliasing Artifacts Produced by Linear and Circular Loudspeaker Arrays used for Wave Field Synthesis.' 120th Convention of the Audio Engineering Society, Paris, France, 20–24 May.

Spors, S., Rabenstein, R. and Ahrens, J. (2008) 'The Theory of Wave Field Synthesis Revisited', 124th Convention of the Audio Engineering Society, Amsterdam, Netherlands, 17–20 May.

Steppuhn, D. (2001) 'Begegnung anderer Art – Neue Audio-DVDs "2+2+2" bei MDG'. *nmz – neue musikzeitung* 2, 17.

Stockhammer, T. (2011) 'Dynamic Adaptive Streaming over HTTP – Standards and Design Principles'. MMSys'11, San Jose, CA, 23–25 February.

Sullivan, G.J., Ohm, J-R., Han, W-J. and Wiegand, T. (2012) 'Overview of the High Efficiency Video Coding (HEVC) Standard', *IEEE Transactions on Circuits and Systems for Video Technology* 22(12), 1649–1668.

Taylor, S. (1998) 'CCD and CMOS Imaging Array Technologies: Technology Review'. Xerox Research Centre Europe Technical Report EPC-1998-106., accessed 12 August 2013 at: http://research.microsoft.com/pubs/80353/ccd.pdf

Teruggi, D. (2009) 'Presto-PrestoSpace-PrestoPRIME'. *International Preservation News* 47(May), 8–12.

Teutsch, H. (2007) *Modal Array Signal Processing: Principles and Applications of Acoustic Wavefield Decomposition*. Berlin: Springer-Verlag.

Thomas, G.A. (1988) 'A Comparison of Motion-Compensated Interlace-to-Progressive Conversion Methods'. *Signal Processing: Image Communication* 12(3), 209–229.

Thomas, G.A. (2011) 'Virtual Graphics for Broadcast Production'. In M. Zelkowitz (Ed.) *Advances in Computers*, vol 82, 165–216.

Ward, P. (2000) *Digital Video Camerawork*. Woburn, MA: Focal Press.

Wiegand, T., Sullivan, G.J., Bjontegaard, G. and Luthra, A. (2003) 'Overview of the H.264/AVC Video Coding Standard', *IEEE Transactions on Circuits and Systems for Video Technology* 13(7), 560–576.

Wikipedia (2010) 'Gnutella', accessed 27 August 2013 at: http://en.wikipedia.org/wiki/Gnutella

Williams, E.G. (1999) *Fourier Acoustics: Sound Radiation and Nearfield Acoustical Holography*. San Diego, CA: Academic Press

Wittek, H. (2007) 'Perceptual Differences between Wavefield Synthesis and Stereophony', unpublished PhD thesis, University of Surrey.

Wood, D. (2007) 'The Development of HDTV in Europe'. EBU Technical Review, July.

Wood, D. and Baron, S. (2005) 'Rec. 601—The Origins of the 4:2:2 DTV Standard'. EBU Technical Review, October, accessed 12 August 2013 at: http://www.ebu.ch/en/technical/trev/trev_304-rec601_wood.pdf

Wright, R. (2011) 'The Real McCoy: What Audiovisual Collections Preserve'. BBC R&D White Paper 211, November, accessed 12 August 2013 at: http://www.bbc.co.uk/rd/publications/whitepaper211

Zotter, F., Pomberger, H. and Frank, M. (2009) 'An Alternative Ambisonics Formulation: Modal Source Strength Matching and the Effects of Spatial Aliasing'. 126th Convention of the Audio Engineering Society, Munich, Germany, 7–10 May.

3

Video Acquisition

Oliver Schreer[1], Ingo Feldmann[1], Richard Salmon[2], Johannes Steurer[3] and Graham Thomas[2]

[1]*Fraunhofer Heinrich-Hertz-Institute, Berlin, Germany*
[2]*BBC Research & Development, London, United Kingdom*
[3]*Arnold & Richter Cine Technik, Munich, Germany*

3.1 Introduction

This chapter presents various aspects of video acquisition that are relevant with regards to the concept of format-agnostic media production. As outlined in Chapter 2, the aim of future media production is to offer the largest possible set of video sensors with the most suitable acquisition capability along various dimensions of a video signal. Most importantly:

- spatial resolution ranging from HD (high definition), via 8k resolution towards ultra-high resolution panoramic video;
- various frame rates ranging from 24/25 frames per second (fps) up to 400 fps; and
- dynamic range from 8 bit per channel up to 16 bit per channel.

Depending on creative aspects at the production side or individual desires by the end user, dedicated facets of visual representation of a scene might be of interest. On the production side, the aim is to capture images with a contrast range chosen to suit the important features in the scene, which can be particularly challenging in scenes containing a large contrast range. In the case of conventional broadcast cameras, the iris is adjusted continuously by the camera man according to the content in the part of the scene being captured. In panoramic video, high dynamic range acquisition will offer the capability to render parts of the scene differently to achieve the best possible contrast. As image resolution increases, it becomes important to increase the temporal resolution (frame rate) to prevent motion blur from negating the benefits of increased spatial resolution. Furthermore, at the end-user side, live video content might be

Media Production, Delivery and Interaction for Platform Independent Systems: Format-Agnostic Media, First Edition. Edited by Oliver Schreer, Jean-François Macq, Omar Aziz Niamut, Javier Ruiz-Hidalgo, Ben Shirley, Georg Thallinger and Graham Thomas.
© 2014 John Wiley & Sons, Ltd. Published 2014 by John Wiley & Sons, Ltd.

replayed in slow motion and only a specific part of the scene may be of interest. This will require interactive navigation in high resolution panoramas and acquisition at higher frame rates to offer acceptable visual quality in slow motion replay. The use of a higher frame rate to provide more natural-looking image reproduction will be hugely beneficial to the immersivity of future television systems. In capture, it would assist in the re-purposing of content for different platforms, enabling the images at the input to a production process to be free of motion blur.

A section is dedicated to each of the dimensions listed above, where the fundamentals and specifics regarding format-agnostic media production are discussed. The aim is to give an insight in future video acquisition technology that improves media experience and supports new forms of user interaction with the content. A flexible concept of media production consists of intelligent mixing and use of various video sources. Hence, a section is dedicated to the use of conventional video content to enhance panoramic video.

3.2   Ultra-High Definition Panoramic Video Acquisition

In the context of format-agnostic video, the acquisition of panoramic video plays an important role. The basic idea behind this is to switch from image capture to scene capture, which allows different players in the media production chain to access their own desired view of the scene. Due to this, very high spatial resolution is required in order to allow the user to zoom-in into desired regions of the scene at acceptable quality. Panoramic video has been a topic of interest for more than a century. This section starts with a short revisit to the history of panoramic imaging and panoramic video. To understand the main challenges of today's technology for acquisition of panoramic video at ultra-high resolution, some fundamentals of camera geometry and the mathematical relations between different camera views are necessary. These are presented in a separate section. Based on these theoretical foundations, the two basic principles for panoramic video acquisition, the star-like and mirror-based approach are discussed. The creation of high quality panoramic video requires some fundamental knowledge of the relationship between the geometry of two views and the depth of the scene, which is discussed in detail. Finally, the different stages of the generation of a complete video panorama are discussed such as view registration, warping, stitching and blending.

3.2.1   History of Panoramic Imaging

The term 'panorama' is composed by the two Greek words *pan = all* and *orama = that, which is seen*. Panoramic imaging dates back to a few centuries ago, when Renaissance painters created paintings of big historical events such as coronations or revolutions. These painters applied the Camera Obscura to draw images in correct perspective and to compose them to a panoramic painting – a technique that is very similar to current digital image stitching for panoramas (Steadman, 2011). In the nineteenth century special buildings or rooms showing panoramic paintings, so-called Cycloramas, designed to provide a viewer standing in the middle of the cylinder with a 360° view of the painting, became very popular (Griffiths, 2004). First experiments with moving images have been performed since the beginning of the twentieth century filmed by omni-directional camera set-ups and re-projected by multiprojection systems. The first installation of an immersive 360° film projection was the Cinéorama system shown at

 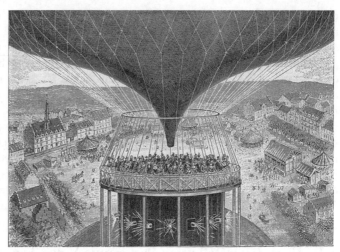

Figure 3.1 The Cinéorama at the Millennium World Exhibition in Paris, 1900: The camera platform with 10 cameras (left), the 360° cinema showing a balloon ride over Paris (right).

the legendary Millennium World Exposition, 1900 in Paris (Wagner, 1955) (see Figure 3.1). Panoramic film projection first became commercial with the invention of Cinerama in 1952 by Fred Waller (Norwood, 1997). The first film in Cinerama technology premiered at Broadway Theatre in New York. The panoramic viewing and immersive sensation was realised by a cylindrical screen with a horizontal viewing angle of 128° and an aspect ratio of almost 3:1, three cinema projectors with overlapping images and a surround sound with six channels (see Figure 3.2). Although the viewing impression was called 'sensational', the projection and the technology had serious drawbacks that hindered a breakthrough. There was a mismatch between acquisition and display geometry that led to the wrong perspective and stitching artefacts on image borders. The use of three individual films was quite inefficient, for example, if one of the films got damaged. Therefore, single-projector solutions like Cinemascope pushed Cinerama from the market again in 1962 and immersive multiprojection systems could only survive in niche markets like theme parks (CircleVision 360, Ub Iwerks, etc.). Even there, they almost disappeared and were replaced over time by single projector systems using either special lenses or spherical mirrors. The IMAX (Image Maximum) technology as one of the most successful is the best example for this development.

During the 1990s, the development of the DVB (Digital Video Broadcasting) standard as well as the DCI (Digital Cinema Initiative) specification of the 2k and 4k standard is considered as the beginning of Digital Television and Digital Cinema. In conjunction, the technological progress in digital video offered many new market opportunities for immersive media. Similar to 3D cinema today, the high quality of digital acquisition and projection technology has caused a renaissance of panoramic multicamera and multiprojection systems for immersive media applications. Digital technology allows precise calibration of omni-directional and multiview cameras, accurate warping and stitching for the generation of panoramic images, exact juxtaposition and blending of multiple projector images, exact adaptation of video panoramas to arbitrarily curved screens and a distortion-free reproduction with very high resolution, brightness and contrast (Raij et al., 2003; Weissig et al., 2005; Bimber, 2006).

Figure 3.2 Cinerama, 1952.

As a result of these advantages several new systems for panoramic imaging have recently been developed and demonstrated successfully. Typical application areas are digital dome projections in planetariums, immersive surround projections for simulation and training centres or the reopening of digital Cinerama theatres like the one in Seattle (Fraunhofer IFF, 2011; Lanz, 2007; HPC Market Watch, 2011). In February 2010, the Fraunhofer Heinrich-Hertz-Institute (HHI) in Berlin, Germany, opened its 'Tomorrow's Immersive Media Experience Laboratory (TiME Lab)', an experimental platform for immersive media and related content creation. The TiME Lab uses up to 14 HD projectors for panoramic 2D and 3D projection at a cylindrical 180° screen as well as a 'Wave Field Synthesis (WFS)' sound system with 128 loudspeakers (Fraunhofer HHI, 2013) (see Figure 3.3, left).

Figure 3.3 TiME Lab at Fraunhofer HHI, Germany, 2010 (left), MegaVision 6k system: The Virtual Stadium, IBC 2005 (right).

As well as the examples given above, several experimental systems covering acquisition, transmission and display have been developed that allow scene capture with ultra-high resolution far beyond HD resolution. First experiments on Super High Definition (SHD) started in early 2000, mainly driven by NTT in Japan. In 2002, NTT demonstrated the first transmission of SHD content with a resolution of 4k × 2k. The content was streamed over the Internet2 (Internet2, 2013) using JPEG 2000 technology (JPEG2000, 2004) and was displayed using one of the first digital 4k-projectors. A similar demonstration was given by the Japanese company MegaVision jointly with NTT during FIFA 2002 (football world championship), (MegaVision, 2004). For this purpose a special relay lens has been developed that could be equipped with one wide-angle lens and three HD cameras. The function of the relay lens is to split a panoramic view through the optical system into three horizontally-adjacent HD images that can then be captured by regular HD cameras. The resulting 6k × 1k images were then displayed by three overlapping HD projectors. The system was tested for the first time during a football game in 2000 and was presented to a small audience during FIFA 2002 in Japan. Since 2003, it has been permanently installed as 'The Virtual Stadium' showcase in the Japan Football Museum and has also been presented at the International Broadcasting Convention (IBC) 2005 at the NTT booth (see Figure 3.3, right). Furthermore, the MegaVision system has been used for capturing content for the Laser Dream Theatre exhibited by Sony at the Expo 2005 in Aichi, Japan (CX News, 2005).

At the same time, the 8k Super Hi-Vision system of NHK has been developed with a resolution of 8k × 4k. For this purpose, a dedicated 8k camera was developed using four 4k CMOS (Complementary metal-oxide semiconductor) sensors (two for green and one each for red and blue) that are mounted with diagonal half-pixel offsets, similar to the well-known Bayer pattern. The resulting 8k image is then screened by a 2-projector system. First live trials for public viewing purposes were carried out in 2005 in Japan. Furthermore the system has been shown by NHK in its 8k-theatre at the Expo 2005 in Aichi, Japan, and was demonstrated at the 2012 London Olympics in collaboration with the BBC (Sugarawa et al., 2013).

In 2006, the German joint project CineVision 2006 experimented with two D-20 D-Cinema cameras from Arnold & Richter Cine Technik (ARRI), (CineVision, 2006; Fehn et al., 2006). The cameras were mounted side-by-side looking in two different directions (see Figure 3.4, left). The two D-20 images were then stitched to one panoramic 5k image that was later shown with five SXGA+ projectors in cinema theatres. Using this system, 5k productions have been shown in a CinemaxX movie theatre in Berlin during FIFA 2006 and in the Brenden Theatre

Figure 3.4   CineVision, 2006 acquisition setup (left), Panoramic screening in Las Vegas, 2007 (right).

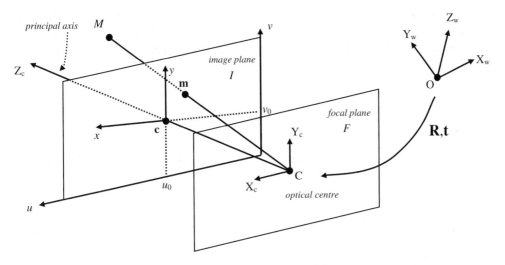

Figure 3.5 The camera model.

in Las Vegas during NAB (National American Broadcasting Convention) 2007 (see Figure 3.4 right).

This review of the history has shown that advances in technology and mainly the advent of digital video leads to systems with ultra-high resolution, resulting in fully immersive video presentations. The application scenario, however, is still quite restricted. All systems address a transmission with a fixed proprietary format, mainly for live public viewing events. Hence, the exploitation of this technology is limited to applications where content is produced in a special format and can only be shown in specialised theatres to small audiences.

3.2.2 The Geometry of Two Views

Before investigating different techniques for video panorama generation, some fundamentals of the geometry of two views are given. A fundamental prerequisite is the definition of how a 3D scene is projected into a single image, the so-called perspective projection. Let us consider a 3D point M in space that is projected onto the image plane I of a camera (see Figure 3.5).

First, a camera coordinate system is defined containing the optical centre C as origin. In order to relate the camera coordinate system to the real world, a Euclidean transformation is used such as:

$$M_c = \mathbf{R} M_w + \mathbf{t} \tag{3.1}$$

The transformation consists of a rotation $\mathbf{R}$ and a translation $\mathbf{t}$. The transformation can be rewritten in homogeneous representation such as:

$$\tilde{M}_c = \mathbf{D}\tilde{M}_w \quad \text{with} \quad \mathbf{D} = \begin{bmatrix} \mathbf{R} & \mathbf{t} \\ \mathbf{0}_3^T & 1 \end{bmatrix} \quad \text{and} \quad \mathbf{0}_3 = \begin{bmatrix} 0 \\ 0 \\ 0 \end{bmatrix} \tag{3.2}$$

The matrix **D** represents the external transformation. The relation between the 3D point M in space and its image in the image plane I can be derived by the fundamental relation of central projection defined by Equation 3.3 relating the 3D point in camera coordinates (X_c, Y_c, Z_c) to the 2D point in sensor coordinates (x, y) considering a normalised distance 1 between the image plane I and the origin C.

$$\frac{x}{X_c} = \frac{y}{Y_c} = \frac{1}{Z_c} \qquad (3.3)$$

Since a 3D point in camera coordinates is considered, the following equation describes the perspective projection from 3D space onto the 2D image plane.

$$\begin{bmatrix} U \\ V \\ S \end{bmatrix} = \begin{bmatrix} 1 & 0 & 0 & 0 \\ 0 & 1 & 0 & 0 \\ 0 & 0 & 1 & 0 \end{bmatrix} \cdot \begin{bmatrix} X_c \\ Y_c \\ Z_c \\ 1 \end{bmatrix}, \quad \text{with} \quad x = U/S, \quad y = V/S \quad \text{for} \quad S \neq 0 \quad (3.4)$$

The equivalent representation in homogeneous form can be written as follows:

$$s\tilde{m}' = \mathbf{P}_N \tilde{M}_c \quad \text{with} \quad \mathbf{P}_N = \begin{bmatrix} 1 & 0 & 0 & 0 \\ 0 & 1 & 0 & 0 \\ 0 & 0 & 1 & 0 \end{bmatrix}, \quad \tilde{m}' = \begin{bmatrix} x \\ y \\ 1 \end{bmatrix} \quad \text{and} \quad s = S \quad (3.5)$$

It has to be noted that the perspective transformation in Equation 3.4 represents the normalised form, where the distance of the image plane I to the optical centre C equals 1. Concatenating the external transformation and the perspective projection into a single term the 3D point in space transforms into its image on the sensor such as:

$$s\tilde{m}' = \mathbf{P}_N \mathbf{D} \tilde{M}_w \qquad (3.6)$$

Finally, an internal transformation is required that maps the 2D point in sensor coordinates (x,y) to its counterpart in image coordinates (u,v). This is basically a shift (u_0, v_0) of the origin from the centre to the lower right corner and a scale containing a change of units in any real metric, for example, *mm*, to *pixels* (k_u, k_v). The focal length f denotes the distance of the image plane to the camera centre.

$$\tilde{m} = \mathbf{A}\tilde{m}' \quad \text{with} \quad \mathbf{A} = \begin{bmatrix} fk_u & 0 & u_0 \\ 0 & fk_v & v_0 \\ 0 & 0 & 1 \end{bmatrix} \quad \text{and} \quad \tilde{m} = \begin{bmatrix} u \\ v \\ 1 \end{bmatrix} \qquad (3.7)$$

The complete projection equation can then be summarised as

$$s\tilde{m} = s \underbrace{\mathbf{A}\tilde{m}'}_{\text{internal}} = \mathbf{A} \cdot \underbrace{\mathbf{P}_N \tilde{M}_c}_{\text{perspective}} = \mathbf{A}\mathbf{P}_N \cdot \underbrace{\mathbf{D}\tilde{M}_w}_{\text{external}} = \mathbf{A} \, [\mathbf{R} \, \mathbf{t}] \, \tilde{M}_w = \mathbf{P}\tilde{M}_w \qquad (3.8)$$

The camera model presented here is called a linear model as the homogeneous point in 3D space is mapped to a 2D point on the image plane by a linear transformation in the projective space.

However, in the case of wide-angle lenses, non-linear distortions must be considered. Current state-of-the-art lenses can be sufficiently modelled by radial distortion following Equation 3.9. The distorted component (u_d, v_d) results from weighting the radial distance r_d of the image point (u,v) from the image centre by first and second order distortion coefficients κ_1 and κ_2. There are more complex distortion models found in the literature but the one below is sufficient for the purpose of this book.

$$
\begin{aligned}
u_d &= u\left(1 + \kappa_1 r_d^2 + \kappa_2 r_d^4\right) \\
v_d &= v\left(1 + \kappa_1 r_d^2 + \kappa_2 r_d^4\right) \quad \text{and} \quad r_d = \sqrt{u_d^2 + v_d^2}
\end{aligned}
\tag{3.9}
$$

More details on camera calibration and non-linear lens distortion models can be found in Stama et al. (1980), Weng et al. (1992) and Salvi et al. (2002).

The stitching of individual camera views can be distinguished in two different cases depending on the geometry of two cameras. If the camera views are resulting from two cameras that are just rotated against each other, the stitching can be performed by a linear projective mapping between both views. Consider two cameras 1 and 2 with identical optical centres C_1 and C_2 (see Figure 3.6, left). A 3D point in space M is then projected in both camera images to $\mathbf{m_1}$ and $\mathbf{m_2}$. As there is no difference in perspective, the two image points are related by the following Equation (3.10).

$$
\tilde{\mathbf{m}}_2 = \mathbf{H}\tilde{\mathbf{m}}_1
\tag{3.10}
$$

The matrix $\mathbf{H}$ is called homography and represents a projective transformation between points on two planes in projective space.

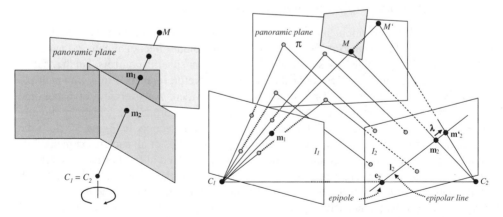

Figure 3.6   Ideal camera configuration: rotated cameras (left), camera configuration with parallax between views (right).

The matrix $\mathbf{H}$ is composed by the rotation matrix $\mathbf{R}$ defining the rotation between both cameras and the two intrinsic matrices $\mathbf{A}_1$ and $\mathbf{A}_2$ describing the internal camera parameters of each camera.

$$\mathbf{H} = \mathbf{A}_2 \mathbf{R}_1 \mathbf{A}_1^{-1} \qquad (3.11)$$

As the panoramic plane is also rotated against the two original cameras, the image points from both cameras can be mapped to the joint plane by a similar projective mapping.

In Figure 3.6, right, the general configuration of two views is presented, where the optical centres of the two cameras do not coincide. If a plane π in space is considered, any image point in I_1 that is a projection of a point on this plane relates to its corresponding image in I_2 by a homography. The homography can be calculated from four corresponding points in camera 1 and 2 (Hartley and Zisserman, 2004). Due to this, different points on I_1 and I_2, which are projections of 3D points on this plane, can be mapped onto a joint panoramic plane. Points in space that are in front or behind that plane contain a parallax represented by λ and oriented along the epipolar line $\mathbf{l}_2$ connecting the epipole $\mathbf{e}_2$ with $\mathbf{m}_2$. Based on a defined plane in space for which the homography is known, any correspondence between the left and right image is calculated as follows:

$$\tilde{\mathbf{m}}_2 = \mathbf{H}_\pi \tilde{\mathbf{m}}_1 + \lambda \tilde{\mathbf{e}}_2 \qquad (3.12)$$

Equation 3.12 is called the generalised disparity equation, as it maps any point in I_1 to its corresponding point in I_2 if the parallax and a common homography between the two images are known. This equation can also be used to map points from two different images onto a single common plane if there is a difference in perspective due to cameras at distant positions. This principle is called *plane + parallax* (Kumar et al., 1994) and has been used for generation of panoramic images from cameras, where objects in the scene are located at relatively large distances from the camera positions (Fehn et al., 2006).

3.2.3   Fundamentals of Panoramic Video Acquisition

In panoramic video acquisition, at any time the full panoramic view must be captured. Hence, any approach based on a rotating camera fails. There are omni-directional cameras available on the market with special lenses, so-called fish eye lenses or spherical mirrors, but they suffer from limited resolution and serious optical distortions. Therefore, camera setups with multiple cameras are considered for high-resolution panoramic video. An overview of panoramic imaging is given in Cledhill et al. (2003). There are two basic principles, which are directly related to the geometrical considerations made in the previous section. In Figure 3.7 (left) the ideal camera arrangement is shown, where all the optical centres of all cameras coincide in a single point. For presentation reasons, the optical centre is moved in front of the lens. Hence, in reality this can only be achieved by using a mirror rig, while the cameras are mounted such that the optical centres are mirrored and virtually meet in a single point behind the mirrors. In such case, there is no parallax between views and this configuration relates to the one displayed in Figure 3.6 (left). An example of such a mirror rig is given in Figure 3.7 (right). This system, the so-called OmniCam, developed by Fraunhofer HHI, is a scalable system, which can be

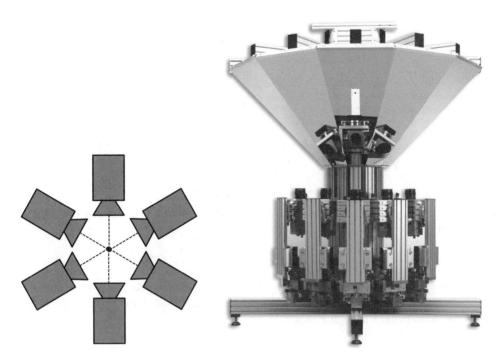

Figure 3.7   Optimal camera arrangement (left), mirror rig by Fraunhofer HHI (right).

equipped with up to 12 HD cameras for 360° shooting. In the configuration presented in Figure 3.7 (right), six HD cameras are used to shoot 180° panoramas. The six cameras generate tiles of 1080 × 1920 pixels each, which are then subsequently stitched to one large panorama with a final resolution of 6984 × 1920 for a 180° panorama. As the cameras are used in a portrait format, the vertical field of view is about 60°, a feature that is important for immersive media.

Since the cameras are mounted exactly below the mirrors, the ideal configuration with all optical centres coinciding in a common point can be achieved (see Figure 3.8, left). However, in such configuration, there is no overlap, which leads to stitching problems at image borders, while creating a common panoramic video.

Hence, the cameras are shifted slightly outwards in a range of 5mm. This radial shift results in an overlap region at image borders of around 20 pixels. Obviously, this modification violates the parallax free condition, but the distortion is minimal and noticeable only very close to the omni-directional camera (below a distance of 2m).

The second principle is depicted in Figure 3.9, the so-called star-like approach. All cameras are mounted in a star-like manner and the cameras are looking inwards or outwards, while all optical centres are located on a circle and the optical axes are perpendicular to the arc. Obviously there is some perspective difference between neighbouring views, which does not allow perfect stitching of views. This configuration relates to the theoretical concept of *plane + parallax* illustrated in Figure 3.6 (right). However, the parallax between the neighbouring views depends on the depth of the scene as derived in more detail in Section 3.2.4. Since objects are far away from the panoramic camera, this parallax distortion can be neglected. In

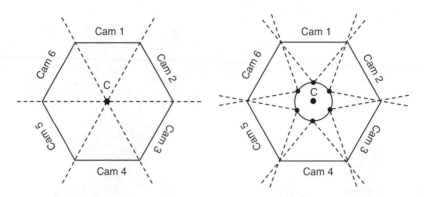

Figure 3.8 Mirror rig approach. Theoretically optimal setup with one common focal point for all cameras (left). Real configuration to achieve overlap at image borders (right).

the case of nearby objects more sophisticated approaches such as *plane + parallax* have to be implemented in order to match pixels from neighbouring views. Commercial products that follow this concept are shown in Figure 3.10.

3.2.4 Geometrical Constraints for Parallax Free Stitching of Two Images

The idea of high resolution video is to combine the images captured by multiple cameras to one common panorama. In this way a much higher resolution can be achieved, which depends on the number of cameras used. In the following, geometrical constraints of this process will be investigated.

Figure 3.11 illustrates the simplest case of two cameras with parallel orientations perpendicular to the line defined by their camera centres C_1 and C_2. The disparity δ is defined by the different positions x_1, x_2 of the projections $\mathbf{m_1}$, $\mathbf{m_2}$ of point M into the two image planes.

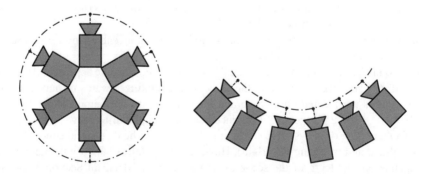

Figure 3.9 Star-like camera arrangement: Outwards oriented (left), inwards oriented (right).

Figure 3.10 Commercial products using star-like approach: Outwards oriented approach from Google street view acquisition device (left), inwards oriented approach by Camargus demonstrated in 2011 (right) (Camargus, 2013).

It can be derived in dependency of depth z_M, the distance between cameras, the so-called baseline B, and the focal length f as follows

$$\delta = x_1 - x_2 = \frac{Bf}{z_M} \tag{3.13}$$

Equation 3.13 shows that the disparity δ is independent of the position of the point M in direction parallel to the baseline (assuming an infinite image plane). Therefore, all points in space which have equal disparity δ define one single depth layer at constant depth z.

In the following, we will determine the constraints for parallax free stitching of two cameras. For this purpose, we will derive the maximal possible depth range that leads to parallax errors of less than one pixel. This depth depends on the distance of the object to the cameras, the

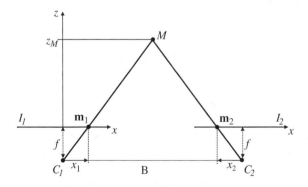

Figure 3.11 Disparity as a function of depth and camera distance.

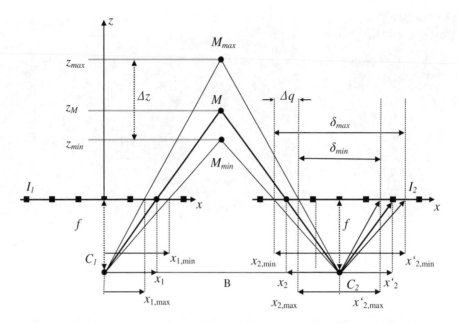

Figure 3.12   Depth layers and optimal depth plane resulting from quantised image plane with pixel size Δq.

baseline between the cameras and the camera sensor resolution, that is, the resulting image resolution.

In Section 3.2.2, we introduced the *panoramic plane* (Figure 3.6). Since we assume that this plane will be located at depth z_M parallel to the image planes of the cameras then it is possible to derive the certain depth range with minimum depth z_{min} and the maximum depth z_{max}, where $z_{min} < z_{max} \leq \infty$ that is still sufficiently suitable for parallax free stitching of the two image planes. For this purpose, it is essential to know how much the depth of a single layer z_M can vary leading to disparities that are still equal or less than the pixel resolution Δq. Figure 3.12 illustrates the problem. The point M is projected onto the image planes of both cameras C_1, C_2. The total depth range Δz can be expressed in terms of the total disparity range $\Delta \delta = \delta_{max} - \delta_{min}$ as

$$\Delta \delta = \delta_{max} - \delta_{min} = Bf \frac{1}{z_{min}} - B \cdot f \frac{1}{z_{max}} = Bf \frac{z_{max} - z_{min}}{z_{max} z_{min}} \tag{3.14}$$

Based on this, we define the total disparity range $\Delta \delta$ to not exceed the physical size of a pixel on the camera sensor $\Delta q = w_h / N_p$ with the sensor width w_h and the image resolution in pixels N_p. The maximal suitable projection error should be half the pixel size in both image planes, that is, $\pm \Delta q/2$. The minimal allowed disparity in this case would be $\delta_{min} = x_{1,max} - x_{2,max}$ where $x_{1,max} = x_1 - \Delta q/2$ and $x_{1,min} = x_1 + \Delta q/2$. Accordingly, the maximal suitable disparity is $\delta_{max} = x_{1,min} - x_{2,min}$. In this case z_M would represent the centre of the projected image points x_1 and x_2, z_{min} the minimal and z_{max} the maximal suitable depth. In this case, the total disparity quantisation error equals the total disparity range and must be lower than twice the pixel size: $\Delta\delta = \delta_{max} - \delta_{min} \leq 2\,\Delta q$. The total disparity quantisation error $\Delta\delta$ is the maximum acceptable layer. We obtain for the maximal suitable disparity

$$\Delta\delta \leq 2\Delta q = 2\frac{w_h}{N_p} \tag{3.15}$$

and according to Equation 3.14

$$\Delta\delta = Bf\left(\frac{1}{z_{min}} - \frac{1}{z_{max}}\right) \leq 2\frac{w_h}{N_p} \tag{3.16}$$

This equation describes the relationship between the camera distance B, the focal length f, the image resolution N_p and the depth range z_{min}, z_{max} with $z_{min} < z_{max} \leq \infty$. As the camera distance and the focal length are constant, the minimal depth can be expressed as

$$z_{min} = \frac{Bfz_{max}}{\Delta\delta\, z_{max} + Bf} \tag{3.17}$$

For a maximal depth at infinity $z'_{max} = \infty$, we obtain for the minimal possible depth

$$z'_{min} = \lim_{z_{max} \to \infty} (z_{min}) = \frac{Bf}{\Delta\delta} \tag{3.18}$$

The derivations in this section can be used to define the available depth range $z_{min} < z_{max} \leq \infty$ for parallax free stitching (i.e., having a disparity variation less than 1 pixel) in dependency of the camera distances. Figure 3.13 illustrates this dependency for a mirror rig configuration, where due to the mirrors the virtual distances of camera centres can be achieved to be very close to the theoretical minimum. In this case, a camera distance $e=B=5$mm will result in the illustrated depth range. Figure 3.13 depicts the depth range of objects, where parallax free stitching is possible. For a linear increase of the minimal object distance the maximal possible object distance increases as well but in a logarithmic fashion. Hence, for some given minimum distance z_{min}, a parallax free stitching until infinity can be achieved.

In practice, camera setups used for stitching usually have arbitrary oriented image planes. In the case of star-like setups (as illustrated in Figure 3.9) the cameras are rotated based on a common midpoint. In order to determine the exact parallax free stitching depth range for all these cases, the camera rotation needs to be considered too. So far, for our derivations we have assumed rectified image planes (see Figure 3.11). Nevertheless, a camera rotation is a depth invariant operation (Hartley, 2004; Schreer, 2005). Any camera pair can be rectified to the setup shown in Figure 3.11. For this case the real pixel width $\Delta q = w_h/N_p$ changes. The rectified virtual camera image will have a different virtual pixel size Δq_v. In order to determine the parallax free depth range for rotated cameras, Equation 3.15 needs to be modified such that the real pixel size $\Delta q = w_h/N_p$ is replaced by this virtual pixel Δq_v which is obtained by taking the camera rotation into account. Nevertheless, in practice in most cases a general approximation of parallax free depth range based on Equation 3.15 is sufficient. Please note, that these values will be further influenced by the aperture of the chosen lenses and other properties.

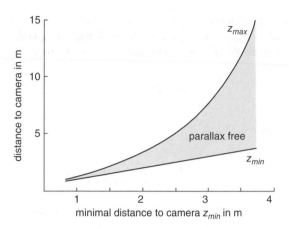

Figure 3.13 Depth range for parallax free stitching in dependency of the minimal camera distance z_{min} based on a fixed camera baseline for a sample mirror rig configuration.

3.2.5   Registration of Views

The generation of panoramic video is basically a question of how different views can be composed into one common view with increased resolution. Depending on the availability of the camera geometry, that is, the orientation and distance between two views, one can distinguish two fundamental approaches. If there is no calibration information available, approximate solutions can be used to relate images from two different cameras observing the same scene. These approaches are called mosaicking techniques that rely on simplified motion models such as translational, Euclidean or affine motions (Capel and Zisserman, 1998). These techniques require a significant overlap of neighbouring views in order to estimate the parameters of the selected motion model, which is also called registration. Well-known approaches are feature-based methods (Brown and Lowe, 2007) and global methods (Shum and Szeliski, 2000). A nice overview on image stitching and registration is given in (Peleg, 1981; Szeliski, 2011). In the case of a mirror rig, no overlap is available between neighbouring views (see Figure 3.8). The registration between views can only be achieved by some dedicated calibration procedure.

In the following, a detailed example for the registration of mirror rig cameras will be given. The registration is based on a large calibration pattern (see Figure 3.14), which is visible in at least two neighbouring cameras (Figure 3.15, left). In Figure 3.15 (right), a contrast enhanced version is shown to allow more robust pattern detection. As explained in Equation 3.10, it is possible to define a homography transformation between two image planes, which can be parameterised purely by related image-based feature points in the two camera views. As the geometry of the pattern is known, it can be used as a reference plane for determining a homography transformation H_{01} between the two neighbouring image views. This transformation can be used as a relationship between the images. A parallax free stitching is possible, if the constraints for a minimal and maximal depth range (as defined in Section 3.2.4) are satisfied. For this case the reference depth plane z_M in Figure 3.12 is equivalent to the *panoramic plane* π in Figure 3.6.

Figure 3.14 Typical pattern design used for mirror rig camera calibration (left) and its application in practice (right).

In order to determine the homography transformation, a standard image based feature detector can be used, as for example, provided by the open source image processing library *OpenCV* (OpenCV, 2013). For robust and accurate feature detection some further pre-processing might be required such as contrast improvement. To support unique pattern registration, special markers can be used to identify the orientation of the calibration pattern. Further on, a unique relationship between the detected features and their counterparts in the reference image plane can be determined. The resulting set of feature pairs can be used to solve Equation 3.10. We obtain for a set of images, points $\mathbf{m}_0$ and $\mathbf{m}_1$ of the two cameras and in both cases a set of reference plane points $\mathbf{m}_{ref0}$ and $\mathbf{m}_{ref1}$ which are related to each other by the homographies H_0 and H_1

$$\mathbf{m}_{ref0} = H_0 \mathbf{m}_0 \tag{3.19}$$

$$\mathbf{m}_{ref1} = H_1 \mathbf{m}_1 \tag{3.20}$$

Figure 3.15 Mirror-rig calibration pattern: (left side) original two neighbouring camera views; (right side) image enhancement to allow more robust pattern detection.

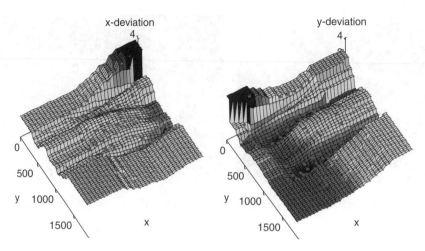

Figure 3.16   Example for deviation compared to the reference in x-direction for the estimated feature points for the two camera views.

The two homographies are used to compute the resulting relationship between the two images defined by the homography matrix H_{01}

$$\mathbf{m}_1 = H_1^{inv} H_0 \mathbf{m}_0 = H_{01} \mathbf{m}_0 \tag{3.21}$$

An important issue is the handling of deviations of this projection and registration process. While the reference pattern provides ideally distributed feature points without any errors, the real image-based features underlie a variety of distortions errors. The standard homography estimation process will minimise this error on a least-square base (Hartley and Zisserman, 2004). Errors caused by the feature detector can be efficiently minimised with this approach. In contrast, systematic errors cannot be handled with this method. An example of these systematic errors is given in Figure 3.16.

It clearly demonstrates the systematic behaviour of the remaining deviations. The main sources for this deviation are the camera lens distortion as well as slightly non-planar mirror surfaces. Note that in the current example a lens distortion correction had already been applied to the input images. However, this process is also based on models that do not perfectly reflect the reality. Nevertheless, the estimated deviation surface shown in Figure 3.16 can be used to fine tune the final stitching result and compensate these systematic errors. The advantage is that the source of deviation is not required to be known. The error will be corrected implicitly.

As well as the geometrical alignment of neighbouring views, photometrical correction must be performed to achieve a common and coherent colour and contrast representation of the final panoramic image. If this photometrical correction is not performed properly, stitching borders will be identified immediately. This photometrical correction can be performed in advance of the acquisition performing a careful joint colour calibration of all cameras. There are several methods available to adjust the colour of two different images based on colour histogram equalisation (Stark, 2000; Grundland and Dodgson, 2005). However, in the setup

Figure 3.17   Result of warping the right camera view to the left image plane.

under consideration, there is almost no overlap between neighbouring views and photometric correction cannot be performed. A solution to this problem might be an additional camera view that provides significant common field of view. Then photometric alignment can be performed via this additional view. Since the format-agnostic scene acquisition already considers scene capture with additional broadcast cameras, a photometric matching must be performed anyhow, as will be discussed in Section 3.3.2. Therefore, the same approach can also be used to align individual views of an omni-directional camera.

3.2.6   Stitching, Warping and Blending of Views

A final step is the selection of the destination camera plane used for stitching. This does not necessarily have to be a real camera plane. Figure 3.17 shows the result of warping the pattern image of the right camera view to the left-hand image plane. Note, that at this stage no blending regions were used.

For panorama cameras that have an overall opening angle close to 180° or more a planar common image plane cannot be used. Therefore, usually cylindrical camera models are used. Figure 3.18 illustrates the difference between the planar and cylindrical case for a real data set. Figure 3.19–3.22 show various results for the stitching of multiple cameras based on the described workflow. Note, that in the overlapping areas of the images a linear blending function was used, which was extracted from the overlapping regions shown in Figure 3.18.

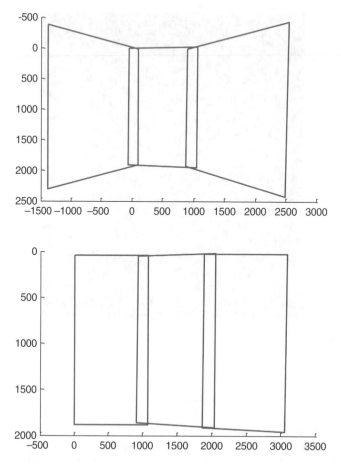

Figure 3.18 Stitching of three neighbouring image planes to planar camera model (left) and cylindrical camera model (right).

Figure 3.19 Stitching results of a cylindrical camera, six sub-cameras and no blending.

Figure 3.20   Stitching result of three cameras to a common planar image plane.

Figure 3.21   Registration example for six sub-cameras on a cylindrical image plane.

Figure 3.22   Stitching result for six sub-cameras on a cylindrical image plane including the blending regions. Reproduced by permission of Berlin Philharmonic Orchestra.

3.3 Use of Conventional Video Content to Enhance Panoramic Video

Ultra-high resolution panoramic video, as described in the previous section, can provide a highly immersive experience of being present at an event, sometimes likened to having 'the best seat in the stadium'. However, sometimes the experience of an event can be significantly enhanced by adding more TV-like coverage, providing views with more detail and/or from different viewpoints. At many live events, large screens are provided for the audience to see such images, and they play an important part in the overall experience.

It is possible to extract a window from a panoramic video image in order to provide a close-up of a particular region-of-interest. This could be done in an automated way, as described later in this book, using content analysis techniques to identify regions-of-interest (Chapter 5) and automated shot selection and framing (Chapter 6). However, there may be insufficient resolution in the panoramic image to allow tight close-ups. Also this does not address the need for shots from many different viewpoints, unless a large number of panoramic cameras are used. It therefore makes sense to consider how conventional video cameras can be used together with a panoramic video feed to provide the best of both worlds: the sense of 'being there' at an event, with the freedom to look around, coupled with close-up views of key parts of the action.

This section considers ways in which this could be achieved, and how the concept of a layered scene representation (introduced in Chapter 2, Section 2.9.4) can be used to represent the relationships between the different views. Section 3.3.1 discusses how the positions and orientations of multiple cameras can be calibrated, to support functionality such as offering a close-up view of a part of the scene from a conventional camera when a viewer selects this area in the panoramic video. Section 3.3.2 discusses colour matching between cam- ⌐s. Section 3.3.3 discusses approaches for blending between views, and how this relates to other work such as free-viewpoint video. In the following sections, we refer to the non-panoramic cameras as *broadcast cameras*, as typically they may have been installed to support a conventional TV broadcast, and are likely to be HD-resolution cameras equipped with zoom lenses.

3.3.1 Calibration of Camera Position and Orientation

In order to make best use of the captured video data, it is necessary to know the relative positions and orientations of the cameras (often referred to as their *extrinsic parameters*, corresponding to the external transformation matrix $\mathbf{D}$ in Equation 3.2), in addition to the field-of-view (which will be time varying in the case of cameras with zoom lenses). This information can be used to relate the views of the panoramic and broadcast cameras, although some assumptions about the depth of the objects being observed are also needed, particularly when the cameras are not sited closely together. To help infer the rough depth of objects-of-interest in a given view, it is helpful to also have some knowledge of the 3D scene structure, such as the location of the ground plane on which people may be standing. Note that the term 'objects-of-interest' is subtly different from 'regions-of-interest': We use the term 'region-of-interest' to refer to a (roughly rectangular) region in the panoramic image, which might correspond to a good area for deriving an interesting region-of-interest, by cropping out a view of the scene. Such a region may contain one or more objects-of-interest (typically people) at different depths, and

Figure 3.23   3D laser scanner and example of scan produced.

knowing the depth of the main object-of-interest is important when registering views between cameras (discussed further in Section 3.3.3).

The same principle applies to knowledge of the microphone positions (see Chapter 4, Section 4.7.2), where knowing their 3D location helps to reconstruct a sound field or control the mixing of microphone levels based on the area of the scene being viewed.

Surveying the 3D Scene

3D information describing the cameras, the microphones and key elements of the scene itself can be obtained directly using surveying equipment. Taking this to the extreme, it would be possible to carry out a complete 3D scan of a scene using a laser scanning tool. Figure 3.23 shows a commercially-available laser scanner, which is capable of producing a 3D point cloud of a venue at the size of a football stadium in a couple of hours.

An alternative approach is to just measure the locations of particular objects-of-interest using a surveying tool such as a 'Total Station', which uses laser distance measurement combined with azimuth and elevation angles to compute 3D coordinates. Both approaches can obtain accuracies of around 1mm over distances of many tens of metres, which is sufficient for this kind of application. We will assume that all cameras and microphones remain fixed in position during a programme, although broadcast cameras may pan, tilt and zoom. It would, however, be possible to use additional sensors to measure the location of moving microphones and cameras, if needed.

Calibration of Panoramic Camera

We will assume that the panoramic camera(s) remain(s) fixed in position, and does not pan, tilt or zoom. The intrinsic parameters of this camera (specifying the mapping between a point with given 3D coordinates with respect to the reference frame of the camera and the coordinates in the panoramic image at which it appears) are assumed to be known from the calibration process described in Section 3.2.5. However, this process did not determine the extrinsic parameters (the position and orientation of the camera in the world reference frame),

Figure 3.24 Calibration of panoramic camera orientation using measured 3D points. Reproduced by permission of The Football Association Premier League Ltd.

as calibration used moveable charts rather than fixed world points. The orientation of the panoramic camera can be determined by identifying the image coordinates of some features in the 3D scene whose positions have been measured by the surveying process. Indeed, this also provides an alternative method to estimating the position of the panoramic camera.

Approaches for camera calibration using 'bundle adjustment' are well-known (Hartley and Zisserman, 2004), allowing camera pose to be computed from a list of 3D feature locations and their corresponding coordinates in the image. However, most implementations will assume a conventional (rather than cylindrical) camera projection. One approach that can work well is to render a conventional image from the central portion of the cylindrical panoramic image, to produce the view that a conventional camera would have seen when looking directly along the centre line of the panoramic camera (that is, having the same orientation as the panoramic camera). Conventional bundle adjustment software (that needs not to know about cylindrical projection) can then be used to compute the pose of the panoramic camera from features in this central area.

Figure 3.24 shows an example of a conventional perspective image rendered from the central portion of a cylindrical image derived from a panoramic camera (shown in Figure 3.25). In this example, a selection of features whose 3D world positions were surveyed and manually identified in the image, are shown with light crosses. These were chosen to be easily-identifiable and distributed across the scene, and included the two ends and mid-point of the football pitch centre line, and numbered signs above the exits to the spectator stand on the far side. The 3D coordinates of these points and the corresponding 2D image coordinates were processed in a numerical optimisation process to compute the position and orientation of the omnidirectional camera.

Calibration of Broadcast Cameras

To measure the time-varying pan, tilt and field-of-view of broadcast cameras, it is possible to use mountings that incorporate rotational sensors, and lenses that supply a data stream giving their current zoom and focus settings. However, such systems are not commonly used in TV production, being brought in only for special requirements such as overlay of virtual graphics, for example, world record lines during races. Even for these special requirements, it is preferable where possible to determine the camera parameters by video analysis, as this

Figure 3.25   Panoramic camera and broadcast camera at a football match.

removes the need both for the sensors themselves, and for retrieving the data from the camera and through the production systems. For events such as football matches, it is possible to estimate the camera pose and field-of-view using computer vision techniques that treat the pitch lines as parts of a large calibration object (Thomas, 2007), using similar approaches to computing the camera parameters to those used in more conventional chart-based calibration methods such as those described in Section 3.2.5. For other kinds of events that do not contain such readily-identifiable markings, approaches such as tracking corner-like features that are learned from the camera image can be used (Dawes et al., 2009).

In order to perform image-based calibration of the broadcast cameras, it is tempting to consider using the image from the panoramic camera to provide reference features that could then be identified and tracked in the broadcast cameras. Knowledge of the scene geometry (for example, from a full laser scan) could enable the 3D location of image features to be determined, allowing broadcast cameras to be calibrated even if they were at some distance from the panoramic camera. In practice, however, this is difficult: the whole point of using broadcast cameras in addition to one or more panoramic cameras is because they can provide images with much more detail, or from significantly different viewpoints, compared to the panoramic camera. The images are therefore generally very different.

Figure 3.25 shows a panoramic camera of the type shown on the right of Figure 3.7 mounted close to a broadcast camera with a large zoom lens at a football match. Figure 3.26 shows an example from a 6k × 2k panoramic image captured with the panoramic camera, compared to an image from the broadcast camera when zoomed in to capture action on the far side of the pitch. It can be seen that there is a significant difference in the level of detail (indeed, there is yet more detail in the broadcast camera that is not visible when the image is reproduced at this small size). Furthermore, there are significant parallax differences between object positions in the two images due to the spacing between the two cameras; for example the blue player 18 appears much further away from orange player 2 in the panoramic image than in the broadcast camera image. These effects make it very difficult to apply conventional feature matching between these images.

Figure 3.26 Comparison of panoramic and broadcast camera images for a football match. Reproduced by permission of The Football Association Premier League Ltd.

The possibility of using the panoramic camera image to help with calibration of broadcast cameras is even slimmer in situations where the views are very different. Figure 3.27 shows a panoramic image captured at a classical concert from the perspective of the audience, compared to four views available from broadcast cameras. These images illustrate the very different nature of the images that are captured. Calibration of the broadcast cameras in this kind of situation, particularly for the close-up shots, would only be practical by the use of sensors on the cameras.

3.3.2   Photometric Matching of Panoramic and Broadcast Cameras

In conventional TV coverage, the brightness, contrast and colour rendition of each camera is usually set before a programme by reference to a test chart. Adjustments are then made to each camera by an operator during a production as the lighting changes, or as cameras change their view between dark and light areas of the scene, as discussed in Chapter 2, Section 2.2.2. A panoramic camera that is constructed from multiple individual cameras will be more difficult to adjust in this way, as changes to parameters such as iris and gain would have to be carefully matched between the cameras, so they are typically left at fixed values. Potential problems with the panoramic camera becoming over- or underexposed can be mitigated by the use of high dynamic range cameras, which will be discussed in Section 3.5. However, there remains the challenge of matching the colour rendition between the panoramic and broadcast cameras.

One possible approach that could allow the matching of broadcast and panoramic cameras to be automated is the well-known method of histogram matching. In this process, two images

Figure 3.27 Comparison of panoramic and broadcast camera images from a classical concert.

showing a similar part of a scene have their colour histograms computed, and look-up tables are then calculated with the aim of making the histograms of the two images match as closely as possible. Let the histogram of the colour components of the reference image be H_{R1}, H_{G1}, H_{B1}, and let those for the image to adjust (in this case the panoramic image) be H_{R2}, H_{G2}, H_{B2}. The cumulative distribution function for the red colour component of each image is computed thus:

$$CDF_{R1}(x) = \int\limits_{0}^{x} H_{R1}(x)\, dx \tag{3.22}$$

Then for each red component level L_{R2} in image 2, we find the level L_{R1} for which $CDF_{R2}(L_{R2}) = CDF_{R1}(L_{R1})$. This gives the mapping value to put at location L_{R2} in the look-up table to apply to values in image 2: $LUT_R(L_{R2}) = L_{R1}$. The process is repeated for the

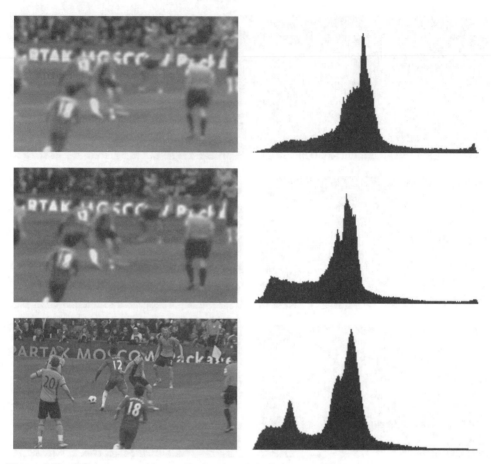

Figure 3.28 Histogram matching applied to the original segment from the panoramic camera (top) to make a colour-corrected version (middle) designed to match the colours in the corresponding broadcast camera (bottom). Luminance histograms are shown on the right of each image. Reproduced by permission of The Football Association Premier League Ltd.

green and blue components. The look-up tables are used to compute the values in a 'corrected' version of image 2, from the original values in image 2.

Figure 3.28 shows how this can be applied to a portion of the panoramic image in order to obtain a better match (Thomas, 2011). Despite the players being in different positions due to parallax, and the differences in resolution, the colour histogram approach provides a good match.

It is possible to use the RGB mapping function generated by the histogram matching process over this small window to process the whole panoramic image. This might not always be expected to give such a good result, as colours that are not visible in the small window will not necessarily be adjusted appropriately. However, encouraging results have been obtained with this approach, particularly when using of the widest-angle broadcast camera together with the corresponding area from the panoramic camera to compute the mapping function.

3.3.3   Blending and Switching Between Camera Views

Let us assume that we now have one or more panoramic cameras and one or more broadcast cameras, whose relative positions, orientations and fields-of-view are known, and that they have been colour-balanced to give very similar colour renditions.

A functionality that this content could support would be allowing a viewer of the panoramic image to select a particular area to zoom in to, and for the system to realise that there is a close-up of this region being provided by a broadcast camera. The system could then switch to this view. However, performing a simple cut between a wide-angle view and a close-up could leave the viewer disorientated (see discussion on avoiding cuts between shots with similar compositions in Chapter 2, Section 2.4.3). Ideally we would like to be able to perform a continuous zoom into the area of interest, starting with an image derived entirely from the panoramic camera, and ending on an image derived entirely from the broadcast camera.

One approach to this problem would be to attempt to introduce detail from the broadcast camera into the panoramic image. This would avoid a sudden switch between images, and also ensure that any colour mismatch remaining after photometric matching had a minimal effect, as low frequency information would always come from the panoramic image. An arrangement of filters that could be used for this is shown in Figure 3.29. The low-pass filters would generally be two-dimensional, filtering out high-frequency detail in all directions. If both low-pass filters have the same characteristic, and cut off below the frequency at which the response of the panoramic camera starts to reduce, then the resulting image should have a flat frequency response.

A fundamental problem with this approach is that it requires perfect alignment between both images for all objects, which is only possible if the cameras have exactly the same viewpoint. This can only be achieved using optical beam-splitters and very precisely aligned cameras. The baseline between the cameras shown in Figure 3.25 is sufficient to cause parallax shifts so large that only a small part of the scene can be aligned at any one time. Figure 3.30 shows an example where the two images are aligned correctly for player no. 12 (providing a useful increase in resolution), but all other players and the background are significantly misaligned. Using a more sophisticated approach, such as depth-based warping, could help eliminate these gross misalignments, but would still leave parts of the panoramic image without additional detail due to them being obscured by foreground objects in the broadcast camera image. Depth-based warping would in any case be difficult to implement, due to the need to match dense features between images of very different resolutions, as discussed earlier.

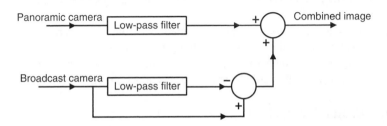

Figure 3.29 Combining low frequencies from the panoramic image with high frequencies from a broadcast camera.

Figure 3.30   Attempting to enhance the detail in a panoramic image using an image from an adjacent broadcast camera.

An alternative approach is to simply generate a zoom into the panoramic image, followed by a cross-fade to the image from the broadcast camera. Again, the issue of the cameras being spatially offset causes problems, but they are easier to overcome: if the portion of the panoramic image is framed such that the main object of interest is at the same location in the image as it is in the broadcast camera image, then the attention of the viewer stays in the right place during the cross-fade. This amounts to needing to know the depth of the object of interest in the image as well as the calibration data for each camera. This approach is essentially the same as that used in applications such as Microsoft's Photosynth, based on the 'Photo Tourism' project (Snavely et al., 2006), where the user can browse between multiple calibrated images of a scene, and a selected object may be chosen to remain at the centre of the rendered image as the virtual viewpoint moves.

Where the broadcast cameras have very different viewpoints from the panoramic camera, transitioning between them with a simple cut would probably work well. Knowledge of the calibration data for each broadcast camera, together with an approximate indication of the depth of the main objects-of-interest in the scene, would allow the approximate field-of-view of each broadcast camera to be indicated in the panoramic image, for example, by overlaying a trapezium corresponding to the area being viewed by each. This could help users identify and select different shots, and may even be helpful to the programme production team (see discussion in Section 9.3.2).

The panoramic video feed could also be presented to a 'companion screen' device, for use alongside a conventional TV broadcast. The ability to interactively explore the whole scene mimics the ability of a spectator to freely look around, while the broadcast 'curated' view replicates that which would typically be shown to spectators on big screens in a live venue.

3.4   High Frame Rate Video

The frame rates used for film and television have been fixed for the best part of a century. A belief has arisen (Ferguson and Schultz, 2008) that the frame rates chosen are close to an upper limit of the human visual system, and that little improvement can be expected from an increase. This section challenges that view, reporting on some experimental work that shows that the use of higher frame rates for capture, storage, transmission and display offers clear advantages at the resolutions associated with SD and HDTV. We will also explain why the frame rates currently in use will increasingly limit the quality of television pictures, if the size of displays and/or the resolution of television systems continue to grow.

In the days of silent cinema, frame rates were not standardised, and were largely in the range of around 16 to 24fps (frames per second). Cameramen and projectionists sometimes varied the speed according to the subject matter portrayed. Thomas Edison however, for a time, recommended 46fps, possibly to prevent flicker (Hendricks, 1961). With the development of sound-on-film in the 1920s, film speeds and hence frame rates standardised at the now ubiquitous 24fps.

To avoid visible flicker, a double or triple-bladed shutter was used to display each image two or three times in quick succession. A downside of this technique is that moving objects being tracked by the eye appear as two or three overlapping images or appear to jump backwards and forwards along their line of motion: an effect also known as 'film judder' (Roberts, 2002). One exception within the film industry is Douglas Trumbull's Showscan system (Trumbull and Jackson, 2013) that uses 65 or 70mm film running at 60fps with a single-bladed shutter. This system is sometimes used for high-speed action films and ride simulations to provide more realistic motion.

The Marconi-EMI television system (now known as '405-line') was described contemporaneously as 'high-definition television'. This and all subsequent TV standards have used a field rate that is the same as the mains frequency (50Hz in Europe).

The reasons given at the time (BBC, 1939) for synchronising the frame rate of television to the mains frequency were to avoid 'beating' against the 100Hz brightness fluctuation in AC-driven studio lights and the 50Hz fluctuation induced by poor ripple-suppression in the HT generation circuitry of early CRT televisions (Engstrom, 1935). The 60Hz mains frequency used in the US similarly led to a 60Hz field rate in their television systems (Kell et al., 1936). Subsequently, on the introduction of colour TV systems, a 1000/1001 adjustment was found necessary in the NTSC system, which means that TV systems described as '60Hz' are generally running at 59.94fps. In addition, these rates are slightly above the 40Hz minimum that was found necessary to avoid visible flicker in the displayed image on contemporary television screens (Engstrom, 1935).

At that time, it was considered sufficient (Zworykin and Morton, 1940) for the frame rate to be high enough merely to exceed the threshold for 'apparent motion' – the boundary above which a sequence of recorded images appear to the eye as containing moving objects rather than being a succession of still photographs. Priority was not given to the elimination of motion artefacts such as smearing and jerkiness. Contemporary tube cameras suffered from image retention, which may have limited the benefits of a higher rate anyway.

A final benefit of choosing a field rate equal to the mains frequency is simple interoperability with cinematic film recording. In 50Hz countries, since the speed difference between 24fps and 25fps is generally imperceptible, a frame of film can be represented as two successive fields of video. In 60Hz countries alternate frames of film have to be represented as three successive fields (a frame and a half) of video, a process known as '3:2 pull-down', which introduces further judder artefacts.

In summary, it appears that the field rates originally determined for television (and kept ever since) were chosen to meet the following criteria:

- greater than the perceptual threshold for apparent motion;
- high enough that flicker was imperceptible on contemporary televisions; and
- simple conversion to and from cinematic film.

3.4.1 Early Work on HDTV Frame Rates

With research into HDTV commencing in the 1970s, the question of the appropriate frame rate for the new television standard was open for re-evaluation. The Japanese broadcaster NHK was the leader in this field, and the 1982 summary of their HDTV research to date by Fujio (1982) identifies 'frame frequency' as a parameter to be determined. There appears to be no published research from them on the subject, however, and the field rate of NHK's 1125-line interlaced HDTV standard remained essentially unchanged from the NTSC standard it replaced, at 60 fields per second.

The question of frame rate, among other parameters, was also investigated by the BBC's Research Department. Stone (1986) performed a number of experiments with a tube camera and a CRT monitor, both modified to support non-standard field rates and other parameters set by the vertical deflection waveform. The issue of increased flicker perceptibility on increasingly large and bright television sets was well known by the 1980s, and taking a leaf out of cinema's book, the use of higher refresh rates was being considered to compensate (Lord et al., 1980). Stone recognised that increasing the frame rate of television would not only reduce the visibility of flicker, but that it would also improve the portrayal of moving objects. He carried out subjective tests and found that for fast-moving subject material (corresponding to a camera pan at a speed of one picture-width per second) increasing the frame rate to 80Hz resulted in a subjective quality improvement of two points on the CCIR 5-point quality scale (Stone, 1986).

Despite this finding, the eventual standardised HDTV formats retained the 50/60Hz frame/field rate. Childs (1988) attributes this to the increases in transmission bandwidth and storage capacity required for a higher rate.

As CRT televisions grew larger and brighter, manufacturers started using frame-doubling techniques to reduce flicker. However, the simple techniques initially employed made the portrayal of moving objects worse, by introducing a 50/60Hz 'film-judder' effect (Roberts, 2002).

3.4.2 Issues with Conventional Frame Rates

Current television field and frame rates cause problems for motion portrayal. Objects stationary within the video frame are sharp, provided they are in focus, but objects that move with respect to the frame smear due to the integration time of the camera's sensor. Shuttering the camera to shorten the integration time reduces the smearing, but the motion breaks up into a succession of still images, causing jerkiness. The perceptual difference between moving and stationary subjects is increased with the increasingly sharper images due to new television systems with successively higher spatial resolutions, so long as the temporal resolution remains unchanged. We describe the ability of a television system to represent the spatial detail of moving objects as

Trajectory of ball, captured with short shutter

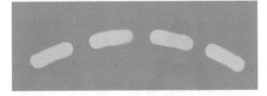

Trajectory of ball, captured with 50% shutter

Trajectory of ball, captured at double frame rate, with 50% shutter

Figure 3.31   Effects of frame rate and shuttering on motion portrayal.

its 'dynamic resolution'. The problems of insufficient dynamic resolution – smearing, jerkiness or a combination of the two – are more noticeable with larger displays where the eye tends to follow the motion across the scene. The problem is illustrated in Figure 3.31, in terms of the movement of a ball across a plain background. In the top illustration, the trajectory of the ball is shown as if captured by a video camera with a very short shutter. Each frame would show the ball 'frozen in time', and the motion would appear jerky when the video sequence was replayed. In the middle illustration, the effect of a (half-) open shutter is depicted. The camera integration smears the motion of the ball out over the background, removing any spatial detail and making it partially transparent. These effects would be clearly visible in the final video sequence. The bottom image shows the effect of doubling the frame rate: both the smearing and jerkiness are reduced. A substantial further increase in frame rate would still be required in this example to eliminate their effects, however.

In cinema, which evolved a high resolution to frame-rate ratio much earlier than television, production techniques have evolved in parallel to deal with the low dynamic resolution of the medium. Tracking shots and camera moves are commonplace, often used in conjunction with short depths of field, which help by softening backgrounds that if moving at different speeds to the tracked subject would otherwise appear to jerk and judder.

The decision to adopt interlaced video for SD television resulted in a lower spatial resolution and a higher image repetition rate than would have been the case in a progressively-scanned

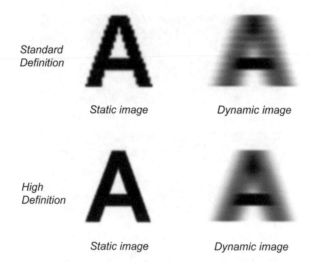

Standard Definition

Static image   *Dynamic image*

High Definition

Static image   *Dynamic image*

Figure 3.32   Static and dynamic resolution at SD and HD.

system of the same frame rate and bandwidth. Hence the dynamic resolution is better matched to the static spatial resolution and so the problems of motion portrayal were considerably ameliorated.

High-definition television (by which we mean television with a vertical resolution of 720 or 1080 lines and a field or frame rate of 50/60Hz) has increased the spatial resolution without, however, altering the frame rates used. Traditional television production techniques have been constrained by this change. For example, during camera pans to follow the action at sports events, HDTV trial viewers reported nausea as the static portion of the scene changed between sharp (when stationary) and smeared (when panning). The implied constraint of reducing the pan rate is not always practical in live coverage, but in practice compromises such as camera shuttering and deliberate softening of the images can help reduce the problem. Regardless of this, simple maths shows that motion of the camera or of objects within the scene at speeds higher than three pixels per field/frame eliminates all of the additional detail gained by the use of high definition, in the direction of motion. This effect is illustrated in Figure 3.32. These problems will be compounded by any future increases in the spatial resolution of television.

Just as shuttering in the camera reduces the extent of smearing, a sample-and-hold characteristic in the final display increases it in a directly comparable fashion. This smearing arises with trackable motion in the displayed video where the eye is following the object across the screen, but where within each displayed image the object remains stationary for duration of the frame or field. This characteristic is to be found in the LCD televisions that currently have a dominant share of the market, and is the reason why these displays have a reputation for representing fast-moving material, such as sport, poorly.

Manufacturers have added processing inside LCD televisions to perform a motion-compensated frame rate doubling, which ameliorates the problem to some extent at the cost of introducing other artefacts when the motion becomes too hard to predict, and during cuts and cross-fades.

In the light of these issues, we propose that higher frame rates be part of any future video format standard, tracking or exceeding any future increases in spatial resolution. This would help redress the imbalance between dynamic and spatial resolutions that exists in current television standards, and is a necessary precursor to further increases in spatial resolution if further undesirable constraints on production techniques are to be avoided.

3.4.3 Practical Investigations into the Effects of High Frame Rates

To investigate the theoretical advantages of high frame-rate capture and display, an intensive week of experiments was undertaken at the BBC in summer 2007. Using a Vision Research Phantom V5.1 camera, a series of 25-second sequences were captured at a resolution of 1024 × 576 and a rate of 300 frames per second. This camera is capable of capturing video at up to 1200fps, and at resolutions of up to 1024 × 1024 pixels, but has only sufficient memory to capture four seconds of video at that resolution and rate. To obtain a TV-standard 16:9 aspect ratio the vertical image was cropped to 576 lines. The Bayer-pattern sensor implies a lower luminance resolution than this, similar in magnitude to the reduction in vertical resolution associated with the use of interlace in standard definition television. A shooting frame rate of 300fps was chosen to allow for shots in excess of 20 seconds long, and to facilitate down-conversion to 25, 50 and 100fps video. Besides, 300fps also has the advantage of simple down-conversion to 60fps. Each 25-second sequence took around 10 minutes to download from the camera.

A variety of subjects was chosen to explore the advantages of high frame-rate capture and display. These included a roulette wheel and a rotating bicycle wheel, for rotational motion; bouncing balls, table-tennis and juggling, as examples of fast-moving 'sports' material; and a fast-panning camera shot with and without a tracked subject. There were few displays that could accept and display video at frame rates higher than around 60fps. CRT computer monitors can in some cases be driven at up to 200fps at reduced resolution, but with a display size much smaller than is normal for HD televisions. For the purposes of these experiments, a projector designed for frame-interleaved stereoscopy applications was selected, which could be driven at 100fps at a sufficiently high resolution: the Christie Mirage S+4K. The material was sent to the display over a Digital Visual Interface (DVI) from a dedicated playout PC, reading uncompressed YUV video from a high-speed RAID array. To create 100fps material, every three successive frames of the 300fps original were averaged (in the linear RGB domain) to simulate an unshuttered 100fps camera. For comparison purposes, six successive frames were averaged to simulate an unshuttered 50fps camera, and finally a third condition alternating between averaging six and dropping six successive frames to simulate a 25fps camera with film-style 50 per cent shuttering.

Further material was computer generated by taking a still image and simulating a sinusoidal pan across it, with camera integration to match the frame rates and shuttering choices described above. The still image chosen was the well-known 'Kiel Harbour' photograph shown in Figure 3.33. The video sequence was rendered at a resolution of 1280 × 720.

The observations were as follows. The most striking differences were seen in the panning shots – real and simulated – where the loss of spatial resolution in the detail of the background was particularly marked in the 720p Kiel Harbour simulated pan sequence, as shown in Figure 3.34. In the standard definition pan shot, lettering that was clearly legible in a static image was unreadable during the pan at frame rates below 100fps. The reduced motion blur on

Figure 3.33 'Kiel Harbour' still.

the tracked pan shot also gave a greater sense of realism and 'three-dimensionality' as the improved dynamic sharpness of both the moving objects and the background improved the quality of the occlusion depth cue.

The table-tennis sequence demonstrated that even 100fps was manifestly insufficient for coverage of this and similar sports when viewed perpendicular to the action. Motion blur was also still in evidence in the juggling sequence at 300fps, played back at 1/3 speed.

It is striking that significant improvements were discernible even at resolutions similar to standard definition television. This implies that high frame-rate capture and display is a

Figure 3.34 Portion of Kiel Harbour, with motion blur representing a pan across the scene taking 4.25 seconds, as captured with a 1/300 s shutter, a 1/100 s shutter and a 1/50 s shutter.

technique that can improve the quality of television in its own right, as well as a necessary consideration as the spatial resolutions of proposed television standards continue to increase.

Further Experiments and BBC Demonstration at IBC 2008

Following the success of the first experiments, a formal project was initiated, with material captured again at 300fps, this time using an Arri Hi-Motion camera (TVB Europe, 2010), providing better video quality, at full HD resolution (1920 × 1080 pixels) (Armstrong et al., 2008).

A wider variety of material was shot, and edited to produce a demonstration shown on the EBU stand at IBC 2008 (International Broadcasting Convention, Amsterdam). Again the main constraint was in the display, which was a projector again running at 100fps, and at sub-HD resolution. The results obtained were however quite dramatic, and the demonstration drew crowds of interested viewers, many of whom were most interested to see with their own eyes something they had long suspected would yield a dramatic improvement in the realism of video presentation. Figure 3.35 shows a pair of stills from the experiment, the top image being that representing the scene (a moving model railway locomotive) as it would have been captured in a conventional 50Hz TV system. The lower image shows the same image, as actually captured with a shutter opening of 1/300 second.

The experiment also confirmed the results observed in the first experiment, but the resulting video material was of significantly higher quality, and covered a wider variety of content types, enabling convincing demonstrations to be given.

Experiments Conducted by NHK in Japan

The Japanese state broadcaster, NHK, has subsequently conducted experiments intended to explore the frame rate requirements for Ultra High Definition TV (UHDTV).

Sugawara (2011), reported the results as indicating that future TV systems would require a display rate above about 80fps to prevent large-area flicker (at the screen sizes and brightness expected), that a frame rate of greater than 100fps was required to prevent a stroboscopic (judder) effect with motion for tracked motion, and a capture time of below 1/320 of a second was required to prevent motion blur in the capture from detracting from the video quality. Sugawara's paper also posed the question as to whether 300 or even 600fps was really feasible, however desirable it might be as a common multiple of current frame rates, and expressed the hope that we might nonetheless be able to agree a single worldwide standard for a higher frame rate. He also posed the question as to whether the 1000/1001 issue might be less of a problem in the future, and concluded by suggesting that the camera, transmission and display technologies to enable the use of a higher frame rate than 60Hz will be developed in the near future.

Work at Sony Corporation

Kuroki at Sony Corporation (Kuroki, 2010) reported the development of a 240fps 4k by 2k camera and display system intended for stereoscopic 3D reproduction. His psychophysical evaluations of sequences from 60 to 480fps found that a frame rate of around 250fps is close

Figure 3.35   Still from BBC test shoot, above as it would have been captured with 50 Hz frame rate, and below, at 1/300 second shutter opening.

to the perception limit for both blur and jerkiness. He suggests that 240fps is 'ideal' because it is a common multiple of both 24 and 60fps.

He noted that his first (2D) prototype gave a better impression of depth at a higher frame rate, which he suggests might be due to more effective pictorial depth cues, such as occlusion, shading and texture gradient and so on. He also noted that a flying ball in open space was harder to visualise. We wonder whether this might possibly be because of the loss of motion blur in the background behind a tracked object. It was this observation that stimulated his development of a single-lens stereoscopic 3D system.

Evolving Frame Rates for Stereoscopic 3D in the Cinema

The world of cinema is considering higher frame rates. In particular, renowned filmmakers Peter Jackson (*The Hobbit*) and James Cameron (*Avatar 2*) are challenging the convention of 24fps as a result of their experiences of stereoscopic 3D film production. Initially James Cameron shot a couple of test scenes at 24, 48 and 60fps in stereoscopic 3D. He used these as the basis of a technical demonstration at CinemaCon, the official convention of the National

Association of Theatre Owners in Las Vegas in March 2011. Cameron is reported (The Hollywood Reporter, 2011) as saying that he fully intends to use higher frame rates and is looking seriously at 60fps.

Peter Jackson shot 'The Hobbit' at 48fps, having tested both 48 and 60fps. He used a 270 degree shutter angle (The Hobbit Movie website, 2011). He reports that it looks more lifelike and easier to watch, especially in stereoscopic 3D and enough theatres were capable of projecting the film at 48fps by the time of the film's release in December 2012 for there to be a significant impact, certainly in critical circles. It would be fair to say that the effect had a mixed reception. As a first, the techniques used clearly developed as the film was shot. The 270-degree shutter opening was designed to enable a reasonable 24fps presentation by dropping alternate frames, giving the appearance of film shot with a 135-degree rather than 180 degree shutter.

The comments made have shown a lack of consensus, with views expressed that 48 is 'too fast' compared to the eye's ability to track motion to the more predictable response that 'it looks like video'.

3.4.4   Future Frame Rates for TV Production and Distribution

The experiments and work described above have shown a very strong indication that a capture, transport and display frame rate higher than 100fps will result in a much more realistic, and hence immersive experience for the viewer. The work at NHK has shown that at higher resolutions (again required for a truly immersive experience) a shorter capture time is required to freeze the motion in the captured scene.

Frame rates for capture and production of 300 frames per second or even 600fps have been suggested, as both are suitable for easy down-conversion to conventional 50 and 60Hz transmission (very useful for international events), or even 24fps for theatrical/film presentation. Such frame rates also enable advantage to be taken of greater artistic freedom for the producer of the content to adaptively select different temporal windows to reduce or eliminate temporal alias effects.

There is obviously a corresponding increase in raw data rate of material captured at such higher frame rates, but it is highly likely that the advantages of sharper images, a lower bit-depth requirement, and freedom from temporal aliases, would enable such material to be compressed with greater efficiency. Indeed, with video compression systems using a long-GOP (group of pictures) (inter-frame) compression mode, where the limit on GOP-length is for example a duration of half a second, a faster frame rate implies more frames in each GOP, and hence more efficient coding. It is thus possible that the higher raw data rate could nonetheless result in little or no overhead once that video is compressed. This is an area currently receiving further study. Another area for further study is the relation between measured noise levels on a video signal, and the visibility of that random noise to the observer, as the frame rate increases.

3.4.5   Consideration of Frame Rates and Motion Portrayal in Synthetic Production

Where an output image is created synthetically by panning a window within a larger captured frame, there are some interesting potential effects that must be considered carefully. If the visual appearance of a panned shot is to match that of a conventional camera, artificial motion blur must be added. However, when that pan is in fact following a moving object, the opposite (and more difficult) effect of removing the motion blur of that moving object as captured by the static camera must be implemented.

It is evident that both shorter shutters (to reduce captured blur) and higher frame rates (to enable the eye to fuse motion and avoid the perception of judder) will be enablers for these production techniques. A higher frame rate in capture and processing may also be very beneficial in this case.

3.4.6   Conclusions on Frame Rates

The spatial resolution of broadcast television cameras and displays has reached the point where the temporal resolution afforded by current frame rates has become a significant limitation on the realism of video reproduction, particularly for fast moving genres such as sport. BBC Research & Development has successfully demonstrated that increasing the frame rate can significantly improve the portrayal of motion even at standard definition, leading to a more immersive experience. If the spatial resolution of television standards continues to increase, the importance of raising the frame rate to maintain the balance between static and dynamic resolution will only increase. Even at the spatial resolutions of SDTV and HDTV, the motion artefacts associated with 50/60Hz capture and distribution rates will become increasingly apparent as television display sizes continue to grow.

Even for television pictures transmitted and displayed at conventional frame rates, capturing at high frame rates can offer some improvement to picture quality through temporal oversampling, giving better control over temporal aliasing artefacts and offering a choice of 'looks' to the director at the post-production stage. It also offers improved compatibility with the different conventional frame rates adopted internationally.

It is evident that a higher capture and display frame rate leads to a step change in picture quality regardless of the spatial resolution. It is thus an important factor in making the presentation of video material more immersive. The current proposal for next generation (Ultra-High Definition TV – ITU-R BT.2020) proposes a single worldwide standard of 120fps for a higher frame rate. Further work is required to decide whether 100, 120 or 150fps is a suitable frame rate for use in the current 50Hz countries, particularly with regard to lighting flicker, and down-conversion for delivery in 50Hz (interlace field rate), 25Hz (progressive), and there is also some uncertainty over compatibility of 120fps with 59.94-based legacy formats. The question of whether acquisition and production might be undertaken at still higher frame rates than these (e.g., 600fps) is not one that has to be settled at present, and can be left to develop, as a way to further improve the delivered immersive content.

3.5   High Dynamic Range Video

The concept of format-agnostic video enables a high degree of interaction with the captured media content at later stages in the production chain. This leads to the requirement that during scene acquisition no decisions should be made that limit the possibilities of interacting with the captured content afterwards. The illumination and absorption characteristics of natural scenes feature higher contrast than conventional video systems and displays can handle. First of all this requires capturing devices with high dynamic range capabilities. High dynamic

Figure 3.36 Two examples of high dynamic range scenes: 'inside a car' (left), 'street at night' (right).

range (HDR) tone-mapping is a well-known technique to adapt the high dynamic range of the captured scene to the much lower dynamic range and brightness of existing displays. In the case of format-agnostic video the editor or even the end user should be able to interactively select the view within the panorama. As a second requirement the tone-mapping process has to happen at the stage of displaying the selected media content rather than at the stage of acquisition like in conventional video production. As a result the tone-mapping process is highly adaptive and dynamic, dependent not only on the type of display but also on the selected view. This section deals with the basics of HDR capturing and tone-mapping and their specific design in format-agnostic video.

3.5.1 The Human Visual System in Natural Environments

In natural scenes the dynamic range of illumination is by principle not limited. For example, an indoor scene, such as inside a car, with bright windows features very high contrast; a night scene on a dark street viewing headlights and taillights of cars is another example of extreme contrasts (see Figure 3.36). The key question in these situations is: what is the relevant dynamic range of the scene? In video content production the human visual system is the ultimate reference. Hence it is important to understand the perception of illumination and contrast in human vision.

The Dynamic Range of Human Visual Perception

Two basic types of photosensitive cells exist in the human eye: *rod* and *cone* cells. Each type is specialised for different illumination conditions. *Scotopic* vision is the vision under low light conditions. In this case only rod cells are sensitive. The luminance level for scotopic vision ranges from 10^{-6} to 10^{-2} cd/m^2. The cone cells are less sensitive and therefore not active at this low illumination level. *Photopic* vision occurs when the luminance level is above 10 cd/m^2. In this case only cones are active. Three types of cone cells exist in normal human vision. Each type is sensitive to a specific wavelength of visible light. Hence cone cells allow the perception of colour. At an illumination level of about 10^6 cd/m^2 *glare* occurs, which is defined as a visual condition that disturbs the observer or limits the ability to distinguish details and objects (CIE, 1999).

The intermediate illumination range between 10^{-2} cd/m^2 and 10 cd/m^2 is called *mesopic* vision (Stockman and Sharpe, 2006). Both the rod and cone cells are sensitive in this range. From the lowest illumination level, where the human vision begins, to the brightest illumination level, where dazzling occurs, the human vision covers a dynamic range of 10^{12}:1 or in photographic terms 40 stops, whereby one stop is a factor of two in brightness (Smith, 2005).

Several mechanisms in the human eye exist, which facilitate this huge dynamic range. The cone and rod cells are sensitive at different illumination levels as mentioned above. Depending on the illumination level the size of the pupil will change. In dark environments the pupil will automatically open to about 8 mm in diameter. In bright environments the pupil will narrow to a diameter of about 2mm. Thus, an adaptation factor of about 16:1 (4 stops) is realised. The biological pigment Rhodopsin is involved in extremely low light conditions. It takes about 45 minutes for human vision to fully adapt to these conditions (Litmann and Mitchell, 1996; Stuart and Brige, 1996).

Another characteristic has been observed, which increases the sensitivity of the human eye by either integrating the stimuli spatially over a larger area or temporally over time. Neighbouring rod cells can be grouped and merged to larger perception elements. Long term fixation can bring a very low stimulus above-threshold (Encyclopaedia Britannica, 1987).

Several of the above mentioned mechanisms require longer adaptation times. Hence the full dynamic range cannot be utilised simultaneously. According to the rules of evolution the purpose of the large dynamic range might be allowing good vision in all sensible natural situations from bright sunlight to a moonless night. The transition between these extreme conditions is typically not immediate, but takes some time in the order of several minutes to an hour. In natural environments the human visual system does not require a faster adaptation rate. In addition, lens flare in the human eye is an important factor, which limits the contrast range.

Just Noticeable Difference

As well as the dynamic range of the illumination another important factor exists, which characterises the human visual system. The Weber-Fechner-law (Fechner, 1889) describes the ability to distinguish two stimuli with slightly different levels. In psychophysics the term *just-noticeable difference* is used for this threshold. The eye sees brightness approximately logarithmically over a moderate range but more like a power law over a wider range. This is the reason why image data are often represented in logarithmic or power-law quantisation. In the realm of digital cinema, the contrast sensitivity has been experimentally evaluated by Cowan and colleagues (2004). It was found that the threshold modulation of the human observer can be as small as 0.002, which leads to the requirement of 12-bit representation for optimum digital cinema representation. This will be discussed in more detail in Section 3.5.4.

Immersion and HDR Imaging

In virtual environments like games, artificial scenes, and remote video broadcasting the term *immersion* describes the extent to which the user has the sensation of being there like they usually feel in conscious reality (Nechvatal, 2009). Hence, it is obvious that the visual stimulus in a virtual environment, being presented by a technical display device, should differ as little as possible from the natural stimulus. This poses a challenge to displays with respect to dynamic range and quantisation of illumination. In order to be indistinguishable from real life, a display would need to serve a logarithmic quantisation of 12 bits over a range of 40 stops of illumination from a moonless night to the direct sunlight.

3.5.2   Conventional and HDR Video Cameras

Conventional video acquisition, distribution and display systems are limited to a defined dynamic range of brightness and contrast. In conventional video production various methods have been established to control the brightness and contrast of the captured scene. Some techniques are applied on set. Examples are lighting up dark areas in the scene by additional lighting fixtures or dimming bright areas like a window by a darkening foil. Other techniques are applied in the camera during capturing, for example, control of the iris, exposure time and electronic gain. As a result the contrast range of the captured scene fits the low dynamic range of the conventional video system. As long as the scene contrast is limited, the above mentioned controls are sufficient to render beautiful images. This approach, however, is not sufficient, if higher scene contrasts exist. In many outdoor productions brightness and contrast cannot be tuned accordingly. In this case information is lost in the extremes of the dynamic range, either through clipped highlights, or fading details in the shadow. An additional requirement comes into play in the case of format-agnostic video productions. All the above mentioned techniques – reducing the dynamic range of the scene to the limited range of the utilised displays – cannot be applied during capturing but only at a later stage in the production chain. As mentioned in the introduction to this section only an interactively selected region within the panorama will be displayed in the end-user device.

Figure 3.37 illustrates the different techniques of controlling brightness and contrast comparing conventional video acquisition with the format-agnostic approach using a panoramic

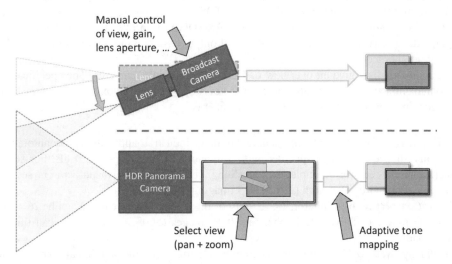

Figure 3.37   Control of brightness and contrast in conventional broadcast production (upper part) and format-agnostic panorama production (lower part).

camera. In the upper part of the diagram the conventional approach is illustrated. Selecting the field of view is achieved by physically panning and zooming the camera to the desired region within the scene. Brightness and contrast are controlled at the stage of capturing, that is, in most cases within the camera. The transmitted signal is limited to the allowed range of the broadcast chain, which typically reflects the constrained capabilities of standard TV displays. In the lower part of the diagram the format-agnostic approach is illustrated. Here, an HDR video panorama is captured. The field of view is selected by defining a region-of-interest (ROI) within the captured image. In this case the brightness and contrast settings have to be adapted to the currently selected ROI. Hence the control of the dynamic range is an adaptive process, which is called *adaptive tone mapping* and will be explained in more detail in Section 3.5.5.

Overview of HDR Video Cameras

For many years photographic film was the only media capable of capturing higher dynamic range. Electronic video cameras and early digital cameras suffered from low sensitivity and contrast. For several years, however, techniques for high dynamic range acquisition have been researched and established in selected digital still and video cameras.

All electronic image sensors feature a limited full-well capacity and non-disappearing read-out noise. These parameters describe the upper limit (how many photons can be collected before clipping) and the lower limit (where the signal disappears below the noise floor) of the achievable dynamic range. While chip designers try to stretch these limits, HDR methods further exceed them. The most common method is taking multiple pictures at different exposure levels and intelligently stitching them together to produce a picture that is representative in both dark and bright areas. One drawback of this technique is the fact that multiple pictures are typically taken at different times or with different exposure durations. Hence, fast moving objects appear at different locations in each exposure and can show up as ghost images in the stitched output image. A novel technique for CMOS sensors was first reported by Li (2008). The so called *Dual Conversion Gain Architecture* (DGA) improves the dynamic range by changing the electrical amplification of the pixel output based on the illumination level. (For a survey of other techniques see Li, 2008.)

The following lists give an overview of HDR still and video cameras. The specifications are from manufacturers' information and hence not fully suitable for comparative evaluations. HDR techniques were first established in still cameras.

- CMOS sensors from several chip manufacturers for industrial applications like automotive and surveillance have been specified with a dynamic range of up to 120 dB. For sensor manufacturers see, for example, Aptina, Sony, NIT, for camera manufacturers see, for example, Point Grey Research (2012), IDS Imaging, Lumenera.
- *SpheroCam HDR* by SpheronVR is based on a line scan camera with 26 stops dynamic range. It delivers full spherical HDR images (360° × 180°) in a scan time of 20 seconds.
- *Civetta* by Weiss AG is another panoramic camera, capturing a full panorama within 40 second scan time with 28 stops dynamic range.
- *Leica S and S2* DSLR cameras for professional photography, dynamic range 12 stops.

In recent years video and motion picture cameras have been developed with a dynamic range higher than regular broadcast cameras. For digital motion picture production this was an essential feature convincing film maker to switch from analogue film to digital capture.

- Pioneering in the digital film camera business, the *Viper-Filmstream LDK 7500* by Grass Valley featured a dynamic range of three decades.
- *Alexa* by ARRI. This digital camera for professional motion picture production captures a dynamic range of 14 stops. The sensor utilises the above mentioned DGA principle, featuring seamless image quality throughout the complete dynamic range. The in-camera ARRIRAW recording (Alexa XT models) allows uncompressed and unencrypted recording up to 120fps.
- *Epic-X* by RED offers HDRx as a special recording mode. HDRx is an option for extending dynamic range up to 6 stops over the base dynamic range offered by the camera. When enabled HDRx simultaneously shoots two images of identical resolution and frame rate – a normally exposed primary track and an underexposed secondary track whose exposure value reflects the additional stops of highlight protection desired. The two tracks are synchronised, meaning there is no gap in time between the two exposures. The exposure times for these two tracks, however, are distinguished from each other reflecting the different sensitivity in both capture modes (Red, 2013).
- *F65* by Sony. Among other models, the F65 is Sony's flagship for digital motion picture acquisition with similar manufacturer's specification: 14 stops of dynamic range, 4k resolution, 120fps.

3.5.3 Conventional and HDR Displays

In order to determine the contrast of a display the following procedure can be used. Take two luminance measurements of the display, the first one with the highest possible input signal, the second with the minimal input signal. The ratio between these values defines the dynamic range. It is also important to consider the ambient illumination while this test is conducted. The second measurement will typically be affected by the ambient illumination. The screen of a display will reflect some light, even when it is completely switched off. This arises from back scattering of the ambient illumination. Hence the available dynamic range is not only given by the design of the display but further limited by the conditions of use. While display manufacturers tend to take these measurements under idealised conditions – namely without any ambient illumination – the ITU recommends illuminating the background at about 15 per cent of the display's peak luminance during any subjective quality assessments of television pictures (International Telecommunication Union, 2002). Under these conditions standard TV sets with a typical peak luminance of 200 cd/m^2 feature a dynamic range of about 50:1, which is equivalent to less than 6 stops.

In the case of cinema projectors a slightly more sophisticated approach is established. Here the peak luminance is set to 48 cd/m^2 and hence lower than in the TV world (SMPTE, 2006), and the dynamic range is specified in SMPTE (2011). Two different values are given for contrast. 'The *sequential contrast* ratio shall be computed by dividing the white luminance by the black luminance, with the measurements made in-situ including the contributions of ambient light' (SMPTE, 2011, p. 5). The value for sequential contrast is between 2000:1 (for reference projectors) and 1200:1 (for commercial theatres). These values represent 11 resp. 10 stops. 'Using a 4×4 checkerboard target, *intra-frame contrast* shall be computed by summing the luminance of the white patches and dividing by the sum of the luminance of the black patches' (SMPTE, 2011, p. 5).[1] The value for the intra-frame contrast is targeted at 150:1 (7.2 stops), resp. 100:1 (6.6 stops), measured with a 4×4 checkerboard. The situation is even worse for smartphones and tablets. The reason is the high ambient lighting, which is common for the use of these devices. Typical values for the maximum brightness are around 400 to 500 cd/m² and for the contrast for high ambient lighting[2] between 15:1 and 80:1. Keep in mind that the definition for contrast is different from the ANSI contrast and hence not comparable in numbers with the contrast values for cinema screens and TV sets (Soneira, 2012). We have to assume that the ANSI contrast will be worse than the quoted numbers.

Compared to the dynamic range of the human eye all these contrast values are quite poor. Figure 3.38 illustrates the comparison of the visual system against natural stimuli (sun, moon, night-sky) and against the capabilities of standard displays and novel HDR displays. For a list of specific values refer to Myszkowsi et al. (2008).

A common technique to implement high dynamic range displays is called *local dimming*. This technique is applied in combination with regular LCD displays. In this case a matrix of LEDs is used as backlight rather than a full-screen Cold Cathode Fluorescent Lamp (CCFL)

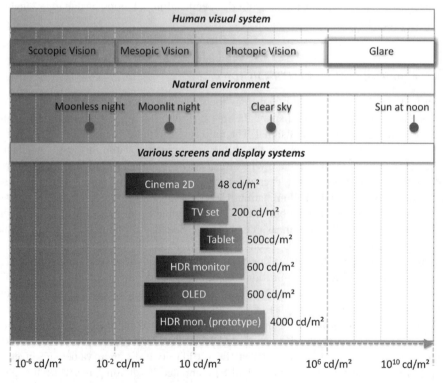

Figure 3.38 Dynamic range of the human visual system, luminance levels of natural stimuli, and dynamic range of common displays.

backlight. Each LED illuminates a small portion of the whole image. By modifying the amount of light the brightness of the related image area can be changed. This enables deeper blacks and higher contrast. The limit of this approach is given by the fact, that each LED covers an area of several pixels of the image. Hence the contrast from pixel to pixel cannot exceed the limits of the LCD's dynamic range. The benefit lies in improving the sequential contrast and the intra-frame contrast, if the scene brightness changes gradually within the frame. Prototype systems based on patented local dimming technology by Dolby feature brightness of up to 4,000 cd/m^2 and ANSI contrast ratio of 20,000:1 under ideal room conditions (Free Library, 2009).

Another approach to develop displays with higher dynamic range is the *organic light emitting diode* (OLED). An OLED display works without a backlight. Thus, it can display deep black levels and a higher contrast ratio than an LCD. OLED displays have been announced by several TV set manufacturers including Samsung and LG. Typical values for brightness are 600 cd/m^2 at a contrast ratio of 5,000:1 (OLED display, 2013). OLED displays are already in use in smartphones and tablet PCs by Samsung and other manufacturers.

3.5.4   HDR Video Formats

The basic considerations for the definition of high dynamic range video formats are very similar to single HDR images and guided by human visual perception. One important parameter is the dynamic range, which has to be covered. From the previous sections a range of 40 stops (12 decades) would be adequate for the human visual system. A second parameter is the quantisation step size between adjacent values (Reinhard et al., 2006). The human eye can distinguish two similar grey patches, if their brightness differs by more than about 0.2%. Many early image formats like, for example, the video broadcasting system, use 8 bit per colour channel. In this case the grey values are gamma compressed, which is a remnant of cathode ray tubes (CRT). The nonlinear relationship in an electron gun between applied video voltage and light intensity follows a power law with a gamma of about 2.2 to 2.8, see also Chapter 2, Section 2.2.2, where the gamma characteristics of video signals are discussed. Nevertheless, the quantisation steps are larger than the relative sensitivity threshold of the human eye. Hence the quantisation can be seen as luminance banding. Data representations of this type are called Low-Dynamic Range (LDR) image formats.

Driven by the capability of human vision and by advancements in image capture, rendering and display, various HDR image formats have been introduced over recent years. The most common formats are listed in Table 3.1.

The DPX format was developed in the 1990s in the realm of photographic film, representing the capabilities of negative film emulsion in those days. For many years DPX served as the common format for digital film production (SMPTE, 2003). The formats HDR/RGBE, TIFF/LogLuv and OpenEXR/Half RGB have been established by computer graphics experts and companies. For a more detailed description refer to Reinhard et al. (2006).

In recent years many Hollywood Studios and Visual Effects Houses have adopted the OpenEXR format. Originally developed by Industrial Light & Magic, it is released under a free software license (OpenEXR, 2013). The Academy of Motion Picture Arts and Sciences (AMPAS) recently developed the ACES format, which among others covers the dynamic range definition for cinema production beyond the days of film (IIF, 2010). In respect of

Table 3.1 Common HDR image file formats, adapted from Reinhard et al. (2006)

Format	Encoding	Bit/pixel	Dynamic range ($\log_{10}$)	Relative step %
DPX	Log	30	2	0.5
HDR	RGBE	32	76	1
	XYZE	32	76	1
TIFF	IEEE RGB	96	79	0.000003
	LogLuv24	24	4.8	1.1
	LogLuv32	32	38	0.3
OpenEXR	Half RGB	48	9	0.1
ACES	Half RGB	48	7.5	0.1

dynamic range ACES utilises the concepts of EXR / OpenEXR to represent image data. Many camera manufacturers including Arri, Red and Sony developed specific raw formats that accommodate the capabilities of their capturing devices. For more details about these formats see the companies' web pages.

It is common to all formats that they describe single image files. In a world of file-based workflows the importance of streaming formats for the production domain is continually declining.

3.5.5   Adaptive Tone-mapping for Format-Agnostic Video

Content captured with an HDR sensor features a dynamic range, which is larger than conventional display systems. Section 3.5.3 illustrated that not only the design of the display devices but also the ambient conditions of viewing affects the noticeable dynamic range. Various techniques have been established for dealing with the enormous dynamic range of natural scenes. During the acquisition process, the camera man can set the iris aperture of the lens, the exposure time of the camera and select the sensitivity of the film emulsion or use neutral density filters in front of the camera for reducing the exposure. In video cameras, the gain of the image sensor can also be controlled electronically. All these modifications will change the overall exposure of the captured image. In feature film productions as well as in still photography additional lighting fixtures are used for lightening dark sectors in the scene in order to reduce the scene contrast. With the advent of higher dynamic range cameras, novel techniques endemic to the image processing domain have been developed and established for compressing the dynamic range to the limited capabilities of the distribution channel and the display device. *Tone mapping* is a technique to map one set of colours to another in order to approximate the appearance of high dynamic range images in a medium that has a more limited dynamic range. Tone mapping addresses the problem of strong contrast reduction from the scene radiance to the displayable range while preserving the image details and colour appearance important to appreciate the original scene content (Mantiuk et al., 2006). For an overview of tone reproduction algorithms refer to DiCarlo and Wandell (2001).

This dynamic range compression is considered a creative process, as it is driven by the visual perception. A simple and steady characteristic curve compressing the dynamic range will not be sufficient to approximate the appearance of the original scene. The tone-mapping will often be controlled by a human operator during the colour grading process in post-production. In general three options exist, from which the operator can choose on a shot by shot base:

1. Concentrate on the highlights in the scene. Define the colour conversion curve such that the highlights are maintained. Shadows will be lost, meaning there are no details left in the shadow areas. The available contrast of the display will represent nice highlights, the shadows will be compromised. If the main action in the scene happens in the brighter regions and the shadow areas are of less importance this might be the setting to choose.
2. Concentrate on the shadows in the scene. Define the colour conversion curve such that the shadows are maintained. Highlights will be clipped, meaning that no detail is left in the highlights. Sometimes the highlights will also turn white, even if they used to be a specific colour. The available contrast of the display will represent nice shadows, the highlights will be clipped. This setting might be used in a scene with very few bright specular highlight spots, where it is less important to maintain the brightness level of these singular spots; compare 'street at night' scene in Figure 3.36.
3. Try to maintain both the shadows and highlights by compressing the contrast. Such the scene will appear 'washed out', if the dynamic range is just linearly compressed. Hence it is more common to utilise an s-shaped transformation curve, maintaining the slope of the curve in the mid-tone range, while compressing more towards the low and high ends. This would render a more pleasing image without harsh clipping on either end and would give the feeling of a more natural look. However, the detail reproduction both in the shadows and highlights will be reduced.

Consider the following example. In Figure 3.39 the tone mapping process is illustrated using the example from Figure 3.36 (left)). The original image (full dynamic range) is shown in Figure 3.39a, whereas the associated luminance histogram is drafted in Figure 3.39b. The chart shows the distribution of darker pixels in the image (to the left) and brighter pixels (to the right). The figures show an elevation in the mid-dark luminance, reflecting mainly the interior of the car, and a second elevation in the mid-bright luminance, reflecting the exterior of the scene. The scene contrast is larger than the available display contrast. In Figure 3.39c tone mapping for the highlight range was applied, which can be seen as lost details in the shadow area, whereas the details in the highlights have been maintained or even emphasised. The draft diagram 3.39d illustrates which range of the luminance histogram can be reproduced on the display and which is lost. The lost range is shaded and marked 'crushed shadows'. In image 3.39e tone mapping for the shadow range was applied, which can be seen as clipped highlight, whereas the details in the shadow range have been maintained or even emphasised. Diagram 3.39f illustrates the parts that can be maintained; the lost range is marked as 'clipped highlights'.

Tone Mapping for Interactive Viewing

In the case of interactive viewing of panoramic HDR content, the tone mapping cannot be carried out up front at the production site as is the case for feature films or regular TV distribution. The tone mapping curves should be based on the current viewing condition of the interactive user. Several parameters influence these viewing conditions. The end user may choose from a list of different display devices for viewing the content, where each device

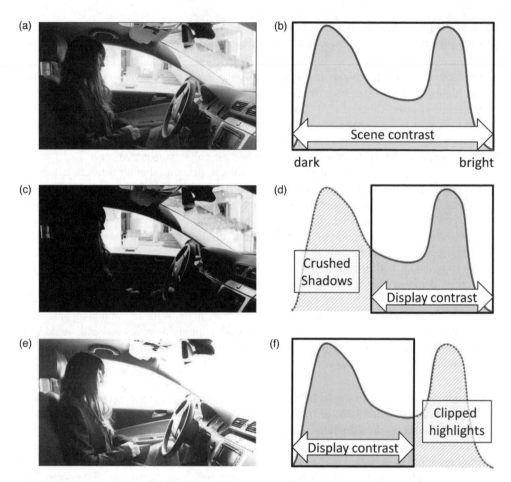

Figure 3.39 The image (a) is 'inside a car' and (b) the associated luminance histogram; (c) and (d) images and histogram, after tone mapping for highlights has been applied; (e) and (f) image and histogram, after tone mapping for shadows has been applied.

may vary in screen brightness and contrast. One of the main features for interactive viewing is choosing a smaller window within the panorama by continuously zooming and panning and watching this ROI on the individual display. Hence the tone mapping has to be content-adaptive. If the ROI of the scene gets brighter or darker over time, the tone mapping should follow in order to match the most important part of the scene contrast to the display contrast. If the user changes the viewing window by zooming and panning, the tone mapping should adapt the dynamic range within the selected window.

Up-front Generating of Two Sets of Transformation Curves

Figure 3.40 illustrates the approach as an example.

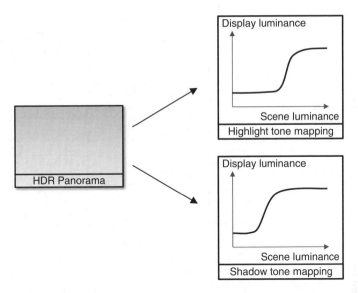

Figure 3.40 Generation of two tone mapping curves adapted for the highlight and the shadow range of the luminance in the panorama.

Colour grading defined by an expert colourist for shadow range: The colour grading curve is defined such that it renders a good representation of the shadow range. In order to display on a device with limited contrast, everything will be made a little brighter – which means that any highlights will be clipped. A special trick is reducing the colour saturation and slightly shifting the colour hue to the blue. The human eye expects less colour saturation in a dark environment as the human perception changes from photopic to scotopic vision. Therefore a scene with less colour saturation and a hue shift to blue appears darker to the viewer, even if the overall brightness level is higher than the real value of the natural scene.

Colour grading for highlight range: Like for the dark range another colour grading can been performed by the colourist. In this case the colour grading curve is designed to maintain a clear representation of the bright areas of the scene. In total the image appears brighter to the viewer, even if the overall brightness level is lower than the real value of the natural scene.

Both tone mappings are designed for a specific display type, for example, a regular TV monitor following the ITU-R BT.709 colour space. Different display types will require an adapted version of these tone mapping curves.

After generating both tone mapping curves they are a priori applied to the HDR content, rendering two low-dynamic-range (LDR) versions of it, one adapted to the brightest areas in the image and the other one adapted to the darkest areas, see Figure 3.41 for illustration. This step happens at the production site. If a viewer wants to watch a dark image area, the shadow tone mapping is selected for a pleasant picture; if he wants to watch a bright area, the highlight tone mapping is selected.

In order to smooth the transition between shadow and highlight regions, a weighting-map is provided. In Reinhard et al. (2006, Chapter 7) several methods are reported as to how such a weighting map can be generated. The rationale behind the weighting map is to calculate

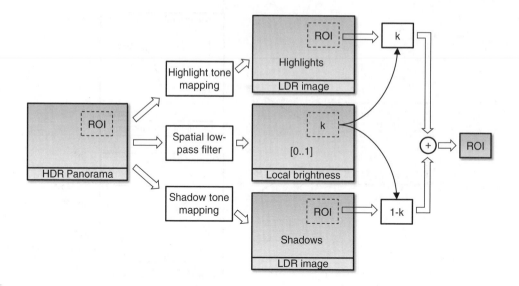

Figure 3.41 Workflow for generating two LDR versions of the panorama in advance. Calculation of the weighted mix of both LDR versions within a region-of-interest (ROI) to be displayed on the users viewing device.

the average brightness of an area. It ranges from 0 for completely dark to 1 for completely bright. If the weight is 0, the shadow-grading is appropriate, if the weight is 1, the highlight grading is appropriate, if the weight is somewhere between 0 and 1 both images are blended. The weighting function within the user-selected region can simply be defined according to the following rules:

- Calculate the scaled average brightness in a range from 0 to 1. This is achieved by low-pass filtering the selected region. This leads to a single factor k for the selected region-of-interest.
- Subsequently blend the shadow-image and the highlight-image according to the following:
- Displayed image $= (1 - k) \cdot$ shadow-image $+ k \cdot$ highlight-image

Interactive Control of the Tone Mapping

While the user interactively zooms and pans within the panorama changing the region-of-interest, the local brightness within the ROI is continuously being calculated. This leads to an ongoing update of the value k and as a consequence to a modified mix of the highlight and shadow image within the ROI. When the viewer moves from a bright area in the panorama to a shadow area, the value of k will gradually change from 1 to 0. As a consequence the displayed image will gradually be lightened revealing more and more shadow details. Moving in the opposite direction from a shadow area to a bright area, the display will continuously be toned down, avoiding clipping in the highlight areas. This built-in behaviour of the system is similar to the biophysical adaptation process of the human eye. In regards of dynamic range,

Highlight tone mapping   **Shadow tone mapping**   **Highlight tone mapping**   Shadow tone mapping

Figure 3.42 Panorama image with illustration of two interactively selected ROIs and the appropriate tone mapping.

the display will always make the best of the content, showing a picture with the best possible contrast of the selected region within the panorama.

Consider the following example. For a typical 180° panorama (captured during a classical concert in the Royal Albert Hall in August 2012) the tone mapping curves were designed for the shadow-image and for the highlight-image to be displayed on a standard TV display featuring the ITU-R BT.709 colour space. The different visual appearance of these versions can be observed in Figure 3.42. A small white box in the upper left corner of the panorama (a) illustrates the ROI, which at first the user has interactively selected on his display. While he watches the people in the box, the shadow-tone mapping is selected. The user can clearly see the content on his display (b), whereas the highlight tone mapping would appear much too dark. The user then moves the ROI over to the orchestra (white box in the centre of the panorama (a). Now the tone mapping is gradually changed to the highlight tone mapping. The orchestra can clearly be seen on his display (c), whereas the shadow tone mapping would clip the highlights.

3.6 Conclusion

This chapter has given an overview of future trends in terms of enhancing the video signal in its main dimensions such as spatial, temporal and dynamic range. Starting from a detailed review of history, the upcoming advances have been discussed in each of the areas.

In terms of ultra-high resolution panoramic video, the theory of panoramic imaging has been presented, which finally leads to two different concepts, the star-like camera arrangement and a mirror-based acquisition system. If high quality ultra-high resolution imaging is required, the mirror-based approach is the most promising. Therefore, the complete process of view registration, warping, stitching and blending has been discussed to explain the challenges and solutions for such kinds of visual scene representation.

The format-agnostic scene representation will contain panoramic video and several views from broadcast cameras in the best case. Hence, merging and blending of such additional views into the panoramic video becomes an important issue. Several approaches have been presented showing the capability for seamless merging of various video sources. This will offer more detail and image quality in specific parts of the captured scene and increase the immersive experience.

Regarding higher frame rate, it has been shown that an increase of frame rate is not just a matter of technological progress per se, but it is an essential property to increase video quality and immersiveness, especially in the case of significantly increased spatial resolution. Recent experiments performed across the world have shown that for some genres like sports, higher frame rate is a particularly important and relevant issue.

The last dimension of a video signal discussed in this chapter is the dynamic range, that is, the available contrast per pixel and colour channel. It was explained that the introduction of higher dynamic range video in production and on the display side faces plenty of challenges. At the end-user side, displays with significantly higher contrast are not yet, as of 2013, in a mature state. However, progress on OLED displays will clearly push product development in this direction. Nevertheless, acquisition with higher dynamic range can still offer increased video quality by introducing tone mapping approaches during media production.

Future media production and delivery will face the challenge that a large variety of video formats will be available. Each of the formats will be able to provide special information that may enhance the visual experience. Hence, it is necessary to keep the available information as long as possible in the media production and delivery chain in order to allow the user to select the desired visual experience. Intelligent processing, coding and conversion between legacy and future video formats will be necessary.

Notes

1. This method is also known as *ANSI contrast ratio* (http://carltonbale.com/the-truth-about-inflated-hdtv-contrast-ratios/).
2. Defined as maximum or peak brightness divided by average reflectance.

References

Armstrong, M., Flynn, D., Hammond, M., Jolly, S. and Salmon, R. (2008) 'High Frame-Rate Television'. BBC Research White Paper WHP 169.
BBC (1939) *Technical Manual, Marconi-EMI System of Television*. London: London Television Centre.
Bimber, O. (2006) 'Multi-projector Techniques for Real-time Visualizations in Everyday Environments', *Proceedings of IEEE Virtual Reality Conference/Workshop Emerging Display Technology*, accessed 13 August 2013 at: http://www.emergingdisplays.org/edt2006/media/oliver/EDT06-noanim-compress.pdf

Brown, M. and Lowe, D.G. (2007) 'Automatic Panoramic Image Stitching using Invariant Features', *International Journal of Computer Vision* 74(1), 59–73.

Capel, D. and Zisserman, A. (1998) 'Automated Mosaicing with Super-resolution Zoom', *IEEE Computer Society Conference on Computer Vision and Pattern Recognition (CVPR'98)*, pp. 885–891, Santa Barbara, 23–25 June.

Carmagus (2013) 'Endzone', accessed 13 August 2013 at: http://www.camargus.com/maxx-zoom.html

Childs, I. (1988) 'HDTV: Putting You in the Picture', *IEE Review*, July/August, 261.

CIE (1999) 'Vision and Colour, Physical Measurement of Light and Radiation', publication no. 135-1999.

CineVision (2006) 'Technologies and Systems for Digital Cinema', accessed 27 August 2013 at: http://www.tomshardware.com/news/cinevision2066-demonstration,2969.html

Cowan, M., Kennel, G., Maier, T. and Walker, B. (2004) 'Contrast Sensitivity Experiment to Determine the Bit Depth for Digital Cinema'. *SMPTE Motion Imaging Journal*, September, 281–292.

CX News (2005) 'Laser Dream Theater at the Aichi Expo', Sony Semiconductor and LCD News, Vol. 40, accessed 13 August 2013 at: www.sony.net/Products/SC-HP/cx_news/vol40/sideview.html

Dawes, R., Chandaria, J. and Thomas, G.A. (2009) 'Image-based Camera Tracking for Athletics'. *IEEE International Symposium on Broadband Multimedia Systems and Broadcasting (BMSB 2009) Bilbao*, 13–15 May.

DiCarlo, J.M. and Wandell, B.A. (2001) 'Rendering High Dynamic Range Images'. *Proceedings of the SPIE: Image Sensors*, Vol. 3965, 392–401.

Encyclopaedia Britannica (1987) 'Sensory Reception: Human Vision: Structure and Function of the Human Eye', volume 27.

Engstrom, E.W. (1935) 'A Study of Television Image Characteristics: Part Two, Determination of Frame Frequency for Television in Terms of Flicker Characteristics'. *Proceedings of the Institute of Radio Engineers*, 23 (4), 295–309.

Fechner, G.Th. (1889) *Elemente der Psychophysik*, 2nd edn. Leipzig: Druck und Verlag, Breitkopf & Härtel.

Fehn, Ch., Weissig, Ch., Feldmann, I., Müller, M., Eisert, P., Kauff, P. and Bloss, H. (2006) 'Creation of High Resolution Video Panoramas of Sports Events', *Proceedings of IEEE International Symposium on Multimedia (ISM2006)*, San Diego, 11–13 December, pp. 291–298.

Ferguson, K. and Schultz, W. (2008) 'Predicting Subjective Video Quality'. *Broadcast Engineering World*, February.

Fraunhofer HHI (2013) 'The Fraunhofer HHI TiME Lab', accessed 13 August 2013 at: http://www.timelab-hhi.com/en/home-page.html

Fraunhofer IFF (2011) 'The Elbe Dome: Immerse in Virtual Worlds', accessed 13 August 2013 at: http://www.vdtc.de/allg/elbe-dom-eng-fraunhofer-iff.pdf

Free Library (2009) 'New SIM2 Solar Series With Dolby Vision LED Backlight Technology Scheduled to Be Available Q209', accessed 13 August 2013 at: http://www.thefreelibrary.com/New+SIM2+Solar+Series+With+Dolby+Vision+LED+Backlight+Technology-a0193038242

Fujio, T. (1982) *High Definition Television*. Tokyo : Nippon Hōsō Kyōkai (NHK) and Technical Research Laboratories, Technical Monograph 32.

Gledhill, D., Tian, G.Y., Taylor, D., and Clarke, D. (2003) 'Panoramic Imaging – A Review'. *Computers & Graphics* 27(3), 435–445.

Griffiths, A. (2004) 'The Largest Picture Ever Executed by Man: Panoramas and the Emergence of Large-Screen and 360 Degree Internet Technologies', In J. Fullerton (Ed.) *Screen Culture: History and Textuality*. London: John Libbey Press, pp. 199–220.

Grundland, M. and Dodgson, N.A. (2005) 'Color Histogram Specification by Histogram Warping', *Proceedings SPIE 5667*, Color Imaging X: Processing, Hardcopy, and Applications, 610, 28 January.

Hartley, R.I. and Zisserman, A. (2004) *Multiple View Geometry*. Cambridge: Cambridge University Press.

Hendricks, G. (1961) *The Edison Motion Picture Myth*, Berkeley: University of California Press.

HPC Market Watch (2011) 'Seattle Cinerama Grand Reopening', accessed 13 August 2013 at: http://markets.hpcwire.com/taborcomm.hpcwire/news/read?GUID=15456683&ChannelID=3197

IIF (2010) 'The Image Interchange Framework,' accessed 13 August 2013 at: http://www.digitalpreservation.gov/partners/documents/IIF_Overview_August_2010.pdf

International Telecommunication Union (2002) *Rec. ITU-R BT.500-11. Methodology for the Subjective Assessment of the Quality of Television Pictures*. Geneva: ITU.

Internet 2 (2013) Accessed 13 August 2013 at: http://www.internet2.edu/

JPEG2000 (2004) *ISO/IEC 15444-1:2004, Information Technology – JPEG 2000 Image Coding System: Core Coding System*. Geneva: ISO.

Kell, R.D., Bedford, A.V. and Trainer, M.A. (1936) 'Scanning Sequence and Repetition rate of Television Images'. *Proceedings of the Institute of Radio Engineers* 24(4), 559–576.

Kumar, R., Anandan, P. and Hanna, K. (1994) 'Direct Recovery of Shape from Multiple Views: A Parallax Based Approach'. 12th International Conference on Pattern Recognition (ICPR'94), Jerusalem, Israel, 9–13 October, pp. 685–688.

Kuroki, Y. (2010) 'Development of the High Frame Rate 3D System', *Proceedings of the 1st Brazil/Japan Symposium on advances in Digital Television*, December 2010, accessed 13 August 2013 at:
http://www.sbjtvd.org.br/2010/anais/development_of_the_high_frame_rate_3d_system.pdf

Lantz, E. (2007) 'A Survey of Large-Scale Immersive Displays', *Proceedings of the Emerging Display Technology Conference*, ACM SIGGRAPH.

Li, X. (2008) 'MOSFET Modulated Dual Conversion Gain CMOS Image Sensors,' unpublished PhD thesis, Boise State University, Idaho.

Litmann, B.J. and Mitchell, D.C. (1996) 'Rhodopsin structure and function'. In A.G. Lee (Ed.) *Rhodopsin and G-Protein Linked Receptors*, Part A, vol. 2. Greenwich, CT: JAI Press, pp. 1–32.

Lord, A.V. et al (1980) 'Television Display System', UK Patent GB2050109, 8 May 1980.

Mantiuk, R., Myszkowski, K. and Seidel, H.P. (2006) 'A Perceptual Framework for Contrast Processing of High Dynamic Range Images', *ACM Transactions on Applied Perception* 3(3), 286–308.

MegaVision (2004) 'The Mega Vision System Overview', October, accessed 13 August 2013 at:
http://www.megavision.co.jp/eng/solution/index.html

Myszkowsi, K., Mantiuk, R. and Krawczyk, R. (2008) *High Dynamic Range Video*. San Francisco: Morgan Claypool.

Nechvatal, J. (2009) *Immersive Ideals / Critical Distances*. Koln: LAP Lambert Academic Publishing.

Norwood, S.E. (1997) 'Cinerama Premiere Book', Red Ballon, accessed 13 August 2013 at:
http://www.redballoon.net/~snorwood/book/index.html

OLED display (2013) 'OLED Television', accessed 13 August 2013 at: http://www.oled-display.net/oled-tv-television

OpenCV 2.4.4. (2013) Accessed 13 August 2013 at: http://opencv.org/

OpenEXR (2013) 'OpenEXR Documentation', accessed 13 August 2013 at:
http://www.openexr.com/documentation.html

Peleg, S. (1981) 'Elimination of Seams from Photomosaics'. *Computer Vision, Graphics, and Image Processing*, 16(1), 1206–1210.

Point Grey (2012) 'Spherical Vision', accessed 13 August 2013 at: http://www.ptgrey.com

Raij, A., Gill, G., Majumder, A., Towles, H. and Fuchs, H. (2003) 'PixelFlex2: A Comprehensive, Automatic, Casually-Aligned Multi-Projector Display', *Proceedings of the IEEE International Workshop on Projector-Camera Systems*, San Diego, CA, 20–25 June.

Red (2013) 'Red Epic Operation Guide', accessed 13 August 2013 at:
http://s3.amazonaws.com/red_3/downloads/manuals/955-0002_v4.0_EPIC_Operation_Guide_Rev-A.pdf

Reinhard, E., Ward, G., Pattanaik, S. and Debevec, P. (2006) *High Dynamic Range Imaging*. San Francisco: Morgan Kaufmann Publishers.

Roberts, A. (2002) 'The Film Look: It's Not Just Jerky Motion . . . '. BBC R&D White Paper, WHP 053, p. 7.

Salvi, J., Armangu, X., and Batlle, J. (2002) 'A Comparative Review of Camera Calibrating Methods with Accuracy Evaluation', *Pattern Recognition*, 35(7), 1617–1635.

Schreer, O. (2005) *Stereoanalyse und Bildsynthese* [Stereo Analysis and View Synthesis]. Berlin: Springer-Verlag.

Shum, H.-Y. and Szeliski, R. (2000) 'Construction of Panoramic Mosaics with Global and Local Alignment'. *International Journal of Computer Vision*, 48(2), 151–152.

Smith, W. (2005) *Modern Lens Design*. New York: McGraw-Hill.

SMPTE (2003) *ST 268:2003. File Format for Digital Moving-Picture Exchange (DPX), Version 2.0*. White Plains, NY: SMPTE.

SMPTE (2006) *ST 431.1:2006. D-cinema Quality – Screen Luminance Level, Chromaticity and Uniformity*. White Plains, NY: SMPTE.

SMPTE (2011) *RP 431-2:2011. D-Cinema Quality – Reference Projector and Environment*. White Plains, NY: SMPTE.

Snavely, N., Seitz, S.M. and Szeliski, R. (2006) 'Photo Tourism: Exploring Photo Collections in 3D'. *ACM Transactions on Graphics (SIGGRAPH Proceedings)* 25(3), 835–846.

Soneira, R.M. (2012) 'Smartphone Displays Under Bright Ambient Lighting Shoot-Out', accessed 13 August 2013 at: http://www.displaymate.com/Smartphone_Brightness_ShootOut_1.htm

Stama, C.C., Theurer, C. and Henriksen, S.W. (1980) *Manual of Photogrammetry*, 4th edn. Falls Church, VA: American Society of Photogrammetry.

Stark, J.A. (2000) 'Adaptive Image Contrast Enhancement Using Generalizations of Histogram Equalization'. *IEEE Transactions on Image Processing* 9(5), 889–896.

Steadman, P. (2011) 'Vermeer and the Camera Obscura', accessed 13 August 2013 at: http://www.bbc.co.uk/history/british/empire_seapower/vermeer_camera_01.shtml

Stockman, A. and Sharpe, L.T. (2006) 'Into the Twilight Zone: The Complexities of Mesopic Vision and Luminous Efficiency'. *Ophthalmic Physiological Optics* 26(3), 225–239.

Stone, M.A. (1986) 'A Variable-Standards Camera and its Use to Investigate the Parameters of an HDTV Studio Standard', BBC Research Department Report 1986/14.

Stuart, J.A. and Brige, R.R. (1996) 'Characterization of the Primary Photochemical Events in Bacteriorhodopsin and Rhodopsin'. In A.G. Lee (Ed.) *Rhodopsin and G-Protein Linked Receptors*, Part A, vol. 2. Greenwich, CT: JAI Press, pp. 33–140.

Sugawara, M. (2011) 'Psychophisical Requirements for Higher Frame Rates in Future Television', DCS 2011, SMPTE Conference at NAB Show, Las Vegas, 6–7 April.

Sugawara, M., Sawada, S., Fujinuma, H., Shishikui, Y., Zubrzycki, J., Weerakkody, R. and Quested, A. (2013) 'Super Hi-Vision at the London 2012 Olympics', *SMPTE Motion, Imaging Journal* 122(1), 29–39.

Szeliski, R. (2011) *Computer Vision: Algorithms and Applications.* New York: Springer.

The Hobbit Movie website (2011) 'Peter Jackson Discusses New Filming Standard for Hobbit', accessed 13 August 2013 at: http://the-hobbitmovie.com/peter-jackson-discusses-new-filming-standard

The Hollywood Reporter (2011) 'James Cameron "Fully Intends" to Make "Avatar 2 and 3" at Higher Frame Rates'. 30 March, accessed 13 August 2013 at: http://www.hollywoodreporter.com/news/james-cameron-fully-intends-make-172916

Thomas, G.A. (2007) 'Real-Time Camera Tracking using Sports Pitch Markings'. *Journal of Real Time Image Processing* 2(2–3), 117–132.

Thomas, G.A., Schreer, O., Shirley, B. and Spille, J. (2011) 'Combining Panoramic Image and 3D Audio Capture with Conventional Coverage for Immersive and Interactive Content Production', International Broadcasting Convention, Amsterdam, 8–13 September.

Trumbull, D. and Jackson, B. (2013) US Patent No.8363117, 'Method and apparatus for photographing and projecting moving images', (for Showscan Digital LLC), issued January.

TVB Europe (2010) 'NAC/Arri Hi-Motion next on Emmy list', TVB Europe, 3 December.

Wagner, R.W. (1955) 'The Spectator and the Spectacle'. *Audiovisual Communication Review* 3(4), 294–300.

Weissig, Ch., Feldmann, I., Schüssler, J., Höfker, U., Eisert, P. and Kauff, P. (2005) 'A Modular High-resolution Multi-projection System', *Proceedings of the 2nd Workshop Immersive Communication Broadcast Systems*, accessed 13 August 2013 at: http://www.informatik.hu-berlin.de/forschung/gebiete/viscom/papers/icob05b.pdf

Weng, J., Cohen, P. and Herniou, M. (1992) 'Camera Calibration with Distortion Models and Accuracy Evaluation', *IEEE Transactions on Pattern Analysis and Machine Intelligence* 14(10), 965–980.

Zworykin, V.K. and Morton, G.A. (1940) *Television: The Electronics of Image Transmission.* New York: Wiley.

4

Platform Independent Audio

Ben Shirley[1], Rob Oldfield[1], Frank Melchior[2] and Johann-Markus Batke[3]

[1]*University of Salford, Manchester, United Kingdom*
[2]*BBC Research & Development, Salford, United Kingdom*
[3]*Technicolor, Research and Innovation, Hannover, Germany*

4.1 Introduction

Over recent years the rapidly changing broadcast landscape has led to a new set of issues and opportunities for audio developments in the audio-visual broadcast chain. Foremost among the drivers for change are ever-developing user requirements and consequent changes in reproduction systems and delivery formats. Current industry practice is proving insufficient or inefficient in meeting the demand for new and improved services across the full range of audio-visual consumption from traditional push based television broadcast and large screen events to video on demand (VOD) and interactive IP based delivery to mobile devices.

From an audio perspective the move from analogue to digital television broadcast was a key milestone of the changes ahead as it introduced surround sound broadcast to the home for the first time. In addition to providing new options and a more immersive experience for the user this relatively incremental change of adding extra audio channels introduced new challenges for audio production. These new challenges included changes in acquisition methods and also new creative decisions to be made by production teams. More channels and greater bandwidth needed new technologies for delivery and this has led to an array of new transmission formats and codecs, each with different capabilities, associated metadata and interfacing requirements. Often the varied number of channels broadcast and the variety of reproduction systems on which the programme may be viewed has also meant that it has been necessary to produce separately for each reproduction format adding a cost burden to broadcasters.

Over the past few years further change has taken place in broadcast. Developments in internet connected or smart TVs have led to increased demand for further consumer choice,

Media Production, Delivery and Interaction for Platform Independent Systems: Format-Agnostic Media, First Edition. Edited by Oliver Schreer, Jean-François Macq, Omar Aziz Niamut, Javier Ruiz-Hidalgo, Ben Shirley, Georg Thallinger and Graham Thomas.
© 2014 John Wiley & Sons, Ltd. Published 2014 by John Wiley & Sons, Ltd.

interactive content and a social element to television viewing across a broader range of receiving devices. These developments are likely to drive demand for additional capability and services across the whole user experience including demands on audio content for reproduction on any receiving equipment. Current audio options that could be extended further include extended implementation of extra specialist commentary channels (containing alternate language or targeted content for special interest or demographic groups), accessibility for visually impaired and hearing impaired users such as audio description (Snyder, 2005) or hearing impaired (HI) audio channels (Shirley and Kendrick, 2006). It could also implement new services and new control over various other aspects of the programme and enable rendering on reproduction systems not currently serviced. These could include personalised binaural reproduction, Higher Order Ambisonics (HOA) and Wave Field Synthesis (WFS) systems for multi-user scenarios in the home or accompanying large screen produced and live events.

To accommodate such a variety of systems it is no longer viable to produce content with reference to the expected reproduction system of the viewer. As has been described in Chapter 2, a channel-based methodology (such as is currently in place, where a sound is panned to a specific loudspeaker, or between two specified loudspeakers), must give way to a methodology where the reference point is the acoustic scene to be captured, instead of a specific loudspeaker layout.

Social TV is beginning to take hold commercially and will provide further challenges beyond additional channel content (Vaishnavi et al., 2010). Some research into Social TV indicates that audio may be the users' preferred communication medium where individuals are sharing a viewing experience (Harboe et al., 2008) creating new demands on audio capability beyond broadcast audio. There is of course a question as to whether this side channel communication is the responsibility of the broadcast engineer, however, where media and communication or additional side channel content has to be synchronised the broadcast engineer may be involved.

Audio-visual media consumption over mobile devices has increased dramatically and will only increase further as broadband speed and mobile device capability increase at a fast rate. Additionally, where mobile devices are used as a second screen to enhance primary screen content, further demand-led development can be anticipated and this is certain to have some impact on audio broadcast work flows and methods. Although, so far, second screen devices have largely provided only extra visual and text content for programmes it seems likely that this will expand to delivering audio content, which must be synchronised and consistent with the primary screen's content. As well as being part of the second screen content the broadcast audio may also be part of the synchronisation between primary and second screen using audio fingerprinting techniques (Duong, Howson and Legallais, 2012).

Moves toward interactive experience of mobile TV content also have the potential to introduce game-technology based paradigms into broadcast content. Most game consoles implement surround sound to provide an immersive experience and binaurally presented mobile games with 3D sound are available. In the broadcast area researchers have utilised IP-based delivery for binaural headphone reproduction as an additional service for radio listeners (Brun, 2011). It is now reasonable to ask how long it will be before a headphone delivered surround sound mix is a standard part of the end users' expectation for broadcast to mobile devices rather than a novelty. Using current methods, this would require an additional production resource and additional delivery channels for each type of device. Once game-based paradigms are introduced there can also be the potential for viewers to choose their own view of media productions and to navigate through the audio-visual scene of a programme and at this point the channel-based production work flow breaks down completely.

The challenge to the audio broadcast engineer is to capture, or create, an acoustic scene in such a way as to enable a single production to cater for channel-based and non-channel-based reproduction devices. An object-based audio system may be used in order that the scene is no longer defined from a specific location but instead with a representation that includes ambient and diffuse sound, audio objects that are either directly picked up and tracked or that are derived from microphone arrays, as well as additional components such as commentary. The scene representation must define locations for these audio objects and events in a 3D coordinate system such that they can be manipulated to match any visual viewpoint chosen by either a production decision or by user choice and rendered accurately on any given reproduction system.

This chapter defines this problem space from an audio perspective reviewing some of the current challenges faced in channel-based audio broadcast (exemplified by stereo, 5.1 surround etc.) and looks at the approaches that will be required in order to deliver to non-channel based sound scene reproduction techniques such as Wave Field Synthesis and Ambisonics. The problems of having many competing audio formats are addressed at both production and reproduction (user) end. The implications of audience interaction with the content is discussed and how this impacts on current techniques. Finally the chapter discusses the use of a format-agnostic/platform independent approach and how this may offer a solution to the aforementioned problems.

4.2   Terms and Definitions

4.2.1   Auditory Event and Sound Event

Following the structure given by Spors et al. (2013) a few definitions have to be considered first in order to understand the relationship between the perceptual effects of a reproduction system and its physical output as it is important to define the terms very clearly. It is important to distinguish between sound events which take place in the physical world and auditory events which describe the perception of sound events (Blauert, 1996) and which make up an auditory scene. Auditory events within an auditory scene are perceived by a listener based on both auditory information and auditory attention. However, complex cognition effects are beyond the scope of this book and only auditory events are discussed here (Blauert, 1996; Kubovy and van Valkenburg, 2001; Cariani and Micheyl, 2010).

A single auditory event is characterised, from others, by its loudness, pitch, perceived duration, timbre and spatial features. Timbre is related to the identity of the sound source, and has multidimensional nature and spatial features (Letowski, 1989). Human binaural perception is the main mechanism enabling sound events to be related with spatial auditory features. The basic functions of spatial hearing are given by Blauert (1996) as:

- localisation (formation of positions of auditory events in terms of azimuth, elevation, and distance);
- extraction of spatial extent;

- suppression of unwanted directional information, for example, suppression of room reflections in terms of the precedence effect (Wallach, Newman and Rosenzweig, 1949), binaural decolouration (Bruggen, 2001) and suppression of noise interferers; and
- identification and segregation of auditory streams into events, such as concurrent talkers (Cocktail-Party-Effect), warning signals.

As a basic example of an auditory event, one can consider the perception of an audio signal radiated by a single loudspeaker. For this scenario, the auditory event is usually associated directly with the sound event (the loudspeaker's radiation) and will most likely be at the same location; the auditory scene will therefore consist of just one auditory event. Consider a more complex scenario where two loudspeakers emit the same audio signal, thus generating two sound events. In this case, there is no longer a one-to-one mapping of sound events and auditory events, that is, the two sound events will be perceived as one auditory event. From a position equidistant from each loudspeaker, a listener will perceive an auditory scene with a single auditory event located between the two loudspeakers.

Similarly, single auditory events can also be created by algorithms employed in multiple loudspeaker playback. In this sense, auditory events are regarded on the basis of auditory cues that are non-concurrent, to facilitate psychoacoustical models employed for their generation using loudspeakers. The *virtual sound event* or *virtual source* is the acoustical reproduction of a computational representation that controls such generation algorithms. A *virtual sound scene* is composed of several virtual sources. Figure 4.1 shows the different scene types and their relationships. If content is produced, this cycle is often followed more than once. For clarity, one can consider the case of several musicians playing in a room and thus creating a sound scene. When the producer walks into the room he will perceive the music as an auditory scene. Going back to the control room he will create a representation of this scene by the means of recording equipment, microphones, digital signal processing and a particular reproduction system. This representation – the virtual scene – is reproduced as a sound scene by the reproduction system and should create the auditory scene the producer desired. Otherwise he will modify the virtual scene again and so forth. Delivery of the content to the audience involves the distribution of a virtual scene and playback on various devices aiming for a comparable or similar auditory scene perceived by the listener.

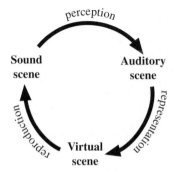

Figure 4.1   Relationship between different scene representations.

4.2.2   Basic Room Acoustics Theory

An additional degree of complexity arises when the perception of sound events that normally take place in reflective environments is taken into consideration. This could be in concert halls, cinemas, living rooms or cars. Such environments influence the perception significantly. In the case of a live concert it delivers the appropriate room impression but in case of audio reproduction over loudspeakers it can significantly alter the perceptual impression and lead to completely different auditory events than originally intended (Toole, 2008). On the other hand a reproduction system must be capable of recreating all important factors of an acoustic environment to deliver the desired impression to the listener. A combination of source and receiver in an acoustic space can be characterised by its spatial room impulse response (Melchior, Michaelis and Steffens, 2011). Figure 4.2 shows a time energy representation of a omni-directional room impulse response of a single source in a room.

Figure 4.2 identifies the three main time segments. Each segment requires different capabilities from a reproduction system. The first impulse represents the direct sound emitted by the source, this part defines the timbre and spatial properties of the source. An accurate reproduction of the appropriate binaural cues here is important to enable a correct perception of the associated auditory event. The second part relates to the early reflections of the environment, these are normally distributed in all directions depending on the surrounding surfaces of the room. In view of a reproduction system these can also be considered as sources emitting a modified version of the direct sound in different directions. Each source in a scene creates its own specific early reflection pattern that needs to be recreated by the reproduction system. The corresponding time range of approximately 20ms to 200ms of these early reflections is important for the characterisation of the size and type of environment. The late part of the impulse responses is normally characterised by a diffuse field and needs to be isotropic and homogeneous to deliver envelopment and immersion for the desired acoustic environment (Kuttruff, 2000).

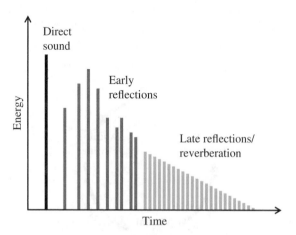

Figure 4.2   Diagram of a typical room impulse response showing, direct, early reflection and late reflection energy.

4.3   Definition of the Problem Space

The aim of sound reproduction is the delivery of a creative idea from the content creator to its audience that evokes the intended experience for the listeners. Callet, Miller and Perkis (2012) note 'An experience is an individual's stream of perception and interpretation of one or multiple events'. Since it is impossible to deliver exactly the same experience to every listener it is important to separate the most relevant variables that influence the experience of the listener. It is also important to distinguish between the quality of experience (QoE), defined in Callet, Miller and Perkis (2012) as 'the degree of delight or annoyance of the user of an application or service. It results from the fulfilment of his or her expectations with respect to the utility and or enjoyment of the application or service in the light of the user's personality and current state'. Since it is beyond the scope of this book to study the quality of experience as it heavily involves the users state and the content itself we will concentrate on quality only. The aim of a format-agnostic reproduction is to achieve the highest possible quality for all target platforms. Quality is defined as 'the outcome of an individual's comparison and judgement process. It includes perception, reflection about the perception, and the description of the outcome. In contrast to definitions which see quality as qualities, i.e. a set of inherent characteristics, we consider quality in terms of the evaluated excellence or goodness, of the degree of need fulfilment,' (Callet, Miller and Perkis, 2012). The quality formation process involves two paths: the perception and a reference. The perception is based on the transformation of a sound scene into an auditory scene, but also involves cognitive reflection based on a reference. The reference reflects the contextual nature of quality, which includes memory of previous experience and expectation. This can form an implicit reference that is used by the listeners to judge the quality. This reference potentially results in a weighting of quality elements and can influence the design of a system in order to ensure that the most relevant quality parameters are covered best. For example, in a news programme it is most important to deliver optimal speech intelligibility whereas in a concert broadcast maybe immersion and timbre plays the most important role. Here we concentrate on an overview of technical influencing factors that need to be considered in order to create content representations that are able to be reproduced in a broad variety of conditions and environments.

To study the challenges of platform independent audio the main discussion focuses on the aural scene quality and acoustic quality aspects to illustrate the challenges of such an approach. Following Blauert and Jekosch (2012) factors of aural scene quality are the identification and localisation of sound in a mixture, speaker intelligibly, audio perspective including distance cues, scenic arrangement, tonal balance and aural transparency. These subjective measures can be assessed by subjective evaluation methods (Bech and Zacharov, 2006) like semantic differential, multidimensional scaling or scaling of preference, suitability, and benchmarking against target sounds. Conversely the acoustic quality consists of instrumental measurements with equipment and is directly related to physical quantities. Even if the creative idea of a content creator is the primary focus, such measures will be used to illustrate the challenges further. The investigations of this chapter will exclude the delivery of the content and just focus on specific reproduction system aspects.

4.3.1   Reproduction Environment

The environment in which content is consumed has a significant influence on the perception of the content. Two main aspects are considered here as being most important in the case of platform independent audio; background noise and acoustic properties of the reproduction environment. Whereas the environmental noise is important for loudspeaker-based and headphone-based reproduction the acoustic properties are particularly relevant if the content is reproduced over loudspeakers.

Acoustic Properties

The acoustic properties of the reproduction environment are overlaid on the desired reproduced sound scene created by the reproduction system. This can lead to significant degeneration of the auditory scene (Toole, 2008) perceived by the audience. Whereas in environments like cinemas this can be controlled by following design guidelines, in broadcast applications there are an unlimited number of reproduction environments. Every listener has a specific environment where the content is consumed, therefore, even if there would be some equality in the reproduction system itself the auditory scene will be distorted individually. The importance of this is reflected in the standardisation of the methods for the assessment of audio quality in an international standard ITU-R (1997). This standard provides a method for subjective assessment of small impairments in audio systems along with a dedicated description of the acoustics of the listening environment. The acoustic properties being considered most important are:

- Late energy: Reverberation time over frequency should be flat and within certain bounds depending on the room size, and anomalies in the sound field such as flutter echoes or colouration need to be avoided.
- Energy and the structure of the early reflections, which should below 10dB, compared to the direct sound in a time span of 15ms after the direct sound

The aspect of the loudspeaker will be explained in Section 4.3.2 but it is important to mention that, for example, the operational frequency response curve is a mixture of the loudspeaker properties and the reproduction room acoustics. For a extensive study on this topic see Toole (2008). Controlling these properties is particularly difficult if content delivery to a larger audience is considered. For example, the reflection pattern for a single position in a room can be controlled using absorbers or reflectors but if a larger zone needs to be controlled trade-offs between different parameters need to be found. Nevertheless, within the broadcast applications considered here the environment can be assumed as being highly uncontrolled (e.g., domestic living rooms)

Background Noise

Another important property of the reproduction environment is the background noise. A standardised method to compare noise in rooms is given in ANSI (2008). In this context the usable dynamic range, or more appropriately, the loudness range is defined by the combination of background noise and the maximum loudness reproduced. To illustrate this consider the following example: a classical concert broadcast is consumed at home, in a car and over headphones while using public transport. Every environment will have its own usable loudness range and either the content provider or the reproduction devices needs to adapt to this loudness range. Without this adaptation the quiet passage could be below the environmental noise level or the loud passages could clip or result in unpleasantly loud reproduction. It is beyond the scope of this chapter to go into details about loudness in terms of normalisation and measurement. EBU Recommendation R 128 contains relevant definitions of loudness measurement (EBU Tech 3341, ITU-R BS. 1770) and loudness range (EBU Tech 3342).

4.3.2   Reproduction Method

Audio reproduction methods have been introduced in Chapter 2. There they were described as means for reproducing content based on different underlying concepts, they will now be considered in the context of reproducing the same content over different systems. The aim is to explain the challenges of deriving an optimal reproduction from a single representation.

Loudspeaker Reproduction

All loudspeaker-based reproduction scenarios share their reliance on the physical parameters of the loudspeakers making up the system. These parameters include:

- amplitude/frequency response;
- operational sound pressure level and dynamic range;
- directivity index;
- non-linear distortions; and
- transient fidelity.

Several of these parameters are highly dependent on the interaction between the loudspeaker and the room (see Section 4.3.1). Apart from monophonic reproduction (as it can be found in certain radios) a number of loudspeakers are often combined into a given reproduction system. Figure 4.3 presents an overview of some different principle loudspeaker-based reproduction scenarios. There are two basic settings:

1. A variable number of loudspeakers distributed in a space surrounding the listener.
2. Built in loudspeakers of specific devices like tablets or sound bars.

Considering the case of distributed loudspeakers first; loudspeaker setups can vary in their layout (2D, 3D, different number of loudspeakers etc) that significantly alter the impression a listener will get when consuming the content. Furthermore the loudspeaker signals need to be calculated based on a virtual scene representation. The algorithm used for this will significantly influence the auditory scene perceived by the listener. To render a virtual scene, the desired reproduction setup needs to be taken into account. Therefore the loudspeaker setup has to either be standardised (with the assumption that it will be always setup in the standardised way – compare Section 4.4.2), or the decoding/rendering at the reproduction side should take the current loudspeaker setup into account. A third option is to generate the loudspeaker signals at the production side and do an adaptation based on the knowledge of the actual loudspeaker locations (Bruno, Laborie and Montoya, 2005).

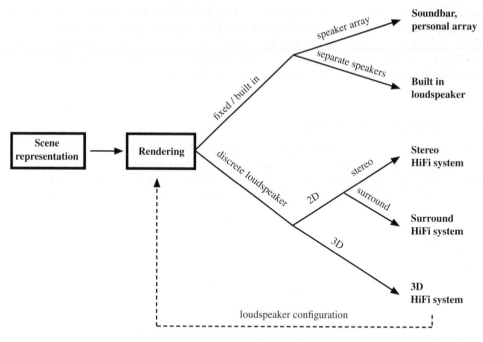

Figure 4.3   Taxonomy of loudspeaker-based reproduction.

All-in-one systems or array-based reproduction integrated in single devices are convenient to setup since they do not require the distribution of a number of discrete loudspeakers in the room. They rely on signal processing algorithms that attempt to distribute the signals as well as utilising room reflections, which shifts some of the problems to the room acoustic (Troughton, 2002).

Depending on the number of listeners, the auditory scene has to be created for portions of the room; ideally for all listening positions. This is a challenge in terms of the acoustic environment since it is more difficult to have control over a larger part of the room but also in terms of the reproduction system itself. Often compromises have to be made and different properties weighted. For example, large live event sound reinforcement systems often concentrate on achieving a good frequency response and a high sound pressure level rather than accurate spatial characteristics. These challenges can be addressed to some extent by using state-of-the-art sound reinforcement systems like line array technology. The scaling of the spatial layout of a sound scene is now discussed.

Scaling of Spatial Audio Scenes
Due to their underlying principles, two-channel stereophonic reproduction systems (see Chapter 2) are able to recreate the correct spatial impression for positions along a line between the loudspeakers. According to ITU (2003) if stereophonic reproduction is extended to surround sound system this line is limited to a point (sweet spot) created by the intersection of the lines of all stereophonic bases. Holophonic systems like Higher-Order Ambisonics or Wave Field Synthesis can enlarge this reproduction area to some extent but in general at the production

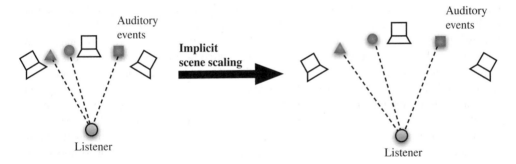

Figure 4.4 Implicit scene scaling.

end the experience for the whole reproduction area needs to be considered and controlled by whatever means possible. If the size of the listening area and the desired number of parallel listeners vary significantly, the scaling of the virtual scene will require consideration.

Implicit scene scaling: Imagine a stereophonic system with three front channels arranged on a circle around the listener (Figure 4.4). Such a system delivers a correct spatial impression only for a limited area (in this example we assume the listener is in the sweet spot). Three auditory events created by such a system are represented by a circle, square and a triangle in the figure. If the loudspeaker positions are arranged on a larger circle the locations of the auditory events are scaled implicitly by the changed loudspeaker setup.

In Figure 4.5 *Explicit scene scaling* is illustrated. A holophonic reproduction system creates auditory events at three different locations (circle and square indicating distant sources and a triangle representing a focused source). First, a small size reproduction system, as shown

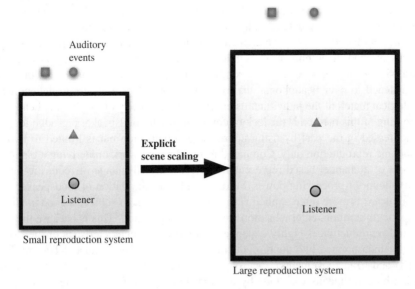

Figure 4.5 Explicit scene scaling.

on the left of the diagram, is considered. Since a holophonic system requires the explicit positioning of the sources, a pure scaling of the loudspeaker layout would result in all sources becoming focused sources due to the larger loudspeaker setup if the source coordinates were unchanged. Therefore explicit recalculation of the source position is required to ensure that only the triangle be a focused source whereas the other two sources are distant sources in locations relative to the loudspeaker array. Furthermore a scaling algorithm has to be applied to ensure that perceptual properties and creative intent of the virtual scenes are kept as much as possible. For example, a source moving around the listening area should do this even if the listening area has a very different shape. The situation can be simplified if some assumptions are made, for example, the listening area is a rectangle. It is also important to notice that in the production situation the virtual scene has to be rendered in a real setup, which is not necessarily the one in which it will be consumed later. A simple concept is to introduce a specific representation of the scene for storage. Such a scene, for example, can be a rectangle with a certain aspect ratio and normalised coordinates. By knowing the real setup it can be scaled by a simple factor in the x, y and z directions. The situation however becomes more complicated if the assumptions need to cover a more general case.

Headphone Reproduction

If only a single listener is considered, headphone reproduction is a potential option. Today headphones are mainly used to play back two-channel stereophonic signals. The benefit is that no additional processing or specific version of the technology is required and it generally enables very high timbral quality. In order to create a new experience compared to two-channel stereo played back over headphones binaural reproduction can be also be considered (see Section 2.7.3). While in the first case the stereo image is spread between the left and right ear inside the head, the latter case enables a externalisation of the experience. In principle one can consider two primary approaches to facilitate a binaural production: either perform a binaural recording, or simulate a loudspeaker setup by binaural means.

Binaural productions have been created for many years by using dummy heads. To implement this method, a dummy head is placed at an appropriate recording position and the binaural signals are recoded directly from the microphones in the ears of the dummy head. When listened to over headphones the externalisation effect is achieved depending on the physiological match of the individual listener and the geometry of the dummy head. A further shortcoming of this method is the lack of compatibility to loudspeaker reproduction. This has been addressed in the past by using appropriate equalisation but is limited to two-channel stereophonic reproduction only. Compared to more recent work on deriving a binaural experience out of a channel-based representations it has the advantage of requiring only a very limited amount of processing power. More recently the simulation of loudspeaker setups by binaural means is more often applied. The main challenge is to achieve a good quality mainly in terms of colouration, externalisation and spatial impression. This is because the quality of a binaural reproduction is highly dependent on individual characteristics of the listener that need to be incorporated into the reproduction system (e.g., Head Related Transfer Functions (HRTFs) and head movements). One of the first attempts to implement virtual loudspeakers using binaural means was done by Karamustafaoglu (2001) to create a virtual listening environment. A five-channel multichannel speaker setup with a dummy head in the reference listening position was used to acquire binaural impulse reposes for a dense grid of dummy head orientations. In the reproduction case each loudspeaker signal is convolved with the binaural room impulse responses and adapted to the tracked head orientation of the listener. This technique requires a dynamic updated convolution and the head tracking of the listener but delivers a high quality simulation of a listening environment including the room characteristics. If no head tracking and/or individualisation is applied the quality that can be achieved is limited compared to a stereo down-mix (Pike and Melchior, 2013). An alternative approach to loudspeaker virtualisation is a direct rendering based on the virtual scene. This field is subject of ongoing research.

4.3.3   Audio-visual Coherence

Besides the inherent aspects of the audio content itself, which is an open field of research, the combination of audio and video highlights some additional challenges that should be considered when designing a platform-independent audio system.

Temporal Coherence

First the audio-visual temporal synchronisation has to be maintained to ensure the delivery of a good experience. An accuracy of between -40ms (early) and +120ms (late) is considered to give a high quality perception of audio-visual synchronisation for average users (Younkin and Corriveau, 2008). Considering different data rates and the parallel delivery of audio video requires careful design of the reproduction system and the transmission path. For more detailed study the reader is referred to van Eijk (2008)

Spatial Coherence

In the case of the combination with picture, the spatial coherence of auditory events and corresponding visual elements needs to be considered (Melchior et al., 2003; Melchior, Fischer and de Vries, 2006). Depending of the capabilities of the visual reproduction (2D or 3D) and the audio system (whether or not it renders a true depth of scene like holophonic systems) misalignment can occur for the listeners. When the visual content is 2D there needs to be a mapping between the modelled 3D acoustic scene, providing depth-of-scene information and the 2D content (de Bruijn and Boone, 2002). This is particularly important for audio systems that provide full depth-of-scene cues such as WFS and HOA and it has been shown that the resulting parallax error introduced when there is such a dimensionality mismatch between the audio and visual scenes is perceptually important (Melchior et al., 2003). It is important to distinguish between the geometrical depth of a sound source and the perceived distance. While only holophonic reproduction systems like WFS or HOA are capable of rendering the former, the latter scenario can be generated using other systems by adding reverberation and controlling the ratio between reverberant sound and direct sound.

Figure 4.6 shows an example of 2D video combined with a virtual source reproduced in true depth. While listener A perceives a match between the virtual source and the visual object on the screen, listener B perceives an angular mismatch, e, between the virtual sound source and

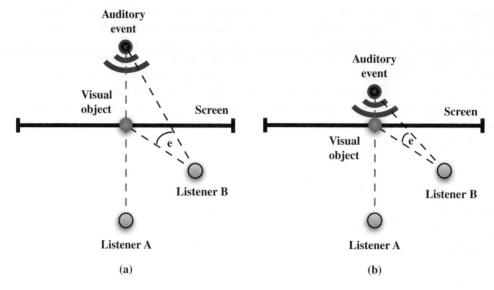

Figure 4.6   Audio-visual coherence using audio reproduction with true depth and 2D video.

the visual object. The amount of error depends on the virtual source and listener positions. If the rendered depth of the virtual sound source is reduced, the error is reduced, but all objects have to be rendered on a single line as close as possible to the screen. The situation gets even more difficult when non-tracked stereoscopic image reproduction is considered since the image will be distorted by the listener position and the two effects overlay. The degree of quality impairments based on limited coherence is also important (e.g., surround sound in cinema with dialogue mixed only to the centre). If audio objects are placed at a specific location within the modelled scene, these positions will only match those reproduced by the system (both visually and audibly) if the reproduction system is scaled appropriately. A smaller screen in this case, for example, requires a narrower audio rendering of the on-screen sources than a larger screen. If the audio reproduction is done using for example, a two-channel stereo system, the loudspeakers can be placed either side of the screen and the phantom source positions will match the visual source positions. If the screen size is reduced, then the loudspeakers could be moved closer together to reduce the width of the acoustic scene accordingly. For more advanced audio systems, the virtual source positions could be produced on the screen to match the visual content location. This will depend on the screen dimensions and thus will require some user input at the render end. As a result the scaling of screen sources and audio only elements have to be implemented separately (see Section 4.3.2). As well as this, in many cases the screen provides a significant acoustic barrier to the reproduction system (e.g., if the loudspeakers are placed behind it), leading to problems with rendering sources to or behind the screen. In some cases it may seem desirable to place an array of loudspeakers above and below the screen in an attempt to achieve a phantom virtual source, but this has proved very difficult in practice. This is due to the different localisation mechanisms of human hearing, which rely on interaural cues in the horizontal plane but on monaural cues in the vertical plane (Blauert, 1996).

4.3.4   User Interaction

Representing the content in new ways, as required for platform independent audio, also opens the field for new interactive applications. The audience can, for example, modify the balance between commentary and background (Fuchs et al., 2012), elements of the content can be personalised (Churnside and Forrester, 2012) (see also Section 4.7.3) or as described for video in Chapter 3 the viewer can choose their viewpoint. All these new forms of interaction have to be considered when designing a system. It is important to note that these bring the system closer to that of interactive gaming application but the important difference is that the aim of broadcasting is still to tell a predefined story and only allow careful interaction rather than completely free navigation and game-like interaction.

Navigation

If applications allow the user to pan or zoom within the visual content, it will be expected that the audio perspective matches the visual perspective. The relative positions of the audio sources then depend upon the dimensions of the screen and on the viewer distance from the screen. As a user zooms in to the visual content, audio sources should move towards and around the listener to match a realistic impression as if the listener were to move their seat within the scene. Zooming in means that some sources that were once on the screen will move off it and move around the viewer until such a point that the source moves behind the listener (Melchior and Brix, 2004). Zoom also raises questions for the best mix between the ambient sound and the audio objects in the scene. Depending on the genre of the recorded material it may be desirable to change the relative level of the audio objects with respect to the ambient sound, giving the listener the impression that they are being physically transported to a different place in the scene. Alternatively, this constant change of the ambient sound level might prove disturbing to the listener. In practice this may be controlled by a set of user preferences. While from a modelling point of view, it is tempting to try and calculate a physical listener position from the pan and zoom data to inform the audio renderer, this may not be practically realisable as the mapping function is non-trivial and even so, it will only be correct for one listener position.

Changing the Balance

A change of balance or an exchange of elements within an audio scene requires the elements to be available separately at the user end. This is an important requirement especially for the transportation of content and demands that the elements of a mix be available separately resulting in a loss of coding efficiency and potentially higher bit rates because masking effects can no longer be employed.

4.3.5   Example Scenario

The example scenario described in this section is an imaginary example of a future system. It will be used to illustrate the potential applications and will use cases of format-agnostic audio. The scenario applies to both traditional and newly developed audio technologies.

A concert of a symphonic orchestra in a traditional concert hall is broadcast, and during the interval an interview is scheduled where the third trumpet player reports on his new mouthpiece. Therefore, two different acquisition tasks for the sound scene have to be performed. First, the orchestral music is captured using a main microphone arrangement and various spot microphones; furthermore, the concert hall acoustics are recorded by some spatially distributed microphones. Finally, the acoustic ambience in the concert hall consisting of room acoustics and the audience is captured by a microphone array. The interview takes place in the foyer of the concert hall. The recording situation here just requires the single microphone of the interviewer.

All signals go to the production stage, where the virtual scene representations are created. The virtual scenes are edited and monitored in the control room and then distributed for broadcast.

The editing in the production stage depends on the addressed rendering systems and their specific requirements on the consumption side. In turn, various virtual scene representations are in use. In this example, four different rendering systems are foreseen:

- Mobile devices using headphones implementing binaural synthesis to play back conventional mixes of channel-based audio using a virtual loudspeaker approach and mobile tracking and individualisation.
- A conventional loudspeaker system in a 5.1 surround sound arrangement plays back hybrid representations consisting of basic channel-based representations. Additional information is provided in different languages, these are delivered in form of additional objects accompanying the channels used for the music signals. The content can be modified with regard to the dialogue level, for example, the interview in order to be adapted to the user's needs and preferences.
- A big screen for public viewing events is equipped with a WFS rendering system. Besides the core audio information being produced as channel-based audio, sound objects can also be reproduced by the system.
- The 3D setup for interactive home application can be realised using artistically designed 3D sound field information. For interaction the audio information is provided in a very generic way. The listener might for example wish to focus his view on the viola player in the middle of the orchestra.

4.4   Scene Representation

4.4.1   *Components of a Virtual Sound Scene*

The sound scene in a recording room can be captured to some extent using a combination of microphones. The captured data provides a representation that enables the storage, play back and modification of the scene in production. The scene representation is then rendered by a reproduction system which ideally renders a virtual sound scene in such a way that the listener perceives the intended auditory scene. In order to understand how this is achieved using different reproduction systems, the various elements that make up a virtual sound scene are discussed in detail below.

The basic component of a virtual sound scene is a virtual sound event, which consists of an audio signal, location information and other parameters. The complexity of these parameters will depend on the capabilities of the reproduction system but location and level are considered as the most basic essential parameters. Virtual sound events may be related to a physical source model such as a point source, plane wave or a source with a more complex directivity (Baalman, 2008). Real sound scenes can be dense and may contain many discrete sources but it is not always necessary or desirable to consider each source as an equivalent virtual source if an effective representation can be achieved with a smaller number of virtual sources. As such there has been work done proposing methods for wide or diffuse virtual sources (Gerzon, 1992b; Kendall, 1995; Potard, 2006; Laitinen et al., 2010; Zotter et al., 2011) which can be used to represent instrument sections, ambiance or reverberation in a virtual sound scene. Using the example of an orchestra; it is not vital that each instrument is recorded individually and represented as discrete, well-localised, virtual sources. Convincing results are achieved by representing sections of the orchestra, such as a choir, using a stereophonic pair of signals from spaced microphones as two virtual sources in the virtual scene. Although the number of individual virtual sources has been reduced considerably the overall auditory scene experienced by the listener is convincingly made up of locatable voices.

The virtual sound scene must then be translated into loudspeaker signals using a specific spatialization method such that it can be reproduced in the listening environment. These loudspeakers emit sound events such that a listener in the room will perceive the desired auditory scene.

The sound scene in a recording room can be captured to some extent using a combination of microphones. The captured data provides a representation that enables the storage, playback and modification of the scene during further steps in production. The scene representation can be fed to a reproduction system which ideally renders a virtual sound scene in such a way that the listener perceives the desired auditory scene. It is possible for this reproduction to be done using different reproduction systems which exploit different physical or psychoacoustical attributes to evoke the intended auditory scene. To understand how this is achieved using the different reproduction systems available, the various elements constituting a virtual sound scene are discussed in detail below.

The basic component of a virtual sound scene is a virtual sound event, which consists of an audio signal, location information and other parameters. The complexity of these parameters will depend on the capabilities of the reproduction system but position and level are considered as the basic parameters.

It is common for a virtual sound event to be related to a physical source model such as a point source, plane wave or a source with a more complex directivity (Baalman, 2008). A wide or diffuse source may well consist of more than one audio signal but as real sound scenes can be dense and may contain a huge number of sources, it is not always necessary or desirable to consider each source as an equivalent virtual source if a plausible representation can be achieved with a smaller number of virtual sources. As such, there has been work done proposing methods for wide or diffuse virtual sources (Gerzon, 1992b; Kendall, 1995; Potard, 2006; Laitinen et al., 2010; Zotter et al., 2011) which are typically used to represent large instruments, reverberation, or ambience in a virtual sound scene. For example, it is not paramount for each instrument of an orchestra to be recorded individually and to be represented as well-localised virtual sources. Plausible results are also achieved when representing bigger parts of the orchestra, or a choir, by a stereophonic pair of signals from spaced microphones as two virtual sources in the virtual scene. Despite there being only two virtual sources in this case, the perceived auditory scene will still consist of individually identifiable and locatable voices.

The virtual sound scene must then be translated into loudspeaker signals using a specific spatialisation method such that it can be reproduced in the listening environment. These loudspeakers emit sound events such that a listener in the room will perceive the desired auditory scene.

4.4.2   Representations of Virtual Sound Scenes

From the recording styles described above, virtual scenes are typically represented in three different ways:

Channel-based Representation

Currently, the most prevalent methods used to represent virtual sound scenes are two-channel stereo sound and 5.1 multichannel surround sound (ITU-R, 1994). These channel-based representations are directly related to a specific layout of loudspeakers with the actual loudspeaker signals being transmitted and stored. The same loudspeaker layout must be in place at every stage of production and reproduction to ensure the optimal quality of the perceived auditory scene. Such an approach has advantages in terms of simplicity and is feasible if only a limited number of loudspeakers is considered. However, in practical domestic applications it is often difficult to achieve the correct loudspeaker setup for even five loudspeakers.

Transformation-based Representation

An alternative method for the storage and transmission of an auditory scene is to use a mathematical description of the sound field capable of creating the desired virtual sound scene such as the spatially orthogonal basis function described by Poletti (2005). This representation method has been implemented with varying degrees of complexity (Cooper and Shiga, 1972; Gerzon, 1992a) and is referred to as transform-domain-based because transformation coefficients are stored rather than loudspeaker signals. To reproduce a transform-domain representation, a decoding stage is required on the reproduction side to transform the coefficients to loudspeaker signals. This methodology affords more flexibility; the loudspeaker layout can be taken into account by decoding so there are less stringent setup requirements. For this representation, there is a trade-off between the desired spatial resolution and required bandwidth for transmitting the coefficients. Currently, the most common concept of transformation-based representations is Higher Order Ambisonics (HOA) (Bertet et al., 2007).

Object-based Representation

The object-based method offers perhaps the most transparency in the mapping of the virtual sound scene. With this method, the signal of each virtual sound source is kept separate and transmitted with some descriptive metadata (e.g., the location of the virtual source). On the reproduction side, a processing stage is required to generate loudspeaker signals that can create the virtual sound scene using the given loudspeaker setup/reproduction method. This could be considered a potential drawback of this method, transferring large amounts of processing to the listener side. An object-based format could also be used to enable interaction and personalisation of the content as it allows a dynamic mixing of the virtual sources (Churnside and Forrester, 2012; Fuchs et al., 2012).

These three principal representations can also be used in combination to derive benefits from more than one method. For example, a channel-based representation can be utilised to store a sub-group of sources within an object-based representation or a 5.1 channel-based production can be represented in a transformation-based representation such as Ambisonics (Gerzon, 1992a). Whatever representation method is chosen, the virtual sound scene needs to be translated to sound events created by the loudspeakers at the reproduction end; this can happen at various levels of complexity.

4.4.3   Implementation Examples

Object-based audio capture and reproduction has been implemented in both commercial and research systems with many overlapping techniques and principles being used with various representations of audio objects applied. The most prevalent and the most relevant of these systems and formats are discussed here.

Current Formats for Channel-based Representation

A common format to store channel-based information is the *Microsoft wave audio* format or one of its extensions like the *Broadcast Wave File* format or the *Wave Format Extendable* (Cross and Puryear, EBU, 1977). For a discussion of further formats like AIFF and different use cases the reader is referred to Dobson (2000).

Current Formats for Transformation-based Representation

Most transformation-based representations are referred to as Ambisonics technologies. The *B-format* for coefficient storage was defined according to requirements of the Soundfield microphone (Craven and Gerzon, 1977). Whereas this traditional Ambisonics approach uses a first order representation, the Furse-Malham set can be used for storage of second and third order representations (Malham, 2005). To cover orders higher than three as well, different format propositions for HOA coefficient exchange exist (Chapman et al., 2009; Nachbar et al., 2011).

Current Formats for Audio Object Representation

Audio objects contain the audio data of the given source accompanied by a set of additional data defining properties such as the position, onset/offset times directivity. Consequently there have been several proposals of potential formats for audio objects containing varying amounts of additional data. This section briefly describes some of the major developments in this area.

VRML/X3D
Mainly used for describing 3D computer graphics the Virtual Reality Modelling Language (VRML) (Carey and Bell, 1997) can also contain descriptions of audio sources that fit into its so-called *scene graph* format. An audio source is in this case defined by adding an audio node to the scene graph. Source directivity can be described in VRML using two ellipsoids that govern the intensity of the source in the respective directions. A disadvantage of the VRML format is that the timing information is sufficiently complicated that it can not be interpreted

simply by inspection and therefore editing source trajectory and other temporal characteristics of the scene can be difficult. Nevertheless the VRML97 standard was introduced in 1997 and was fully incorporated into MPEG-4's AudioBIFs format as described below. The VRML format has since been superseded in 2004 by X3D, which became an ISO standard in 2004. This allows, among other things, an XML syntax.

VRML utilises a local coordinate system allowing the easy translation and rotation of objects. Its hierarchical structure means that objects can be grouped together and these groups can then be grouped together and so forth. The problem with this structure is that it becomes very difficult to infer the movements of audio sources and editing can become a very complicated procedure. For most audio scenes there are not many audio objects present, thus, the hierarchical structure can be considered needlessly complex and highlights some of the weaknesses associated with this format.

MPEG-4

MPEG-4 is a standard for object-based description and distribution of audio visual content (Pereira and Ebrahimi, 2002). It allows not only the streaming of audio-visual data but also includes the ability to add an extensive amount of metadata to the transmission. The scene description element of MPEG-4 is dealt with by the Binary Format for Scenes (BIFS) with the audio scene description defined by AudioBIFS (Schmidt and Schröder, 2004). Subsequently Advanced Audio BIFS (Vaananen and Huopaniemi, 2004) have been proposed that can handle spatial audio description and metadata.

The MPEG-4 format offers all the necessary tools to effectively describe a spatial audio scene, including the relevant room acoustic information, source directivity and level to name but a few characteristics. However, the standard is long and highly complex and is thus very difficult for audio scene designers to use. Consequently spatial encoders and decoders of this format are uncommon and it has not been widely adopted in the spatial audio community.

SAOC

Spatial Audio Object Coding (SAOC) (Herre et al., 2012) is an extension to MPEG Surround and was finalised in 2010 by the ISO/MPEG Audio standardisation group. SAOC utilises the principles of Spatial Audio Coding (SAC) applied in MPEG Surround's channel-based system to an object-based system and is therefore more flexible. Being a description of audio-object coding, the standard focuses more on compression of the data rather than a format for describing the acoustic scene. The primary aim of SAOC is compression; successful encoding and decoding of a stream of audio objects with the minimum amount of data required to represent the scene.

Audio3D

Hoffmann, Dachselt and Meissner (2003) presented the Audio3D format, which is based on an XML description syntax. It is primarily designed for computer gaming and can be used in conjunction with an X3D description.

The Audio 3D format is aimed at being platform and Application Programming Interface (API) independent and is suitable for both real-time and offline sound reproduction. Similar to X3D and other formats that are used in combination with graphics, Audio3D utilises a hierarchical audio scene graph to describe audio objects. The format allows the representation of room acoustic information and even obstacles in the path of the sound such as walls, for example. The size of the radiating sound sources can also be defined in this format. Different types of audio object can be described including sources without a position such as background music and also moving audio objects.

SMIL

The Synchronised Multimedia Integration Language (SMIL) (Rutledge, 2001) is an XML description of multimedia scenes for the web allowing the integration spatially and temporally of multimedia objects. SMIL has since been extended to include the description of spatial audio attributes in three dimensions (Pihkala and Lokki, 2003) using the *Advanced Audio Markup Language* (AAML). The timing of objects within SMIL is dealt with by a central time line, which is a versatile way of placing objects at a specific time or relative to another object. It is also possible to edit this time line, in real-time using, for example, the detection of mouse clicks. The extensions to SMIL mean that the coordinate system for the audio objects are different to the visual elements, the former being in meters and the latter being in pixels. A listener position must also be defined such that the spatial audio scene is rendered and perceived correctly.

ASDF

The Audio Scene Description Format (ASDF) (Geier, Ahrens and Spors, 2010) is used to describe audio objects in a scene is an XML description based on some of the features of SMIL. The description format is easily extensible and readable by humans, making it very easy to edit and to author acoustic scenes without bespoke software tools.

SpatDIF

One strategy for the encoding and transmission of audio objects is the Spatial Sound Interchange Description format (SpatDIF) (Peters, 2008). Like ASDF, SpatDIF is based on an XML description, however, unlike ASDF it cannot describe the trajectory of an audio object but rather uses a series of control messages with time stamps that describe the source position at each point in time. This can result in problems of either very large amounts of data being recorded in the description or insufficient sampling points to describe accurately the movement of the source. Additionally it is not intuitively obvious how sources move simply from inspection of the file and thus once again, any changes to temporal features of the scene are difficult to edit without bespoke software solutions.

4.5   Scene Acquisition

This section will draw upon the example scenario described in Section 4.3.5 to illustrate some of the important capture techniques that are used for the recording of a scene for format-agnostic broadcast/rendering.

Principally there are two practical recording styles employed to record a sound scene:

1. A sound scene normally consists of a mixture of sound events. The virtual/reproduced sound scene in an object-based scenario generates this mixture by building the scene explicitly including each object of the sound scene. In this case the audio signals of the different objects are recorded separately, and by using a set of spatialisation parameters, a complete scene is composed. An example of this would be the individual instruments making up the

orchestral performance. The recording of the room's acoustics using a array of sources can also be considered as an application of this approach; although the sound event consists of many reflecting sources it will be represented as one *diffuse object*.

2. Alternatively, *microphone arrays* use a special set of microphones in order to record several sound events simultaneously. Here only an implicit definition of the spatialisation parameters or scene objects exists. This approach is often used to capture an entire sound scene in concerts particularly in cases where the recording is made in a good, or interesting, acoustic environment. Either the microphone signals drive discrete loudspeakers or they are transformed before playback.

In practice, as for the example of an orchestral concert, these two concepts are often combined into a virtual sound scene that consists of recordings from a main microphone array and a number of individual virtual sound objects. All individual signals are picked up simultaneously by the main microphone array but additionally spot microphones close to specific instruments or instrument groups are used to enhance the reproduction and allow the sound editor to have some creative freedom at the production end including specifying spatialisation parameters, relative levels of the source and the rendering methods with which the scene is generated.

The common stages of such a production are recording (considering the main microphone array design and/or spot microphone placement and the capture of required audio signals either simultaneously or sequentially) (Owsinski, 2009), mixing (creating a virtual sound scene) (Owsinski, 2006) and mastering (adjusting the overall characteristics of the scene, often done in the domain of loudspeaker signals) (Katz, 2002). For an object-based approach these stages need to be adapted since the content is no longer produced for a specific loudspeaker setup. Thus, attention must be paid to the recording of the audio objects and the sound field components, the spatial and level mixing of objects and sound field components, and mastering should be performed based on the scene representation (Melchior, 2011a).

4.5.1   Capturing Discrete Audio Objects

Discrete audio objects correspond to the direct energy shown in the impulse diagram in Figure 4.2. The capture of discrete audio objects depends greatly on the scenario that is being recorded. In the ideal case one could place close, spot microphones on the sources, resulting in a clean recording with little or no ambient sound component from the reproduction environment and other sources affecting the recording. The aforementioned broadcast scenario consists of several audio objects comprising the musicians in the orchestra, these would be recorded using close microphones positioned near the performers. The recording techniques used for this capture require that the sound sources are as acoustically dry as possible with little or no reverberation present so it is desirable to have the microphones as close as possible to the intended source. As part of the performance described in Section 4.3.5 there is also a choir performing Handel's Messiah. The choir consists of 20 individuals. The essence of a choir is the ensemble performance, so it is not desirable to apply close miking techniques to each individual, not only that, but the number of audio objects and microphones required would certainly be too great. To record the choir for an object-based scene description several spaced microphones can be used at a suitable height above the performers. These can be encoded as several discrete objects (Theile, Wittick and Reisinger, 2003) or can be mixed and rendered as one object with a broad radiation characteristic such that the width of the ensemble is more accurately reproduced.

For an object-based scene representation, it is important that the position of the audio objects in the scene are obtained correctly at the production end to enable a correct spatial audio rendering at the reproduction end. This is important for scenarios where the sources need to be rendered in the correct position within the sound scene or relative to the visual content and particularly if the user is permitted to interact with the audio-visual content. Methods for extracting the content and position of sound sources in the scene therefore must be employed. The complexity of the capture of both position and content of audio objects depends on the type of audio source and genre being recorded and also the capture hardware available. In the following, two types of audio objects are defined that need to be considered when capturing an audio scene, namely *explicit objects* and *implicit objects*.

Explicit Audio Objects

Explicit audio objects are objects that directly represent a sound source and have a clearly defined position within the coordinate system. This could include a sound source that is recorded at close proximity either by microphone or by a line audio signal and is either tracked or stationary with defined coordinates. An example of an explicit audio object would be instruments close miked in a performance that are static and have little or no crosstalk from other sources, or the audio feed from an interview (in this case the position of the object is less important).

Obtaining the Object Position
The position of an explicit object can either be determined prior to the performance (e.g., if the positions of the musicians is constant) or be obtained from tracking data if the position varies.

Practical Capture Examples
In the recording of the classical concert each of the performers can have a microphone (or microphones) set up close to their position. The location of these microphones is measured using a laser scanning measurement device or theodolite thus enabling the content and position of the audio objects to be encoded into the desired format for transmission. In the example capture scenario, there is also a soloist singer who will move around the stage as she sings. She is fitted with a wireless tracking device that communicates her position within the pre defined coordinate system in real-time to the production room where tracking data is encoded as metadata with the corresponding audio data as a dynamic audio object.

For other capture scenarios there are cases where close miking is not possible and in such a case the audio objects cannot be recorded directly but instead must be derived from arrays of more distant microphones unless it is intended to render the audio source as a diffuse object.

Implicit Audio Objects

Implicit audio objects represent sound sources in a more indirect manner and generally describe sources that cannot be closely miked as in the explicit case. These could include signals that are picked up by more distant microphone techniques or by microphone arrays where the source of sound may be derived from a combination of several recording devices.

For the example scenario an implicit audio object could come from some dance performers who shout and clap their hands during the performance but cannot be tracked or close miked because they need freedom of movement during the performance. In this case the audio content can be derived from a set of remote microphones positioned in the scene. It may also be possible to derive the positions from these microphones so they can be rendered as audio objects or a group thereof.

Obtaining the Object Position

In the case of implicit audio objects, techniques can be applied that enable the positioning of the source, either from audio data alone such as described in Oldfield, Shirley and Spille (2013) or from additional information such as visual content tracking. In many cases it is easier, and requires less equipment, if the object position is determined using only the audio data, in which case various beamforming techniques or time-difference-of-arrival (TDOA) techniques can be employed to determine the position of the source using triangulation. Beamforming and acoustic holography while being more accurate techniques for source localisation are usually not practical for a broadcast scenario where the number of microphones available and positioning is often limited. If the source position is to be derived from the audio data only the positions of the microphones need to be measured accurately as part of the sound scene description prior to the recording. The location of these sensors can be measured manually with a theodolite or using laser-scanning devices to build up a 3D map of the production environment, this can additionally help in the registration/position synchronisation of audio content with the video content from any cameras in the scene if these are measured as well.

The accuracy of the source positioning depends in many cases on how many microphones detect the source. If only two microphones pick up the source, it is only possible to position the source somewhere along a line of possible locations. If however there are three or more microphones picking up the source, it can be localised with a much higher degree of accuracy as the third sensor allows the inclusion of depth localisation.

There are many methods for automatic source localisation algorithms in the literature (e.g., Benesty, 2000; Silverman et al., 2005; Do, Silverman and Yu, 2007) most of which require complicated statistical methods but with today's powerful computing abilities it is possible to perform some if these techniques in real-time, which is a prerequisite for a live broadcast scenario. Problems occur however, for large-scale events with sparse microphone arrays as the search area is very large and the captured data therefore limited; putting additional strain on the localisation algorithms.

4.5.2   *Capturing the Sound Field Component*

The sound field component of the scene, although capturing the whole sound scene from one location, is most useful for reproducing the early reflection energy shown in the impulse diagram in Figure 4.2. Capturing the sound field component requires a sampling of the incident sound field from all directions at a given location. This is usually done using array-based microphones such as the Soundfield® (with four capsules) or the Eigenmike® (with 32 capsules) microphones. Both of these microphones provide omni-directional information of the sound field, however, they differ in their spatial resolution. After post-processing the capsules' signals, the Soundfield® microphone provides audio data in a first order Ambisonics format, whereas the Eigenmike® signals can be used for a HOA (see also Chapter 2) representation up to fourth order. The post-processing steps include a simple matrixing operation and a specific filtering to equalise the array response (Batke, 2009; Kordon et al., 2011).

Practical Capture Examples

For the example capture scenario an Eigenmike® was installed in the standing crowd at the performance. As such the recorded sound field component contains not only the ambience of the room but also the less diffuse information from the people within close proximity to the microphone array. The results of recording in such a scenario are an increased sense of envelopment in the scene and an intimacy that would not otherwise be possible. Additionally, rotations and directional attenuation by beamforming methods of the sound field component have a tangible effect on the perceived audio scene, making interaction with the visual content more appealing.

4.5.3   Capturing the Diffuse Field

The main characteristic of the late reflected energy shown in the impulse diagram in Figure 4.2 is that it can be considered as a diffuse sound field. Therefore, a representation in a virtual scene must be able to represent a diffuse sound field. This requires a high number of decorrelated signals, which are reproduced over a number of different directions by the reproduction system. The decorrelated signals can either be generated by an appropriate microphone array or by means of signal processing. Since high quality decorrelation is a challenging task in our concert hall example the hall has been measured by a high resolution spherical microphone array before the recording. Directional impulses responses are generated to give the sound designer additional options to modify the late sound field according to the program material (Melchior, 2011a). To capture additional signals including audience noise a space microphone array has been installed above the audience. A modified Hamasaki square (Rumsey, 2001) is used, which is extended by additional four microphones to generate eight decorrelated signals suppressing as much of the direct sound as possible. In the virtual scene the signals are saved together with a direction and reproduced in order to create a homogeneous isotropic sound field.

4.6   Scene Reproduction

In this section we will cover the reproduction of the virtual scene. This will be illustrated by using the concert hall example from Section 4.3.5 and will discuss the different reproduction scenarios with respect to the problem space description given in Section 4.3. We will demonstrate how this approach can be used within current systems and also how it offers a future-proof approach that will allow the possibility of emerging audio formats. With these different scenarios in mind we will discuss the reproduction of the scene in terms of delivery, environment, rendering and interaction. Detailed explanation of the specific reproduction systems mentioned here are discussed in Chapter 2.

4.6.1   Scenario: Mobile Consumption Via Headphones

For the reproduction for single listeners the rendering can be done utilising binaural techniques based on an HRTF approach. Adopting this technique means that audio objects are positioned in the modelled reproduction space and the HRTF corresponding to that source location is then applied to the source content. A summation of the audio objects convolved with their respective HRTFs results in a complete auditory scene when reproduced over headphones. If a sound field recording has been made, this can be decoded for a binaural output. To enable interaction with the scene, user tracking can be employed to determine not only the listener position but also the direction the listener is looking. This can help to update the audio content as described in Section 4.6.4.

4.6.2   Scenario: Interactive Multichannel Reproduction

If the audience consists of a small number of individuals, binaural techniques are not suitable as each user would have to wear a set of headphones and a different tracker and renderer would be needed for each person.

The audio renderer should in this case receive information as to the screen dimensions and the most likely listener position so that the width of the audio scene can be made to match the reproduction infrastructure in order that the audio sources match up with the video sources.

4.6.3   Scenario: Big Screen

When the scene reproduction is being delivered to large audiences, holophonic techniques such as WFS or HOA are often considered the best solutions as they constitute what could be considered a volumetric solution to spatial audio reproduction. This is particularly significant for larger audiences where it is important that everyone experiences an equally impressive reproduction irrespective of their location in the reproduction environment.

Delivery

It is assumed that the content will be delivered to a number of different big screens and therefore an object-based delivery is used. This enables the flexibility of scaling the scene depending on the size and shape of the reproduction environment. Furthermore, some of the content will be played back using a 3D system and some of it on a 2D system.

Environment

The environment for this scenario is a cinema, which can be assumed to be acoustically controlled according to the appropriate specifications. Therefore, no dynamic processing is required and the concert can be played at the original volume as well as loudness range. To ensure a high quality rendering all loudspeakers of the system are equalised for a number of measurement positions in the audience. The concept underlying this is that all loudspeaker locations can reach most of the audience area and then can be successfully used for sound field rendering.

Rendering

Several hybrid reproduction systems have been suggested in the literature that aim to utilise the strengths of the respective systems. For example, it is well known that WFS provides excellent localisation cues in the horizontal plane but that it is very seldom implemented in 3D. Consequently, WFS systems can be perceptually enhanced through hybridisation with other systems to provide additional height cues. In this scenario a WFS-based rendering using a horizontal ring of loudspeakers is used. Furthermore, the ring is adapted to the stadium seating and ascends towards the back. To deliver a more enveloping experience an additional grid of speakers is mounted under the ceiling. This is fed with diffuse signals and reflections that have been acquired during the acquisition of the scene. In the 2D case these signals are rendered into the horizontal layer. The direct sources, which are mainly the spot microphone signals of the orchestra, are rendered to point sources behind the screen. In order to match the visual image on the screen, the depth of the direct sound positions is reduced by an automatic scene scaling algorithm. Further sound field elements and the main microphone signals are rendered as point sources distributed around the audience. This enables a stable position for the whole listening area. Diffuse elements are rendered using specific sources. In this case the scene scaling algorithm optimises the arrival times and distribution of energy over the field. Simply using a number of plane waves would not work in this scenario since they only give a good diffuse field impression on a limited listening area. All diffuse signals of the scene are mixed to a single set of sources to optimise efficiency.

4.6.4   Scenario: Interactive and Free Viewpoint 3D

In this scenario audio objects are mixed into the sound field representation to allow full adaptation of the audio content to match the user's interaction with the audio-visual content.

Environment

It is well known that the audio reproduction will be affected by the reproduction environment (Toole, 2008). Using an object-based approach could potentially allow a user of the system to have some degree of control over the level of the reproduced reverberation to complement the acoustics of their specific listening environment. As the recorded content will essentially be overlaid with the room impulse response at the listener location, being able to control the level of the ambience could reduce the overall effect of the reverberation slightly.

Rendering

The main consideration for the rendering of the audio in this case is the updating of the positions of the rendered sources to match the requirements of the user interaction.

Panning using Audio Objects

When a user pans video content, the visual view angle changes and this coordinate can be communicated to the audio renderer, which can update the positions of the audio objects relative to the listener. If the reproduction is being made using a system capable of rendering depth information such as WFS, there will be no audio sweet spot associated with the rendering, however, if the rendering is done using stereophonic-based systems, the rendering of the audio objects will only be positioned correctly for one listening position, which will usually be the centre of the reproduction room.

Panning using Sound Fields

This update in view vector from the video rendering when the user navigates in the video content can be used as a means of rotating the sound field. The sound field can be rotated easily at the decode stage depending on the loudspeaker setup (Ahrens, 2012).

Zooming using Audio Objects

When zooming into the content the viewer perspective changes and correspondingly the audio perspective changes. Zooming in to the content creates a wider audio scene as the audio objects move around the listener. Practically, as the zoom angle is increased the angle between the audio objects and the listener will increase as if the listener were actually walking into the scene. As the zoom angle increases still further, the audio objects will continue to move around the listener until such a point that they are behind him.

Zooming using Sound Fields

Applying zoom to a sound field is a non-trivial task. Various translations can be applied to the sound field (Gumerov and Duraiswami, 2004; Ahrens, 2012). Practically, the application of zoom to the sound field component may depend both on user preferences and the recording scenario. Some listeners may find it disorientating for the sound field to be zoomed in and out of and may instead prefer simply a change in the relative levels between the sound field content and the audio objects.

4.7   Existing Systems

This section reviews some existing systems found in both the commercial and research spheres that implement the principles of a platform independent approach to audio as described above.

4.7.1   Commercial Systems

Dolby Atmos

The Dolby Atmos system (Robinson, Mehta and Tsingos, 2012) was first introduced to the market in April 2012 as an object-based audio approach to the audio for motion picture. Dolby Atmos is a hybrid channel and object-based system. It has been designed as an addition to the traditional channel-based system and thus maximises backward compatibility allowing mixing and producing using many of the same tools currently used for the channel-based methodology. The concept is primarily aimed at the cinema market and specifies the addition of loudspeakers in the ceiling to provide additional height cues to give a fully 3D listening experience. Other additions to the recommended loudspeaker set up are the inclusion of full range surround loudspeakers such that sounds can be panned off-screen without a change in timbre resulting from the different response of front and side loudspeakers. The standard Dolby Atmos loudspeaker setup therefore forms a 9.1 system augmenting the traditional 7.1 set up with the ceiling loudspeakers (Lts and Rts) such that the loudspeaker set up becomes: left (L), right (R), low frequency effects (LFE), left side surround (Lss), right side surround (Rss), left rear surround (Lrs), right rear surround (Rrs), left top surround (Lts) and right top surround (Rts). The 9.1 channel system is normally realised with arrays of loudspeakers, which allows individual loudspeakers to be addressed by the object-based rendering engine. Within Dolby Atmos, audio objects can be freely and dynamically moved around within the 3D space and can be assigned to specific loudspeakers or groups thereof for a fully immersive rendering.

DTS Systems

DTS, provides many spatial audio tools and codecs. In early 2013 DTS announced their upcoming Ultra High Definition (UHD) audio solution which, similarly to Dolby Atmos, provides a hybrid channel-based and object-based platform for content creators. The development of their UHD solution came from a desire for a format-agnostic rendering whether over headphones, 5.1, a sound bar or larger audio systems. They are also proposing a new format known as Multi-Dimensional Audio (MDA), which is intended to accelerate the next generation of content creation for object-based systems, however, details of this format are not yet publicly available.

Iosono

Iosono is a company developing spatial audio solutions for different applications. Their technology is mainly based on the traditional array concept of WFS but has since been developed further to systems with larger loudspeaker distances while retaining the perceptually advantageous properties of WFS (Melchior, Heusinger and Liebetrau, 2011). Foundational to the reproduction of acoustic scenes in WFS is an object-based representation (Melchior, 2011b). Iosono thus utilises an object-based audio system within their authoring and rendering software (Meltzer et al., 2008), which allows a fully customisable and dynamic audio scene. An object-based approach, such as this, means that the rendering engine deals with the mixing and relative levels of the sound sources, while the creator can concentrate on accurately positioning the sources in the scene to create a realistic impression of being present in the scene. This object-based approach puts no limitations upon the specification of the rendering system, which can be anything from a simple 2-channel stereo set up to a full WFS system with the number of channels only limited by the audio infrastructure and computational processing power.

4.7.2   Research Projects

There have been a number of research projects that have utilised and extended an object-based audio approach, some of which are described briefly in this section.

Soundscape Renderer

The Soundscape Renderer (Geier, Ahrens and Spors, 2008) is a sound field synthesis authoring and presentation tool, written at T-Labs Berlin. The software has been released under an Open Source license and allows rendering of acoustic scenes to several reproduction formats including WFS, HOA and binaural. They have developed and implemented the Audio Scene Description Format (ASDF) (Geier, Ahrens and Spors, 2010) to describe the audio objects in the scene, which is an XML description based on some of the features of SMIL. The description format is easily extensible and human readable, making it very easy to edit and author acoustic scenes without bespoke software tools. The content can then be dynamically reproduced over the available loudspeaker setup using the Soundscape Renderer. The Soundscape Renderer team is an active research group, publishing many papers in the spatial audio field and object-based sound rendering techniques.

FascinatE Project

The FascinatE project is a research project funded under EU FP7, that aimed at developing an end-to-end audio-visual broadcast system that would provide an immersive, interactive experience for live events. The system was designed to allow users to create and navigate their own user-defined audio-visual scene based either on user preferences, or by free pan, tilt and zoom control of the video content.

The FascinatE project recreated the entire sound field at the end-user side corresponding to the navigation within the panorama, including audio zoom and rotation. The challenging task was to achieve this on any loudspeaker setup with respect to the interactively controlled view at the end-user terminal. This requires as much information as possible about the audio objects and the sound fields. To perform individual modifications by the end-user, for example, zooming in a specific region of a football stadium, audio objects and sound fields have to be captured and transmitted separately. On the reproduction side all these audio signals have to be adequately composed and adapted to the existing loudspeaker setup.

As part of the project methods were developed for real-time live audio object detection ? extraction (Oldfield and Shirley, 2011) in order to enable dynamic audio interaction for dynamically changing user-generated scenes. A modelled 3D audio scene was also defined allowing users to pan around and zoom into the high resolution panoramic video with corresponding changes to the audio scene. This was enabled through the use of an object-orientated approach (Batke et al., 2011), combining a set of audio objects with a recorded sound field component.

At each end-user device, a specific FascinatE Rendering Node exists that contains two main blocks as depicted in Figure 4.7: a *scene composer* and a *scene presenter*. The scene composer was used to select and compose single audio objects or recorded sound fields according to the user request. However, it was not an ordinary mixing of different audio channels. It allowed the user for example, to rotate, select/deselect or adjust the level of specific audio objects and sound fields. The audio presenter then received the selected and composed audio objects as well as the adapted sound field signals and facilitated audio reproduction using the desired technique including stereophonic reproduction, HOA or WFS based on information as to the available loudspeaker setup.

As part of the project there were three test capture events: A live football match, a dance performance and an orchestral performance. The following paragraphs briefly describe some of the outcomes from and the techniques applied during these test captures.

Live Football Coverage

The audio recording of a live football match poses problems primarily relating to the capture of the audio objects in the scene. The audio objects are primarily described by the on-pitch sounds such as ball-kicks, whistle-blows and player communications. These sounds constitute

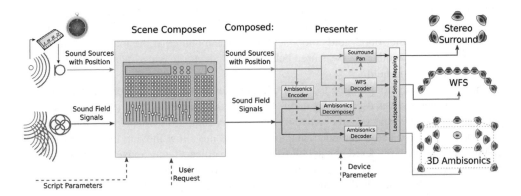

Figure 4.7   The Flexible Rendering Node (FRN) in more detail.

implicit audio objects as tracking devices and close microphones cannot be employed, hence the audio objects have to be derived from the available data. This can be difficult due to the large amount of background noise present in the scene. Consequently, additional capture techniques must be employed to separate only the salient audio content from the available microphones positioned around the pitch, from this information, the position of the detected sounds can then be derived using various time-difference-of-arrival techniques (Oldfield and Shirley, 2011). Other static audio objects such as the commentary are easier to deal with and are generally positioned centrally in the acoustic scene.

For this test capture, a standard broadcast microphone setup was employed consisting of 12 highly directional microphones around the periphery of the pitch with the sound field recorded by a SoundField® microphone and additionally an Eigenmike® that records the sound field up to fourth order Ambisonics. Recording the sound field to a higher spatial resolution in this way enables relative changes in the listener view angle to be represented in the acoustic scene by rotating the sound field around the listener. This introduces some challenges with the implementation of zoom, as the sound field is only recorded at a discrete location, although translations can be applied (Gumerov and Duraiswami, 2004; Ahrens, 2012). Zoom also poses some problems for the rendering of audio objects, as the relative positions of each object should change with user navigation, but the degree to which they change is not obvious as each viewer will have a different size screen and be a different distance from the screen. Hence, the rendering in this case adopted an approach focusing on what would feel perceptually correct/plausible to the listener rather than a geometrically correct approach, which would require tracking of the listener position and a non-trivial coordinate transformation from the 2D panoramic content to the 3D audio scene, which was deemed a subjectively unimportant complication.

Dance Performance
The second test capture took place at the, 'Arena', venue in Berlin in May 2012, during the production of a dance project by the Compagnie Sasha Waltz and Guests and the Education Programme by the Berlin Philharmonic Orchestra conducted by Sir Simon Rattle, performing a choreography of the Carmen-Suite by Rodion Schtschedrin. The audio setup featured a large

number of microphones on a course grid above the performers to pick up the sounds from the dance stage and 62 microphones for the instruments of the orchestra. Two Eigenmikes® were also used for the sound field component. One was mounted at the location of the 180° panoramic camera, another one was used behind the conductor to experiment with the quality of the recorded sound field.

Once again, it was not possible to attach microphones to the dance performers but rather a grid of microphones was used for the audio objects in the scene. Each microphone covered a given zone, which would become active or inactive depending on whether the microphones contained significant audio content. This approach did not allow moving objects but rather, moving sources were rendered as a set of discrete audio objects being active at different times during the performance. The array of microphones above the performers also allowed for experiments with diffuse audio objects that could be used to recreate the ambience of the performance space but not recorded at just one location as was the case with the sound field recordings.

Orchestral Performance

The simplest case for a test capture was a recording of a BBC Proms concert in the Royal Albert Hall, London. The recording team was able to capture three symphonies by Vaughan Williams performed by the BBC Scottish Symphony Orchestra and conducted by Andrew Manze. In this case the audio objects were all static instruments in the scene. Each microphone was closely positioned to an instrument section (brass, woodwind, strings etc.) although ideally each instrument would have had an individual microphone, this set-up did allow the necessary amount of panning and zooming and the user can get a sense of being present in the acoustic scene. An Eigenmike® was positioned on top of the 180° panoramic camera, which was placed among the many standing spectators. Having the Eigenmike® in this location gave an amazing sense of actually being present in the hall during the concert as individual claps and cheers from the corresponding locations in the surrounding crowd are easily discernable. This is one advantage of combining an object-based approach with a sound field representation using the Eigenmike® as it would not be either possible or desirable to use spot microphones to record individuals in the crowd but the increased spatial resolution afforded by the Eigenmike® means that the feeling of *presence* is increased and will adapt as expected with listeners' navigation within the scene.

4.7.3   Perceptive Media

Perceptive media is a term coined by the BBC to describe media content that can be adapted without the user necessarily knowing that the content is being tailored to them. It is not about the user interacting with the content *per se* but rather the content is altered and personalised based upon user profiles gathered from various sources along with any geographic or climate information obtained from, for example, the user's IP address. Practical realisations of the perceptive media concept generally rely on an object-based audio paradigm to facilitate the necessary customisation at the user end thus offering a different application for this format. In this case the object-based approach is not about a spatial audio rendering but rather, having all the necessary audio assets available as individual objects so they can be added in or out of the mix as required by the customising/rendering engine. An example of a perceptive media broadcast was a radio drama produced by the BBC, which utilised an object-based audio paradigm to allow an adaptive storyline. The radio show, entitled *Breaking Out* was designed to have a narrative that would adapt based upon variables such as the geographic location of the listener and the time and date on which they are listening, for example. The drama was composed of several audio assets/objects, corresponding to sound effects, actors voices, music and so forth. recorded at the production end which, along with the adaptive elements rendered using a text-to-speech engine in the listener's browser, made up the content of the audio drama. The drama was delivered over IP, utilising the audio processing ability within the HTML5 standard such that all the audio processing and mixing was done at the user-end, including the addition of any artificial reverberation used to evoke the sensation of the acoustics of the room in which the conversation in the drama is taking place. Details about the first perceptive media radio drama can be found in Churnside and Forrester (2012). This project focused more on the opportunities for creative expression that could be exploited through the use of an object-based audio approach rather than how such a representation of the audio scene could be recorded, transmitted and reproduced. Another application of perceptive media could be in speeding up or slowing down the speed of an audio drama or performance, rather than using temporal stretching principles. For example, it is possible to remove some of the silences between words such that the reading speed increases but there are no intelligibility issues associated with stretching the audio.

4.8 Conclusion

This chapter goes beyond the current state-of-the-art, described in Chapter 2, and has described work defining and implementing a format-agnostic audio system that can deliver both platform independent and interactive audio for TV broadcast.

Section 4.1 introduced some of the drivers for change in the broadcast environment and the need for format-agnostic audio delivery and contextualised enhanced user requirements for broadcast.

In Section 4.2 the terms and definitions used in the chapter were described, in particular setting out the important relationship between the acoustic scene to be represented, the virtual scene used to describe it and the sound scene as presented to the viewer.

Section 4.3 identified the problems of representing an acoustic scene in TV broadcast by describing the important acoustic characteristics and identifying barriers and challenges to be overcome in order to represent this accurately, or effectively, for a viewer. The problem space described here sets out the challenges to be overcome in delivering audio in a format-agnostic manner and discussed using an object-based audio representation in order to tackle these challenges. The section discussed the issues such as listening conditions and limitations of a variety of reproduction systems introduced in Chapter 2, and covered problems such as scene scaling, audio-visual coherence and aspects of user interaction. An example scenario of a broadcast of a music concert was presented, which highlighted the challenges and possibilities for format agnostic audio.

Issues around the make-up and representation of an acoustic scene were discussed in Section 4.4. Channel-based, transformation-based and object-based solutions were described together with examples of formats that may be implemented for each of these approaches.

In Section 4.5 the description of object-based methods was expanded further and the types of audio object that must be represented in an object-based system were defined. This section looked at how different types of audio objects may be captured and how coordinate locations for each object may be derived either from audio signals captured, or by separate tracking.

Format-agnostic reproduction was described in Section 4.6 using example scenarios including for mobile, interactive and large-screen reproduction. Delivery and implementation of panning systems to respond to user interaction were also described with regard to pan and zoom interaction capability.

Finally Section 4.7 provided examples of where object based audio has been implemented to date including recent commercial and research implementations.

4.8.1   Open Issues

While the aforementioned systems have to a large extent been successful in their implementation, there are still many open questions that should be addressed before object-based audio broadcast can become commonplace. For example, recording and reproduction of diffuse audio objects is not covered by any of the example implementations mentioned here (see Section 4.5.3). There is also still much work to be done to finalise a streaming format for an object-based audio description. While it has been shown that file formats such as ASDF and SMIL can provide a useful set of tools to be used in object-based renderers, they are not defined as broadcast formats, which may limit their widespread adoption. The MPEG-4 format however, can be used for broadcast but as mentioned above, it is a rather complicated standard; AudioBIFS are difficult to implement and have therefore not yet been adopted as part of the de facto audio streaming format.

References

Ahrens, J. (2012) *Analytic Methods of Sound Field Synthesis*. Heidelberg: Springer.

ANSI (2008) 'ANSI/ASA S12.2-2008: American National Standard Criteria for Evaluating Room Noise'. Melville (NY).

Baalman, M.A.J. (2008) 'On Wave Field Synthesis and Electro-acoustic Music, with a Particular Focus on the Reproduction of Arbitrarily Shaped Sound Sources', PhD thesis, Technische Universität Berlin.

Batke, J.M. (2009) 'The B-Format Microphone Revised'. *Proceedings of the 1st Ambisonics Symposium*, Graz, 25–27 June.

Batke, J.M., Spille, J., Kropp, H., Abeling, S., Shirley, B. and Oldfield, R.G. (2011) 'Spatial Audio Processing for Interactive TV Services'. 130th Convention of the Audio Engineering Society, London, UK, 13–16 May.

Benesty, J. (2000) 'Adaptive Eigenvalue Decomposition for Passive Acoustic Source Localization'. *Journal of the Acoustic Society of America* 107(1), 384–391.

Bertet, S., Daniel, J., Gros, L., Parizet, E. and Warusfel, O. (2007) ' Investigation of the Perceived Spatial Resolution of Higher Order Ambisonics Sound Fields: A Subjective Evaluation Involving Virtual and Real 3D Microphones'. AES 30th International Conference, Saariselkä, Finland, 15–17 March.

Blauert, J. (1996) *Spatial Hearing: The Psychophysics of Human Sound Localization*. Cambridge, MA: MIT Press.

Blauert, J. and Jekosch, U. (2012) 'A Layer Model of Sound Quality'. *Journal of the Audio Engineering Society* 60(1/2), 4–12.

Bruggen, M. (2001) 'Coloration and Binaural Decoloration in Natural Environments'. *Acta Acustica with Acustica* 87(3), 400–406.

Brun, R. (2011) 'The Festival of Nine Lessons and carols In Surround Sound', accessed 14 August 2013 at: http://www.bbc.co.uk/blogs/radio3/2011/12/the-festival-of-nine-lessons-and-carols-in-surround-sound.shtml

Bruno, R., Laborie, A. and Montoya, S. (2005) 'Reproducing Multichannel Sound on Any Speaker Layout'. 118th Audio Engineering Society Convention, Barcelona 28–31 May.

Callet, P.L., Miller, S. and Perkis, A. (2012) 'Qualinet White Paper on Definitions of Quality of Experience (2012)'. Technical Report Lausanne, Switzerland, Version 1.1, European Network on Quality of Experience in Multimedia Systems and Services (COST Action IC 1003).

Carey, R. and Bell, G. (1997) *The Annotated VRML 97 Reference Manual*. New York: Addison-Wesley Professional.

Cariani, P. and Micheyl, C. (2010) 'Towards a Theory of Information Processing in the Auditory Cortex', in D. Poeppel, T. Overath, A.N. Popper and R. Fay (Eds) *Handbook of Auditory Research: The Human Auditory Cortex*. New York: Springer, pp. 251–390.

Chapman, M., Ritsch, W., Musil, T., Zmölnig, J., Pomberger, H., Zotter, F. and Sontacchi, A. (2009) 'A Standard for Interchange of Ambisonic Signal Sets Including a File Standard with Metadata. *Proceedings of the 1st Ambisonics Symposium Graz*, Austria, 25–27 June.

Churnside, A. and Forrester, I. (2012) 'The Creation of a Perceptive Audio Drama', NEM Summit, Istanbul, Turkey, 16–18 October.

Cooper, D.H. and Shiga, T. (1972) 'Discrete-Matrix Multichannel Stereo.' *Journal of the Audio Engineering Society* 20(5), 346–360.

Craven, P.G. and Gerzon, M.A. (1977) 'Coincident Microphone Simulation Covering Three Dimensional Space and Yielding Various Directional Outputs'. United States Patent, US 4,042,779.

Cross, N., and Puryear, M. (n.d.) 'Enhanced Audio Formats for Multi-Channel Configurations And High Bit Resolution'. Microsoft White Paper.

de Bruijn, W.P.J. and Boone, M.M. (2002) 'Subjective Experiments on the Effects of Combining Spatialized Audio and 2D Video Projection in Audio-Visual Systems'. Presented at 112th AES Convention, Munich, 10–13 May.

Do, H., Silverman, H.F. and Yu, Y. (2007) 'A Real-Time SRP-PHAT Source Location Implementation Using Stochastic Region Contraction (SRC) on a Large-Aperture Microphone Array'. *Presented at IEEE International Conference on Acoustics, Speech and Signal Processing*, 15–20 April.

Dobson, R.W. (2000) 'Developments in Audio File Formats'. *Proceedings of the International Computer Music Conference (ICMC)*, Berlin, 27 August–1 September, pp. 178–181.

Duong, N., Howson, C. and Legallais, Y. (2012) 'Fast Second Screen TV Synchronization Combining Audio Fingerprint Technique and Generalized Cross Correlation'. Consumer Electronics-Berlin (ICCE-Berlin), *IEEE International Conference*, Malaysia, 2–5 May, pp. 241–244.

EBU (2011) Specification of the broadcast wave format (bwf), version 2.0 Tech Doc 3285.

Fuchs, H., Hellmuth, O., Meltzer, S., Ridderbusch, F. and Tuff, S. (2012) 'Dialog enhancement: Enabling user interactivity with audio'. Presented at the NAB Show, Las Vegas, 14–19 April.

Geier, M., Ahrens, J. and Spors, S. (2008) 'The Soundscape Renderer: A Unified Spatial Audio Reproduction Framework for Arbitrary Rendering Methods'. 124th Convention of the Audio Engineering Society, Amsterdam, 17–20 May.

Geier, M., Ahrens, J. and Spors, S. (2010) 'Object-Based Audio Reproduction and the Audio Scene Description Format'. *Organised Sound* 15(3), 219–227.

Gerzon, M.A. (1992a) 'Hierarchical Transmission System for Multispeaker Stereo'. *Journal of the Audio Engineering Society* 40(9), 692–705.

Gerzon, M.A. (1992b) 'Signal Processing for Simulating Realistic Stereo Images'. Preprint 3423, 93rd Convention of the Audio Engineering Society, San Francisco, 1–4 October.

Gumerov, N.A. and Duraiswami, R. (2004) *Fast Multipole Methods for the Helmholtz Equation in Three Dimensions*, 1st edn. Amsterdam: Elsevier.

Harboe, G., Massey, N., Metcalf, C., Wheatley, D. and Romano, G. (2008) 'The Uses of Social Television'. *Computers in Entertainment (CIE)* 6(1), 8.

Herre, J., Purnhagen, H., Koppens, J., Hellmuth, O., Engdegård, J., Hilper, J. et al. (2012) 'MPEG Spatial Audio Object Coding – The ISO/MPEG Standard for Efficient Coding of Interactive Audio Scenes'. *Journal of the Audio Engineering Society* 60(9), 655–673.

Hoffmann, H., Dachselt, R. and Meissner, K. (2003) 'An Independent Declarative 3D Audio Format on the Basis of XML2003'. International Conference on Auditory Display, Boston, MA, 6–9 July.

ITU (2003) 'ITU-R BS.1284-1: General methods for the subjective assessment of sound quality', International Telecommunications Union, Geneva.

ITU-R (1994) 'BS.775: Multichannel Stereophonic Sound System With and Without Accompanying Picture'. International Telecommunications Union, Geneva.

ITU-R (1997) 'BS.1116-1: Methods for Subjective Assessement of Small Impairments in Audio Systems including Multichannel Sound Systems'. International Telecommunications Union, Geneva.

Karamustafaoglu, A. and Spikofski, G. (2001) 'Binaural Room Scanning and Binaural Room Modeling'. 19th International Audio Engineering Society Conference, Schloss Elmau, Germany, 21–24 June.

Katz, B. (2002) *Mastering Audio*. Boston, MA: Focal Press.

Kendall, G.S. (1995) 'The Decorrelation of Audio Signals and its Impact on Spatial Imagery'. *Computer Music Journal* 14(9), 17–87.

Kordon, S., Krüger, A., Batke, J.M. and Kropp, H. (2011) 'Optimization of Spherical Microphone Array Recordings'. International Conference on Spatial Audio, Detmold, Germany, 10–13 November.

Kubovy, M. and van Valkenburg, D. (2001) 'Auditory and Visual Objects'. *Cognition* 80(1), 97–126.

Kuttruff, H. (2000) *Room Acoustics*. New York: Taylor & Francis.

Laitinen, M.V., Philajamäki, T., Erkut, C. and Pulkki, V. (2010) 'Parametric Time-Frequency Representation of Spatial Sound in Virtual Worlds'. ACM Transactions on Applied Perceptions.

Letowski, T. (1989) 'Sound Quality Assessment: Cardinal Concepts'. 87th Convention of the Audio Engineering Society, New York, 18–21 October.

Malham, D. (2005) 'Second and Third Order Ambisonics – The Furse-Malham Set', accessed 14 August at: http://www.york.ac.uk/inst/mustech/3d_audio/secondor.html

Melchior, F. (2011a) 'Investigations on Spatial Sound Design Based on measured Room Impulse Responses', PhD thesis, Delft University of Technology.

Melchior, F. (2011b) 'Wave Field Synthesis and Object-Based Mixing for Motion Picture Sound'. *SMPTE Motion Imaging Journal* 119(3), 53–57.

Melchior, F. and Brix, S. (2004) 'Device and Method for Determining a Reproduction Position'. Patent EP 1518443 B1.

Melchior, F., Brix, S., Sporer, T., Röder, T. and Klehs, B. (2003) 'Wavefield Synthesis in Combination with 2D Video Projection'. Presented at the 24th AES Conference, Banff, 26–28 June.

Melchior, F., Fischer, J.O. and de Vries, D. (2006) 'Audiovisual Perception using Wave Field Synthesis in Combination with Augmented Reality Systems: Horizontal Positioning'. 28th AES Conference, Piteå, Swede, 30 June–2 July.

Melchior, F., Heusinger, U. and Liebetrau, J. (2011) 'Perceptual Evaluation of a Spatial Audio Algorithm based on Wave Field Synthesis using a Reduced Number of Loudspeakers', 131th Convention of the Audio Engineering Society, New York, 20–23 October.

Melchior, F., Michaelis, U. and Steffens, R. (2011) 'Spatial Mastering – A New Concept for Spatial Sound Design in Object-Based Audio Scenes'. Presented at the ICMC, Huddersfield, UK, 30 June–5 August.

Meltzer, S., Altmann, L., Gräfe, A. and Fischer, J.O. (2008) 'An Object Oriented Mixing Approach for the Design of Spatial Audio Scenes'. Presented at the 25th VDT Convention, Leipzig, 13–16 November.

Nachbar, C., Zotter, F., Deleflie, E. and Sontacchi, A. (2011) 'Ambix – A Suggested Ambisonics Format'. 3rd International Symposium on Ambisonics & Spherical Acoustics, Lexington, KT, 2–3 June.

Oldfield, R.G. and Shirley, B.G. (2011) 'Automatic Mixing and Tracking of On-Pitch Football Action for Television Broadcasts'. 130th Convention of the Audio Engineering Society, London, UK, 13–16 May.

Oldfield, R.G., Shirley, B.G. and Spille, J. (2013) 'Object-Based Audio for Interactive Football Broadcast'. Multimedia Tools and Applications, DOI: 10.1007/s11042-013-1472-2.

Owsinski, B. (2006) *The Mixing Engineer's Handbook*. Boston, MA: Thomson Course Technology.

Owsinski, B. (2009) *The Recording Engineer's Handbook*. Boston, MA: Thomson Course Technology.

Pereira, F. and Ebrahimi, T. (2002) *The MPEG-4 Book*. Upper Saddle River, NJ: Prentice Hall.

Peters, N. (2008) 'Proposing SpatDIF – The Spatial Sound Description Interchange Format (2008)'. Ann Arbor, MI: Scholarly Publishing Office, University of Michigan Library.

Pihkala, K. and Lokki, T. (2003) 'Extending SMIL with 3D Audio (2003)'. International Conference on Auditory Display, Boston, MA, 6–9 July.

Pike, C. and Melchior, F. (2013) 'An Assessment of Virtual Surround Sound Systems for Headphone Listening of 5.1 Multichannel Audio'. 134th Audio Engineering Society Convention, Rome, 4–7 May.

Poletti, M. (2005) 'Three-Dimensional Surround Sound Systems Based on Spherical Harmonics'. *Journal of the Audio Engineering Society* 53(11), 1004–1025.

Potard, G. (2006) '3D-Audio Object Oriented Coding', PhD thesis, University of Wollongong.

Robinson, C., Mehta, S. and Tsingos, N. (2012) 'Scalable Format and Tools to Extend the Possibilities of Cinema Audio'. *SMPTE Motion Imaging Journal* 121(8), 63–69.

Rumsey, F. (2001) *Spatial Audio*. Amsterdam: Focal Press.

Rutledge, L. (2001) 'Smil 2.0: Xml for Web Multimedia'. *Internet Computing, IEEE* 5(5), 78–84.

Schmidt, J. and Schröder, E.F. (2004) 'New and Advanced Features for Audio Presentation in the MPEG-4 Standard'. 116th Convention of the Audio Engineering Society, Berlin, 8–11 May.

Shirley, B. and Kendrick, P. (2006) 'The Clean Audio Project: Digital TV as Assistive Technology'. *Technology and Disability* 18(1), 31–41.

Silverman, H.F., Yu, Y., Sachar, J.M. and Patterson III, W.R. (2005) 'Performance Of Real-Time Source-Location Estimators for a Large-Aperture Microphone Array'. *IEEE Transactions on Speech and Audio Processing* 13(4), 593–606.

Snyder, J. (2005) 'Audio Description: The Visual Made Verbal', *International Congress Series* 1282, 935–939.

Spors, S., Wierstorf, H., Raake, A., Melchior, F., Frank, M. and Zotter, F. (2013) 'Spatial Sound with Loudspeakers and its Perception: A Review of the Current State'. *IEEE Proceedings*, 1–19.

Theile, G., Wittek, H. and Reisinger, M. (2003) 'Potential Wavefield Synthesis Applications in the Multichannel Stereophonic World'. Presented at the 24th AES Conference, Banff, 26–28 June.

Toole, F.E. (2008) *Sound Reproduction: The Acoustics and Psychoacoustics of Loudspeakers and Rooms*. Amsterdam: Focal Press.

Troughton, P. (2002) 'Convenient Multi-Channel Sound in the Home'. 17th UK Audio Engineering Society Conference, London, 9–10 April.

Vaananen, R. and Huopaniemi, J. (2004) 'Advanced Audiobifs: Virtual Acoustics Modeling in Mpeg-4 Scene Description'. *IEEE Transactions on Multimedia* 6(5), 661–675.

Vaishnavi, I., Cesar, P., Bulterman, D. and Friedrich, O. (2010) 'From IPTV Services to Shared Experiences: Challenges in Architecture Design'. Multimedia and Expo (ICME), *IEEE International Conference*, Singapore, 19–23 July, pp. 1511–1516.

van Eijk, R.L.J. (2008) 'Audio-Visual Synchrony Perception', PhD thesis, Technische Universiteit Eindhoven.

Wallach, H., Newman, E.B. and Rosenzweig, M.R. (1949) 'The Precedence Effect in Sound Localization'. *American Journal of Psychology* 57, 315–336.

Younkin, A. and Corriveau, P. (2008) 'Determining the Amount of Audio-Video Synchronization Errors Perceptible to the Average End-User'. *IEEE Transactions* 54(3), 623–627.

Zotter, F., Frank, M., Marentakis, G. and Sontacchi, A. (2011) 'Phantom Source Widening with Deterministic Frequency Dependent Time-Delays'. *DAFx-11 Proceedings*, Paris, 19–23 September.

5

Semi-Automatic Content Annotation

Werner Bailer[1], Marco Masetti[2], Goranka Zorić[3], Marcus Thaler[1] and
Georg Thallinger[1]

[1] *JOANNEUM RESEARCH, Graz, Austria*
[2] *Softeco Sismat Srl, Genoa, Italy*
[3] *Interactive Institute, Kista, Sweden*

5.1 Introduction

In recent years, we have witnessed significant changes in the way people consume media, in particular in digital form. These trends include increasing levels of interactivity, increasing screen resolutions, spatial sound with higher fidelity and a steep growth of media consumption on mobile devices. The TV industry has responded to these trends in various ways. Content is often produced at different fidelity levels, for instance with SD and HD (and soon 4K) offerings of the same TV channel, and reframed versions are sometimes created for viewing on small mobile displays. From a high-level perspective, one can observe that these approaches are actually contributing to a multiplication of formats, being deployed over several media delivery chains. Relatively modest interactivity or personalisation features are offered, and most of the time these are confined to the second screen only. In contrast, the format-agnostic paradigm advocated in this book (cf. Section 2.9) has the objective of changing the way content is produced at the source, so that the highest levels of quality and flexibility are maintained as far through the delivery chain as possible, giving more possibilities for the content to be customised based on the preferences of the end user and the functionality offered by their device.

In traditional broadcast production, a single (or at most a few) output streams are produced from a limited number of input streams. The task of selection from the different streams in real-time is carried out by a well-trained production crew, by viewing the inputs on an array of

Media Production, Delivery and Interaction for Platform Independent Systems: Format-Agnostic Media, First Edition. Edited by Oliver Schreer, Jean-François Macq, Omar Aziz Niamut, Javier Ruiz-Hidalgo, Ben Shirley, Georg Thallinger and Graham Thomas.
© 2014 John Wiley & Sons, Ltd. Published 2014 by John Wiley & Sons, Ltd.

screens and using specialised tools and interfaces that support them in this task, as described in Section 2.4. When moving to format-agnostic production, the production of a range of different output streams is required, each tailored to specific user groups or individuals, and to specific devices. In addition, the output may not only be a finalised stream, but a scene in which a user can navigate and interact. This means that it is no longer feasible for the production team to manually create each of the output options. They need to be supported by automation, for example, the *Virtual Director* tools described in Chapter 6. In short, a Virtual Director in the context of a format-agnostic production system is an intelligent software component that automatically frames and cuts between multiple camera views. Its purpose is to present the most relevant action in the scene in an aesthetically pleasing style that is appropriate and specific to the production domain.

In order to enable the Virtual Director to take appropriate and editorially pleasing decisions, it needs information about what is going on in the scene, that is, descriptive metadata of the audio-visual content being captured. For content selection, this includes metadata about objects and people, specific actions and "interesting things" happening. Additionally, in order to perform framing of a virtual camera, this includes the regions where these objects and events are visible, as well as their trajectories. Adding the required annotations manually would of course ensure the highest possible quality, however, just as performing content selection, providing all these annotations in real-time is not feasible with a reasonably small number of people. Automatic metadata extraction is an alternative and scalable option. The drawback of automatic methods is a lower reliability of the results. Thus, in practical systems, a semi-automatic approach is applied: automatic tools provide a large set of low-, mid- and high-level metadata, and operators review the metadata with the content, being able to make corrections and add further relevant annotations. It depends on the specific type of metadata to be generated, whether an automatic, semi-automatic or manual approach is preferable. For example, tracking objects automatically is relatively reliable, while drawing and updating person tracks in real-time is hardly feasible for an operator. In contrast, recognising a complex and rare event automatically might not be feasible with acceptable reliability, while a human could easily set in- and out-time points for the event in order to highlight it as important to the Virtual Director.

The remainder of this section discusses general requirements for the semi-automatic analysis tools as well as the required types of metadata. The following section, Section 5.2, presents architectures of analysis subsystems and metadata models that fulfil these requirements. Subsequent sections discuss different crucial types of analysis tools: domain-independent saliency analysis tools in Section 5.3; person detection and tracking in Section 5.4; and online concept/action detection in Section 5.5. User interfaces that enable the production team to view and update metadata are described in Section 5.6 and Section 5.7 concludes the chapter.

5.1.1   Requirements on Semi-automatic Annotation Tools

The requirements on the semi-automatic annotation tools are strongly driven by the need for real-time operation. As the Virtual Director needs to make decisions in real-time and with low latency, the automatic metadata extraction tools need to be capable of working under the same constraints. Automatic tools for offline analysis typically report results for segments in retrospect, while in the online case preliminary results must be reported in order to meet the latency constraints. For the manual annotation tools, the real-time requirement mainly poses challenges for user interface design in order to enable efficient inputs and interactions.

Another requirement concerns the applicability in a broad range of content domains. This means that both the automatic and manual tools shall be either domain agnostic or it needs to be possible to adapt or train the tools easily to a new domain of interest. For example, there is a large amount of work on automatic sports video analysis (see e.g., Wang and Parameswaran, 2004), which can extract relevant events, but targeted to a specific kind of sport. Similarly, there are highly specialised sports logging tools. However, a practically usable set of format-agnostic production tools should not be restricted to certain content domains.

The temporal granularity of the annotations is an important aspect. On the one hand, the temporal granularity has to be sufficiently fine to meet the latency constraints; on the other hand the extracted metadata must be sufficiently coarse so that the Virtual Director can still process all the information it receives. Finding the right tradeoff in terms of granularity and avoiding redundant updates is very important.

Finally, the usefulness of the extracted information for the Virtual Director tools is a key criterion. The metadata needed by the Virtual Director tools is partly different from that used in typical content search applications. The result of the annotation should provide mid- or high-level descriptions of the content that are semantically useful for making shot framing and selection decisions. Rather than describing entire segments after they have ended, start/end events and changes of specific metadata values are relevant. In addition, localisation in the scene is of crucial importance for framing decisions, in particular when processing panoramic video.

5.1.2   Requirements on Metadata

In the following, the requirements related to metadata, and the different types of metadata that are relevant in the format-agnostic production process, are discussed.

The system has to deal with both real-time and non-real-time metadata, that is, metadata created, extracted, processed and transmitted online, or created offline, such as information about the venue, ontologies of objects of interest, and visual grammar rules. Metadata come with different temporal granularity, from dynamic metadata varying by frame up to metadata valid for the whole production. In a live scenario, metadata also have a temporal scope, that is, it can be discarded after some interval of validity. Most metadata in the system flow downstream, such as all metadata extracted from the content. However, there are also upstream metadata, for example, user interactions or information from the distribution network back to the Virtual Director.

Sensor Parameters

In order to allow a format-agnostic representation of the scene, a large variety of metadata from production needs to be made available in different succeeding modules of the processing chain. Each scene can be represented by an audio scene and a video scene. The video scene usually consists of several spatially distributed camera clusters, and the audio scene consists of individual audio objects and one or more sound field representations (cf. Section 2.9 on format-agnostic media). Audio objects are recorded with close-up microphones or estimated using an array of microphones including shotgun microphones. The metadata can be classified in two main groups, either static or dynamic. Metadata describing the scene geometry will be defined during the setup of the complete capturing framework, and remain fixed during the whole lifetime of the live event coverage (as discussed in Section 3.3.1). On the other hand, some sensors such as broadcast cameras will pan and zoom during capture. Therefore, internal and external parameters will change accordingly. These metadata are dynamic and have to be updated on a frame by frame basis. Dynamic metadata are also needed to represent the fact that a sensor can be switched on and off during the event. Positions and parameters of all these sensors are available and are needed to map annotations of audio-visual content to positions in the scene.

Content Description

This encompasses metadata describing low-, mid- or high-level features of audio-visual content, which are typically extracted by automatic tools or added by the production team. Metadata produced by automatic content analysis with no or only limited semantic information are called low- and mid-level metadata (e.g., colour features, blobs and trajectories of moving objects). High-level metadata related to regions or objects of interest, available audio/video streams, and so forth are often the product of relations between mid-level attributes or manual annotation and bear semantic information, that is, represent objects, actions or events.

In particular, the following metadata for content description are of interest: the precise description of object regions and their trajectories over time, the detection results of specific classes of objects (e.g., people), consisting of candidate region and confidence, and if applicable, centre position and/or orientation. Other metadata include the identification of specific object instances, the classification of actions, the amount of activity/saliency/relevance in a certain area of the depicted/recorded scene and the movement of audio sources (explicit audio objects). In the manual annotation process, the production team will mainly deal with regions of interest (ROIs) within the scene, time codes of the start and end of key actions, and references to objects, people and actions from a controlled set customised for the production.

Domain/Scene Knowledge

Domain, or scene, knowledge includes knowledge about the content domain or type of event, static and dynamic external metadata about the event and static description of the scene setup. For different use cases and scenarios, basic and additional knowledge about the domain and the scene needs to be made available. An example is the provision of additional information on players in a football game, game statistics and so on which can be requested from the user. This additional knowledge can be classified into general knowledge about the domain of the production (e.g., structure of a football game, relevant types of events), the scene setup (e.g., area of field, stage, audience, reverberation time and limits of acoustic space) and dynamic event information (e.g., sports scores, statistics, of the current or related simultaneous events, music concert set lists).

Production Rules and Visual Grammar

There are a number of general principles that cameramen are trained to use when framing a shot, and that a programme director will use when positioning the cameras, instructing the cameramen and selecting shots. Some of these are common across many programme genres, while others are specific to particular kinds of programmes, or may vary depending on the screen size and aspect ratio of the target viewing device. Furthermore, some of these principles will be applied more flexibly by some directors and cameramen than others, as each has a personal style. The Virtual Director needs to capture these general principles so that the kinds of shots offered by the automated shot framing and selection decisions adhere to the expected look and feel of a professional TV production.

5.2   Metadata Models and Analysis Architectures

This section discusses the representation and storage of metadata from audio-visual content annotation. The discussion includes models for the representation of metadata, architectures for automatic content analysis and metadata storage, in particular, for providing time-based indexing and access of content annotations.

5.2.1   Metadata Models

A metadata model is a data model for representing technical descriptions of the content items and manual annotations as well as results of automatic analysis. A number of different metadata models are in use for different types of multimedia data and in different application domains.

The EBUCore Metadata set described in EBU (2012) aims to define a minimum set of metadata properties for video and audio content for a wide range of applications in the broadcast domain. It is based on Dublin Core to facilitate interoperability beyond the broadcast and audio-visual media domain, for example, with the digital library world (DCMI, 2003). EBUCore adds a number of metadata properties that are crucial for audio-visual media, as well as qualifiers (allowing more precise definition of the properties' semantics) to the Dublin Core elements. Wherever possible, the use of controlled vocabularies (e.g., for people, places, genres) is supported. An important advantage over Dublin Core is also the support for parts (e.g., fragments, segments) of media resources, which enables annotation of time-based metadata. A representation of EBUCore in Web Ontology Language (OWL) has been published.

The Society of Motion Picture and Television Engineers (SMPTE) has issued standards for technical and descriptive metadata of audio-visual content, including a container format, metadata scheme and structured list of metadata elements. The SMPTE Metadata Dictionary is a structured list of 1,500 metadata elements in seven distinct classes: identification, administration, interpretation, parametric, process, relational and spatio-temporal (SMPTE, 2010). The elements can be represented in binary format and the standard defines the key, length and value (KLV) format and semantics for each of the elements. It is used for all metadata embedded in Material eXchange Format (MXF) files, but the elements defined in the dictionary are also used outside MXF. The context of the elements and their semantics in relation to their parent elements are not precisely specified.

MPEG-7, formally named *Multimedia Content Description Interface* is a standard developed by the Moving Picture Experts Group (MPEG, 2001). The standard targets the description of multimedia content in a wide range of applications. MPEG-7 defines a set of description tools, called description schemes and descriptors. Descriptors represent a single metadata property (e.g., title) of the content description, while description schemes are containers for descriptors and other description schemes. It is important to highlight that MPEG-7 standardises the description format, but not the methods for feature extraction and comparison. MPEG-7 Systems specifies mechanisms for supporting synchronisation between content and their descriptions, as well as for partial updates of metadata documents. This is of interest in real-time use cases.

Four MPEG-7 profiles have been defined in part 9 of the standard. Among them, the Audio-Visual Description Profile (AVDP) described in MPEG (2012) is the only one also including low-level visual and audio descriptors, which are needed for representing the results of automatic audio-visual content analysis. Thus, the profile includes functionalities for representing results of several – (semi-)automatic or manual – feature extraction processes with confidence levels in multiple parallel timelines, as well as for describing feature extraction tools, version, contributors, and the date/time the tool was applied. The top-level structure for content description supported by AVDP considers three cases. The most basic case is used to describe a single item of audio-visual content. The second case is used to describe several content items and their relations (e.g., similar segments or copies). The third describes the result of summarisation and the related audio-visual content.

A shortcoming of current metadata standards and formats is the capability for representing detailed calibration metadata, which are not sufficiently supported by any of the formats discussed above. This group of technical metadata properties is of high relevance in multisensor capture environments in order to represent position and orientation metadata of sensors, as well as further device characteristics (e.g., lens parameters) that are needed for fusing information captured by different sensors.

Due to the need for fine-grained spatio-temporal metadata, MPEG-7 is a good candidate for the representation of automatic analysis results. For other types of metadata, it can be complemented by other formats, or extensions when there is no appropriate format available.

5.2.2 Architectures for Audio-visual Analysis

This section outlines the major issues and solutions needed to fully exploit metadata annotation and semantic feature extraction from audio-visual content. Extracted features and annotation manually (or semi-automatically) added to the media content are useless if they cannot be used to add value to end-user services. In this sense extracted features have to be encoded in a proper format and distributed to the other system components in the chain. Independent of the specific format used, the metadata life cycle covers all the steps depicted in Figure 5.1.

Metadata extraction is carried out by audio-visual analysis components and annotation tools and is described later in this Chapter. Metadata filtering/preprocessing usually involves some domain knowledge and aims to reject spurious or superfluous metadata. Metadata storage, fusion and enrichment is usually the scope of a metadata storage component, needed to provide persistence and to avoid collected metadata being lost. Regarding metadata usage, traditionally either a pull or a push model can be adopted. In a pull model the user (or

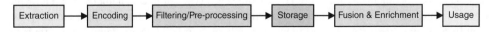

Figure 5.1 Metadata life cycle.

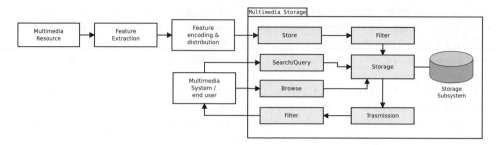

Figure 5.2   Data store architecture according to Kosch (2004).

a multimedia component that needs audio-visual metadata) submits a query to the metadata storage component and gets a single response back (e.g., the Virtual Director when it queries the semantic data store for the area of the scene with the most objects of interest). In a push model instead, the client (either an end user or another system component) establishes a connection with the metadata storage component and is fed by the metadata stream. A publish/subscribe mechanism can be adopted as an evolution of the push model. Usually a client is interested in a selected group of descriptors, hence, metadata are filtered before being streamed. In a format-agnostic production system this model is used to continuously stream metadata of the objects of interest found (e.g., where bounding boxes representing these objects are grouped) to the Virtual Director. An architecture that implements all these functionalities (outlined in Kosch, 2004) is depicted in Figure 5.2.

5.2.3   Storing MPEG-7 Metadata

As discussed above, MPEG-7 provides a standardised metadata set for the detailed description of audio-visual media that can be used as the basis for a broad range of multimedia applications, including media production. The large number of descriptors in an MPEG-7 document poses a challenge for an effective data store solution. A suitable MPEG-7 storage solution should satisfy several other critical requirements: fine-grained storage, typed representation, multidimensional index structures for efficient access, and path indexing to retain and browse the inherent hierarchical structure of MPEG-7 documents. In addition to these basic features, the use of MPEG-7 in some domains poses extended requirements like the proper management of complex data types like arrays, matrices, time points and duration used by some MPEG-7 low-level descriptors.

Structure of an XML Document

One example of a data model that can be adopted is the data model of XML Path Language (XPath) to represent XML documents (Clark and DeRose, 1999). The XPath data model models XML documents as an ordered tree using seven types of nodes, namely, root, element, text, attribute, namespace, processing-instruction and comment. Without losing generality nor simplifying things too much we will focus here only on four node types: element, text attribute and reference, as depicted in Figure 5.3.

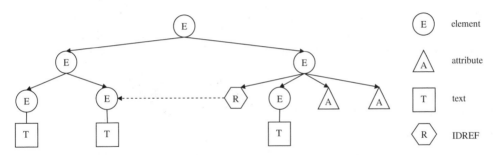

Figure 5.3 Main XPath node types.

Approaches to Provide Persistence to an XML Document

A main drawback in using XML to encode MPEG-7 metadata is the overwhelming number of XML elements and attributes typically generated by audio-visual analysis components. Storage solutions encompass native data stores (databases especially designed to store XML documents), data store extensions for traditional relational databases and ad-hoc database schemas. The following sections will review the relevant solutions for these categories.

Native XML Data Stores

The field of native XML databases is still quite an active research area. A broad range of solutions, commercial and open source, can be found on the market. Here only the most successful and widespread solutions are listed. Among the commercial native XML database products worth mentioning are Excelon-XIS (Excelon, 2013), Infonyte XQL Suite (formerly PDOM) (XQL, 2013), Tamino (Schöning, 2001) and X-Hive/DB (X-Hive, 2013). A comparison between these solutions is beyond the scope of this book. A general remark is that the notion of a "native" data store and what this really implies is somewhat fuzzy with respect to the notion of relational databases and seems to be more of a marketing strategy. Some (but not all) of these solutions store the XML document in its original format focusing on exploiting the most advanced compression techniques to minimise the space occupied. A common feature of all native solutions is the use of the XML standard to model the stored data, express queries and encode query results. Generally these solutions cannot handle data not encoded in XML. Alongside these commercial solutions there is a broad range of open source ones, among them eXistdb (eXistdb, 2013) and Basex (Basex, 2013).

Extensions for Traditional DBMS

Usually three different approaches can be followed to represent XML documents in relational databases. The first approach, sometimes referred to as *unstructured storage* implies storing the XML document as text in a CLOB (Character Large Object) field. The most common commercial DBMS (Data Base Management System) solutions (Oracle, IBM DB2, MS SQL Server) support the unstructured storage of XML documents. In the second approach (generally called *structured storage*) a fine-grained metamodel able to represent the different XML tree nodes and their relationship is built. This approach usually tends to create over-bloated DBMS schemes and is fairly rigid. The third approach (generally called *structure mapping*) implies the definition of a generic DBMS schema to handle an XML document.

Generic DBMS Schemes to Handle XML

Historically traditional DBMS (Data Base Management Systems) data stores have been suggested to provide persistence to XML elements. However, using a relational database to store hierarchically structured data, such as XML elements, poses some issues. To store XML elements in a relational data store two approaches can be followed: the *schema-conscious* (or *structure-mapping*) approach and the *schema-oblivious* (or *model-mapping*) approach. Following a schema-conscious approach basically means creating a data store table for each different node type. This has several drawbacks, the most significant one being the complexity in following XPath queries efficiently and the tendency to generate over-bloated schemas. On the other hand the major drawback of schema-oblivious approaches is the lack of an effective typed representation mechanism, as usually all elements and attribute values are represented as plain text (as formatted in XML) no matter what their nature. Some solutions, like SM3 (Chu et al., 2004), try to take the best from the two approaches. This is due to the fact that an XML document can be viewed as a tree graph. In an XML tree, the internal nodes correspond to the element types with complex element content in an XML document, while the leaf (or evaluated) nodes correspond to the single-value attributes and element types with simple content in the XML document. The idea of SM3 is to use a model-mapping approach to map all internal nodes and use a structure-mapping approach to map all leaf nodes. While mitigating most of the issues related to structure-mapping approaches, a general drawback of this solution is that the MPEG-7 schema has to be fixed in advance and tables for any leaf type (element or attribute) have to be created in advance even if no elements of that type will eventually be generated by the audio-visual analysis modules.

Looking at the inherent hierarchical tree structure of an XML document, another database schema categorisation is between *edge-oriented* approaches and *node-oriented* ones. Among the edge-oriented approaches we can list the Edge schema (Florescu and Kossmann, 1999) that stores XML data graphs as directed graphs (each edge has a source node and a target node). The Edge approach stores all edges in a single table. The Monet approach (Schmidt et al., 2004) can be seen as a variation of the Edge approach and stores edges in multiple tables, one for each possible path. One node-oriented approach found in literature is XRel (Yoshikawa and Amagasa, 2001).

XParent as a Simple Schema to Store XML Data

XParent is an edge-oriented approach that, in contrast to XRel, stores node parent–child relationships in a separate table (XRel is limited to storing node paths). This speeds up query execution, while not consuming much space. XParent can be used to store large XML data collections like the MPEG-7 snippets produced by audio-visual analysis components. The XParent schema is shown in Figure 5.4.

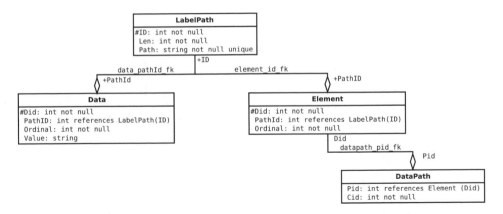

Figure 5.4   XParent schema.

5.2.4   *Bulk Loading Techniques for Massive MPEG-7 Metadata Storage*

To benchmark the execution of queries over a data store solution, some reference data has to be loaded first into the data store. Historically several generic solutions exist. Some of the most widely used are: DBLP bibliography files (DBLP, 2013), SHAKES (Bosak Shakespeare collection) (Bosak, 1999), and BENCH (from the XML benchmark project) (Schmidt et al., 2001).

To maximise the import throughput of metadata, content has to be loaded using bulk-loading techniques when possible. The XParent schema is very simple and, although being highly normalised, it can be seen that records in the Data and DataPath tables are not referenced by anybody, so they do not need to be inserted one at a time but can be loaded all at once. Moreover, using specific DBMS features that enhance the performances in populating the database (like disable auto-commit, use copy methods instead of insert, remove indexes and foreign key constraints and many others) the performance of database loading can be greatly increased. As an example, using commodity hardware, an impressive performance boost (passing from 200 minutes without bulk loading techniques down to less than two minutes with bulk loading techniques to load one million of XML elements) can be achieved.

5.2.5   *An Example Architecture of a Semantic Layer Management System*

Adopting a small, generic and simple schema like XParent can be a solution to keep pace with the MPEG-7 input coming from the analysis components. Tests proved that the semantic data store does not lose any MPEG-7 descriptor even if several ones were produced in parallel for each frame. XParent, however, may not be the best solution for semantic processing of XML content. This is mainly due to its generic nature and lack of domain context. In the audio-visual analysis domain for example, the information of a person's position in the scene, held in an MPEG-7 MovingRegion descriptor, is spread across several records at XParent level, so is not easily viewed as a unique entity. Moreover, encoding any value type as text prevents any possibility of using inference on the data. Another reason that prevents the use of XParent for

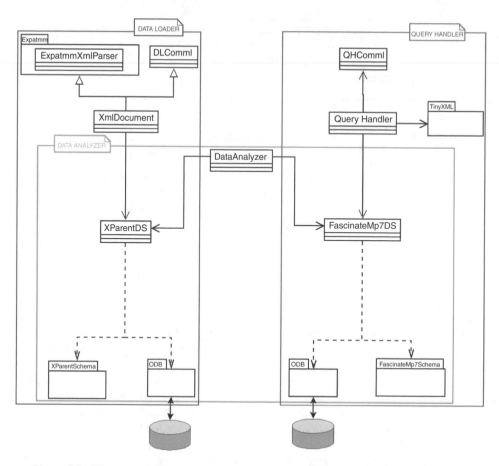

Figure 5.5   The semantic layer management system as implemented in the FascinatE project.

semantic queries is the limitations of the Structured Query Language (SQL) and the difficulty in building an effective SQL engine. For example a query like "track the position of this group of people from frame x to frame y" would be very hard to translate into SQL code and will definitely end up in a long sequence of SQL instructions.

For these reasons usually a semantic layer that leverages upon the raw generic data store layer is envisaged. A two-tier architecture, in spite being more complex to implement, can bring other advantages, for example, the upper semantic layer can be engineered as a pluggable component; in this way a semantic data store can cope with different application domains simply by plugging in the appropriate semantic layer. Another advantage is that content can be filtered while loaded into the semantic layer. The high level architecture of the semantic layer adopted by the FascinatE project is shown in Figure 5.5.

The data store implemented in the FascinatE project supports different types of semantic objects and is based on a DBMS. A detailed analysis of the semantic components implemented for the FascinatE project is beyond the scope of this Chapter. The invocation of the DataAnalyzer is triggered by crucial events at XParent level (for example, the processing of the closing tag for a list of MPEG-7 descriptors for a given domain). In this way only descriptors useful for semantic processing are processed and stored into the semantic layer. The QueryHandler component translates and performs the query requests sent by other components.

5.3 Domain-independent Saliency

As explained in Section 2.9, panoramic images can provide a very useful way of getting an overview of a complete format-agnostic scene. We will therefore focus on the task of saliency detection in a panoramic image, as captured by a 180-degree static camera of the type described in Section 3.2. In particular, we will focus on spatio-temporal saliency as a measure for the relevance of content independent of the domain. Saliency does not provide semantic meaning on its own, but rather indicates where a human observer of the scene might naturally look. This section considers two applications using saliency information that are relevant to format-agnostic production. First, saliency of different areas in the scene can be used in order to adaptively assign bandwidth for encoding of high-resolution video, as discussed by Alface et al. (2013). Second, salient regions can be provided as candidate virtual camera views to the Virtual Director.

5.3.1 Spatio-temporal Visual Saliency

One of the very first works focusing on saliency-based scene analysis is by Itti et al. (1998). The model used in this work is based on the sensitivity of visual neurons of the human brain to local spatial discontinuities. For example, a bright object stands out in a dark environment and vice versa. This so-called centre-surround model is built from difference images of a Gaussian image pyramid. To improve the saliency detection results, feature combination strategies have been proposed by Itti and Koch (2001), which make use of supervised learning to determine the weights of features. In Itti and Baldi (2005) this method is extended by a probabilistic aspect, resulting in a measure called Bayesian surprise. The Bayesian surprise measures the divergence of probability distributions between a learned model and the current observation. Hou and Zhang (2007) propose a saliency detection method in the frequency domain based on the log spectrum. This work also proposes a simple threshold method to detect objects in the images based on the saliency map. Guo et al. (2008) use the Quaternion Fourier Transform to operate in both time and frequency domains. Tong et al. (2011) estimate motion vector fields in order to obtain temporal information. The information about the position of human faces is used as an important feature in this work due to its special focus on surveillance videos.

Most of these methods focus only on still images. When dealing with video, the temporal context cannot be neglected, as there is the need for saliency estimates that are consistent over time, and that do not cause abrupt changes of salient regions. Also the heuristic of focusing on human faces is quite domain-dependent, and in many settings it might draw attention to the audience rather than the actions on the field or stage. Some methods also assume that the most salient objects reside in the centre of the image. This does not hold for the panoramic images used in the format-agnostic production approach proposed in this book. For example, in a soccer scenario, the most important things, the goal shots tend to happen at the image boundaries. Finally, some of the proposed features are costly to compute, and have to be excluded or replaced in order to enable processing of a high-resolution panorama in real-time.

A traditional centre-surround model as proposed by Itti et al. (1998) built from the difference images of a Gaussian image pyramid of the input video, can be used as a basis. The intensity and four broadly-tuned colour channels (red, green, blue and yellow) are calculated. To obtain the pyramid, images are iteratively subsampled by a factor of two (averaging pixel intensities). Following the inter-scale subtraction to build the centre-surround model, each downsampled image is expanded back to the original size using nearest neighbour resampling so that the subtraction can be done on a per pixel basis in the original resolution. The shrinking and expanding has the effect of averaging the intensities per block. To compare the features in a centre-region with its surroundings, the difference between a fine-scale image and a coarser-scale image is calculated. As the human visual cortex is sensitive to complementary colours (e.g., the combination of red and green) the colour chromatic complements red-green and blue-yellow (defined as differences of the corresponding channels) are chosen as visual features instead of taking the colour channels separately. The centre-surround feature of pixel intensity is defined as the absolute difference of the intensities of centre and surround pixels.

5.3.2   Estimating Grid-based Saliency

One approach for estimating grid-based saliency is presented by Alface et al. (2013). In a video sequence, the spatio-temporal saliency histograms are used to distinguish between static and moving salient regions. The video is divided into cells of size $c_x \times c_y$. For each cell a reference histogram $\mathbf{H}$ of saliency values is built from the recent n frames. To compare the current spatio-temporal saliency histogram $\mathbf{H}'$ with the reference histogram $\mathbf{H}$, the Earth Mover's Distance (EMD) is used (Rubner et al., 2000). EMD aims to determine the optimal alignment between bins. As the number of bins and the normalisation are the same for both histograms, the distance calculation can be done efficiently using the cumulative sum of the histogram differences:

$$
d_{\mathrm{EMD}}(\mathbf{H}, \mathbf{H}') = \frac{1}{nc_x c_y} \sum_{i=1}^{|\mathbf{H}|} \left| \sum_{j=1}^{i} (\mathbf{H}_j - \mathbf{H}'_j) \right|,
\tag{5.1}
$$

with $|\mathbf{H}|$ being the number of bins, and $\mathbf{H}_j$ denoting the jth bin.

In order to suppress the effect of reflections on smooth surfaces (e.g., stage floors), post-processing is applied to filter these areas. Those areas are largely unstructured, but get texture only from shadows and reflections. The content in each grid cell is assessed by the number of edges. For this purpose, the Canny edge detector is used to determine the edges and calculate the percentage of pixels belonging to an edge (Canny, 1986). Finally, this percentage is used to reweight the cell saliency. It can be observed that the salient regions are preserved while most of the undesired regions are filtered out.

An example of a resulting grid-based saliency map is shown in Figure 5.6. It can be seen that only the actors in the foreground remain salient, while people in the background and objects like the marks on the floor are suppressed.

Figure 5.6   Frame from the panorama discussed in Section 3.2.5 (top), saliency map calculated from difference of the pyramid levels (middle) and spatio-temporal grid-based saliency (bottom), darker values indicate higher saliency. Reproduced by permission of the Berlin Philharmonic Orchestra.

5.3.3   Salient Regions for Controlling Automated Shot Selection

A saliency map consists of individual saliency values for each image grid cell. From the most salient cells (using for example, a percentile-based threshold), salient blobs can be extracted. The rectangular bounding boxes of these blobs can then be used as alternative virtual camera views in an automated shot selection process. The score for the bounding box is defined by the averaged saliency values within the rectangle. Higher scores indicate higher saliency. A more sophisticated method is to provide further bounding boxes covering multiple blobs. There are two issues to consider. First, the number of all possible blob combinations grows exponentially. Therefore, exhaustively testing blob combinations, in particular those including other blobs, should be avoided. In order to speed up score calculation, the saliency map information might be represented in an *integral image*. An integral image, also known as the summed area table is a data structure and algorithm for quickly and efficiently generating the sum of values in a rectangular region of an image. This enables the calculation of the mean saliency value of an arbitrary rectangular in constant time. A bounding box covering multiple salient blobs typically encloses some regions with lower saliency. This has to be considered (e.g., by reweighting

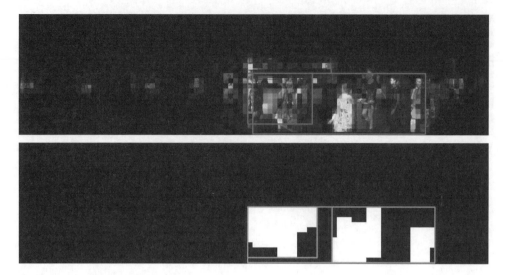

Figure 5.7   Salient regions of a frame: the bounding boxes are used to update the state of the Kalman filter (top), stabilised salient regions smoothed by Kalman filter, slightly different due to the smoothness constraint over time (bottom).

based on the area) in order to avoid decreasing the average saliency value of the larger blob because of the lower saliency of an enclosed blob.

In order to avoid quick jumps of virtual camera positions, the movement of salient rectangles needs to be smoothed. The Kalman filter uses the series of recent measures over time to predict the new state (Kalman, 1960; Bishop, 2006). It tolerates noise and is able to predict the underlying model. The positions of bounding boxes can be modelled using the centre, the size and the movement of the bounding box. At each iteration, that is, between two subsequent frames, the Kalman filter predicts the new state and is corrected by the current measured position. The correction step induces the Kalman filter to update the model. The current measure comes from the best matched bounding boxes between two frames where the Euclidean distance between two measures is calculated. The effect of the Kalman filter is that sequences of matching bounding boxes are stabilised. Current unmatched bounding boxes are added to the buffer and kept for the next n frames after having continuously matched. Old unmatched bounding boxes from the last frame are kept in a buffer, and will be removed if they remain unmatched in the next n frames. This delayed reaction increases the robustness against short-time mismatches.

Figure 5.7 shows an example of the effect of filtering over time. The temporally filtered bounding boxes are slightly different than the ones detected for the frame, as they react to local changes with some inertia. This provides an input to the Virtual Director that is suitable as virtual camera view.

5.4   Person Detection and Tracking

People in the scene are of interest in every content domain and are natural areas of interest for a viewer. Thus, detection and tracking of people throughout the scene is an important feature.

This section discusses approaches that support person detection and tracking in a static camera environment, as would be appropriate in an image from a stationary panoramic camera. Each track represents an individual person and is therefore labelled with a unique ID. As a result of tracking, the position of each person in the image plane is reported for every frame of the image sequence. On the basis of these positions, and the composed person trajectories, behaviours of a single person and/or groups of people can be derived. To infer what is going on in the scene, regions where groups of people are visible or frequently used paths can be provided. Furthermore, on the basis of single person trajectories it is possible to follow a specific person for a certain time segment. Information about a single person or groups of people can be summarised using statistics. For instance, the distance walked and the average velocity of a person during a sports event can be calculated.

In order to support production systems, the methods for detection and tracking of people should be applicable for a wide range of various content domains such as sports events or dance performances. To adjust for the different needs of each content domain the methods have to deal with many challenges, for example, lighting changes, occlusion by other people or objects, motion blur, fixed or a varying number of people to track. Thus, it cannot be expected to detect and track all people at all times and therefore concatenation of stable tracks by reducing discontinuities is of high importance.

5.4.1   Person Detection

Most of the recent state-of-the-art tracking approaches are based on tracking-by-detection algorithms. First, moving objects are detected by specific object detectors using discrimination between the foreground and the background based on different features like colour, edges and texture. Each of the detected object regions is commonly represented by an ellipse or a rectangular bounding box. With statically mounted cameras a particularly suitable method for person detection is to establish models representing the background. These background models, built for example, by averaging of pixel intensities or on the basis of colour histograms, are used to handle varying backgrounds. After subtraction of the background the processed foreground pixels of each frame are clustered and result in sufficiently large regions (blobs) that are then used for further tracking.

Due to its excellent performance, methods for person detection based on Histograms of Oriented Gradients (HOG) proposed by Dalal and Triggs (2005), are widely used in state-of-the-art tracking algorithms. In a nutshell, for each frame of an image sequence the descriptors are extracted using a sliding window. The descriptor, based on histograms representing the orientation of gradients in the window, is calculated for multiple scales at different positions in the image. A scale ratio defines the ratio between the scales of the sliding window in order to detect the appearance of a person for each possible person size (height in pixels) in the image at each location. Each descriptor is then verified by a classifier trained on the basis of a set of samples representing people with different appearances in terms of environment, perspective, light, occlusion and clothes. As a result, each image region (described by a bounding box) that provides a sufficiently large matching score is classified as a person.

A shortcoming of person detection based on HOG, is missed detections due to motion blur. The detector is not able to detect person regions with low gradients caused by rapid motion (a typical situation in sports scenarios). Therefore, simple foreground estimation is used in

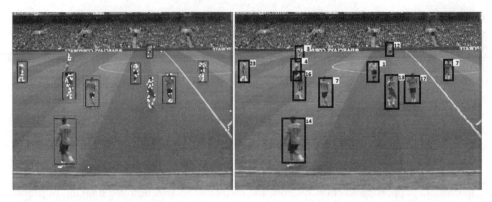

Figure 5.8 The detected person regions provided by the HOG and blob detector including the clustered feature points from tracking (left), the bounding boxes and the assigned person IDs identifying the derived person tracks (right).

Kaiser et al. (2011) in addition to person detection based on HOG descriptors to support the person detection process. Thus, the missed detection rate is significantly reduced by the fusion of the resulting blobs with the extracted person detections using the HOG detector. These merged person regions are then processed in the tracking step of the proposed tracking by detection algorithm (see Figure 5.8). Furthermore, a scene scale calculation is used to provide an estimate of the person's height at a given image location in relation to the camera position. This allows the HOG detector to be applied to a limited set of (the most) probable sliding window scales only. Hence, false detections of both detectors are discarded by a verification of the detected bounding box's height.

In case of missing camera parameters a scene scale calculation can be performed on the basis of manually measuring person sizes in the image. Nevertheless, manual measurement of person sizes is a tedious process especially in a multi-camera environment. To overcome this, Mörzinger and Thaler (2010) proposed a method to automatically calculate the scene scale without the need for any prior knowledge of the camera parameters based on bounding boxes provided by an object detector. For each location in the image the scene scale provides an estimate of the person scale (height) at every image location with respect to the ground plane. The ground plane position is indicated by the bounding box's bottom centre point. This is done by solving the linear correlation of the person's foot positions in form of x and y coordinates to the height of the bounding box. A RANSAC (Random Sample Consensus) algorithm is used to find the best subset out of all available detections providing the optimal linear correspondence on the basis of a linear regression by exclusion of false detections (Fischler and Bolles, 1981).

5.4.2   Person Tracking

In order to match the corresponding detected person regions between two consecutive frames either an appearance and/or motion model is established for the detected object. Each model is established for each single person and is generally updated for every frame. On the basis of these models tracking algorithms are able to discriminate between the detected people assigning them to the appropriate person track.

Motion models are based on the person's position, velocity and direction of movement. For person tracking, the defined motion model and the previous state of the person are used in many state-of-the-art approaches to predict the current position of the person. In Kaiser et al. (2011) the predicted position is compared to the position of the actual detected person to support establishing correspondence between the person detections in the image sequence. To model the motion of an individual track based on its position and velocity Kalman filters are applied in a wide range of tracking algorithms (Kang et al., 2003). By using Kalman filters the prediction of the person's position is corrected by the measurement of the person's position in the current frame.

The appearance model is based on image features derived from the regions representing the detected people. To reach a highly discriminative representation of each person, colour, edge and texture features are used. A survey of different features suitable for appearance models is given in (Yilmaz, Javed and Sha 2006). Establishing a colour histogram of each detected region is a widely used method to represent the appearance of people. For instance in Thaler and Bailer (2013) person tracking is based on a colour histogram comparison of the detected people between consecutive frames. A more sophisticated appearance model invariant to different rotations, translations and scales is proposed by Kang et al. (2003). The model is based on combinations of different colour distributions estimated for each person using a surrounding reference circle. Therefore, colour distributions for multiple arrangements of circle portions based on a Gaussian colour model are estimated. The circle portions are defined by concentric circles with different radii and orientations.

A very popular method for calculating appearance features to support tracking of objects is the KLT feature point tracker (Lucas and Kanade, 1981; Tomasi and Kanade, 1991; Shi and Tomasi, 1994). For feature point selection the KLT algorithm is used to find multiple feature points with sufficient neighbouring texture selected by a cornerness measure in each frame of an image sequence. In the subsequent tracking step the geometric relationship of these salient points to the next frame is established using a tracking window that assigns the best match. The tracking window with a typical size of 7×7 pixels is located around the feature points.

In order to increase the discriminability of each track and to overcome occlusions a combined motion and appearance model is often used. The algorithm proposed in Thaler and Bailer (2013) tracks people on the basis of person regions derived by merging the detection results of a HOG and blob detector. To determine the corresponding detected person regions between consecutive frames a KLT-based feature point tracker is used processing about 1,500 feature points per HD frame. Clustering feature point regions with sufficient height and number of points with respect to a given scene scale completes the person detection process and increases the person detection rate. As the initial step for tracking, person IDs are assigned to each feature point cluster contained by a detected person region. The connection of appropriate person regions between consecutive frames is based on these feature point clusters assumed to correspond to the same person in the image sequence at all times. For each frame the person ID of each feature point cluster is assigned to the surrounding bounding box of the person region representing the track (see Figure 5.8). In the case of an already tracked feature point cluster not included by a detected person region, the track's position is updated by the current corresponding cluster location. If the number of cluster points is too low, the track's new position is estimated from a motion model based on the person movement and a subsequent colour histogram comparison of both the old and the new person region.

In case of occlusion, feature point clusters might be merged and cannot be easily differentiated further, for example, a person region contains feature points with more than one assigned person ID or at least two regions include feature points corresponding to the same cluster. To handle overlapping tracks Thaler and Bailer (2013) suggest a combined motion and appearance model. The motion model is based on the direction of movement and the velocity of the feature points and the corresponding person region as well as the motion history of the track. The appearance model is based on a colour feature verification and the geometric characteristics of the person region. The number of involved feature points is also considered by the model.

Furthermore, a person tracking score calculated for each frame is assigned to each track. The features used for calculation are also used to build the motion and appearance models. To enhance tracking, the person tracking score is used for a correct assignment of person IDs to person regions in ambiguous situations, for example, if a feature point cluster extracted by the KLT point tracker representing one person is included in more than one detection. The tracking score is also used to ensure long-term stability of the tracks. Furthermore, on the basis of the confidence value, the proposed algorithm is able to respond to the needs of different scenarios, such as supplying all found person detections and tracks versus only the very stable ones.

5.4.3 Multicamera and Panoramic Environment

Tracking in a multicamera or panoramic environment is very popular due to consequent benefits like establishing a common 3D ground plane. Using multiple cameras or a high resolution panoramic camera enables the coverage of wide areas at an appropriate resolution from static viewpoints. Furthermore, specific areas in the scene can be captured from different viewpoints by at least two cameras in order to solve ambiguities caused by occlusions.

Multicamera

In Kang et al. (2003) tracking of soccer players captured by a multicamera environment is based on joint probability models: an appearance model and two Kalman filter based motion models are combined. Blobs representing soccer players are detected using background modelling, using a colour appearance model and two motion models. The appearance model is based on a combination of different colour distributions and is therefore invariant to rotation, translation and scale changes. The 2D motion model is based on the players' motion in each image plane defined by the upper left and lower right bounding box position between consecutive frames. Using homographies (see Figure 5.9) projecting the ground-planes of individual image views defined by the players' feet positions in a common 3D ground-plane, a 3D motion model is established. Thus, the detected players' positions can be analysed by a top down view provided by the 3D ground plane representing the soccer field. The prediction of the players' motion for both models is based on a Kalman filter. Together with the appearance model a robust method for person tracking across different camera views is proposed. Nevertheless, with a runtime of 1 second per frame at a resolution of 720×576 for each view, person tracking within real-time constraints is not feasible.

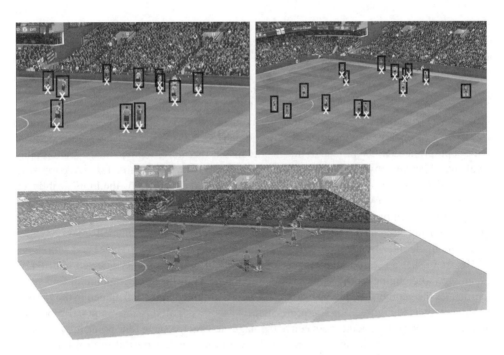

Figure 5.9 The upper right image is projected into the upper left image with respect to a common ground plane using an inter-image homography, calculated on the basis of the corresponding foot positions (indicated by the white crosses) of the detected person regions (indicated by the black bounding boxes).

In the case of tracking across multiple (partly) overlapping views the handover of the assigned unique person IDs between person tracks provided by different camera views representing the same person has to be ensured. As mentioned above camera parameters (in the form of planar homographies) are commonly used to establish a geometric correspondence on the basis of a common ground plane between the different image views. The common ground plane is defined by the detected person's foot positions represented by the bottom centre of the corresponding bounding boxes. Each of these points is assumed to be located on the ground plane (indicated by a white cross in Figure 5.9) and is denoted as foot point in the following. In case of missing camera parameters due to the lack of camera calibration the methods described in the following can be used to automatically establish correspondence between adjacent camera views. Each method is based on person detections located on a common ground plane and the resulting correspondence enables an easy handover of person IDs.

For instance in Khan et al. (2001) field of view lines are automatically established in each camera view. The edges of field of view (EOFOV) indicate the boundaries of the overlapping area covered by adjacent camera views with respect to the ground plane. Furthermore, the correspondence between each EOFOV and the corresponding image border is estimated on the basis of person motion near the image borders. Every time a person appears at the border of one image view, any person detected in the corresponding area of other views is assumed to be the same person. Their foot positions are collected as possible points located on the EOFOVs connecting to the particular image border. At least two points are needed to determine each EOFOV. As a result, every time a person appears at an image border, the same person ID is linked to the detected person closest to the corresponding EOFOV according of the established relationship model.

Calderara et al. (2005) modified the approach in order to solve the ambiguity arising if more than one person is crossing the same EOFOV at the same time. In this approach, straight lines are estimated from foot points of people at the moment when they are fully visible appearing at a border. These lines are built from at least two points located near the image border and linked to the appropriate EOFOV instead of image borders. Furthermore, an inter-image homography is calculated between two camera views based on at least four point pairs. For that purpose at least four corner points of the overlapping area in one view, described by the EOFOV induced by an adjacent camera view and vice versa, are used. The shortcoming of this method is the constraint that each of the EOFOV has to be crossed at least twice by a person, which is not the case in many situations, for example, because of inaccessible or occluded areas.

For automatic inter-image homography calculation the algorithm in Thaler and Mörzinger (2010) establishes a set of point pairs each consisting of two foot points detected in different image views belonging to the same person. On the basis of these corresponding foot point coordinates the geometric relationship of each view's ground plane is calculated in the form of a homography (see Figure 5.9). Therefore, multiple subsets of all possible point pairs, built by combining detected foot points between both image views at the same time point, are randomly chosen. The subset providing the homography with the smallest overall re-projection error (applied to the selected point pairs) is used for further calculation. This selection is performed for multiple samples of overlapping images and the resulting point pairs are collected over time. The collected point pairs are then filtered by a RANSAC algorithm in order to discard outliers and with the resulting point pairs the final homography is calculated. On the basis of this inter-image homography the foot point for each detected person in the other view can be calculated.

As discussed in Section 5.4.1, on person detection, for each image view a scene scale can be automatically calculated in order to determine the person's approximate height in the overlapping image area. This input can be used together with the estimated inter-image homographies to enhance the tracking performance in the overlapping areas, for example, to handle missed detections by processing of appearance models. Additionally, by utilising different perspectives of the same scene occlusion can be solved.

Panoramic Camera

Installation of a multicamera environment is often not feasible due to the lack of appropriate mounting points or limited accessibility to areas suitable for cameras or cabling. Using panoramic cameras enables wide areas to be covered from a single viewpoint. Another advantage of processing a panoramic view instead of using multiple overlapping views is the known geometry within the panorama that makes the handover of person IDs quite simple. Furthermore, different brightness in images of the same scene due to different camera types is not an issue and a tedious camera calibration process is not needed.

An approach for person detection and tracking in a 180° omni-directional panoramic video sequence at a resolution of approximately 7000×2000 is described in Kaiser et al. (2011).

Figure 5.10 Bounding boxes including person IDs of all detected people. Reproduced by permission of the Berlin Philharmonic Orchestra.

The ultra-high definition panorama is acquired by an omni-directional camera capturing live content in a wide range of different domains. This algorithm processes individual parts of the panoramic image distributed on several workstations in order to reach real-time throughput. The ultra HD panorama is therefore divided into separate adjacent image tiles with a maximum resolution of 1080×1920 (FullHD). Each tile is analysed on a separate CPU in parallel using a tracking by detection algorithm. The algorithm combines two basic object detectors and a feature point tracker. For detecting people blobs provided by a simple background subtraction approach and detections provided by a person detector using a Histogram of Oriented Gradient (HOG) based approach are fused. Tracking of people is based on feature points inside the person regions found provided by a KLT-based feature point tracker. The handover of the assigned person IDs between individual image tiles in order to track the people across the whole panoramic scene (see Figure 5.10) is performed on a separate workstation. Therefore, all corresponding person regions located near the image borders within a small time window with approximately the same height and vertical position are merged by updating their IDs and positions. The correspondence is based on the known relative orientation of the tiles to each other and corresponding vertical coordinates in neighbouring regions.

5.4.4 GPU Accelerated Real-time Tracking Beyond HD

Considering the increasing amount of video data at a resolution of at least FullHD, it is a big challenge to carry out online person tracking supporting real-time broadcast systems with minimal latency. Most state-of-the-art methods for person detection and tracking do not meet real-time requirements when analysing high resolution image sequences. For instance, the implementation of the Histograms of Oriented Gradient person detector by Dalal and Triggs (2005) needs almost 1 second to process one frame at a resolution of 320×240. The runtime for tracking 10,000 feature points using the multithreaded CPU implementation of the KLT feature point tracker provided by the OpenCV library for one FullHD frame takes about 130 ms. To achieve a significant speed up, most recently a lot of algorithms have been ported on Graphics Processing Units (GPUs) in order to analyse large amounts of data in parallel. Below approaches are described that have been implemented for NVIDIA's CUDA-enabled GPUs in order to utilise their parallel architecture (NVIDIA, 2013). CUDA is a GPU programming language introduced by NVIDIA and is similar to C. CUDA enables parallelisable functions (so-called kernels) to be processed by several thousands of threads concurrently on the basis of a couple of multiprocessors each containing hundreds of processing cores.

To speed up person detection GPU implementations of Histograms of Oriented Gradient based detectors were presented by Wojek et al. (2008) and Prisacariu and Reid (2009). For the first approach a runtime of 385 ms for one frame is reported at a resolution of 1280×960. The latter implementation, known as *fastHOG* has a runtime of 261 ms at the same resolution for greyscale images and provides as good results as the CPU implementation by Dalal and Triggs (2005). The optimised GPU implementation of a KLT-based feature point tracker in Fassold et al. (2009) detects a large number of feature points simultaneously. As a result, a speed-up factor of four to seven can been achieved compared to a multithreaded CPU implementation provided by the OpenCV library. To benefit from NVIDIA's recent more CPU-like architecture named Fermi (described in NVIDIA, 2009) the GPU implementation has been adapted in Kaiser et al. (2011). Due to the revamped memory hierarchy, the performance of the tracking step previously limited by the shared memory can been improved. Relying on the local memory the algorithm is able to handle bigger tracking window sizes efficiently. As a result, the algorithm is able to track 5–10 k feature points in 20 ms in one FullHD video stream on an NVIDIA Geforce GTX 480.

To enhance the real-time capability of person detection and tracking the algorithm proposed in Thaler and Bailer (2013) uses two GPU accelerated methods: the fastHOG person detector and a KLT-based feature point tracker based on Fermi. Additionally, the original implementation of the fastHOG was modified in order to reduce the processing time of person detection to 70 ms for one HD frame. In order to reach this speed-up the scale step for the sliding window was changed from 1.05 to 1.30. In consideration of the speed-up through this coarser sampling the decrease of the person detection rate by 4 per cent is negligible. Furthermore, using a given scene scale calculation, fastHOG is applied for possible sliding window scales only. Additionally, a simple blob detector is used to support the person detection process if fastHOG is not able to detect the person regions under the presence of motion blur. The missed detection rate is significantly reduced by using the resulting blobs, after verification on the basis of the scene scale calculation.

To establish the correspondence of the resulting person regions between the consecutive frames the Fermi optimised GPU implementation of the KLT feature point tracker described by Kaiser et al. (2011) is used. Unique person IDs are assigned to clustered featured points and their covering person region representing the people in the scene. The runtime of the proposed algorithm for analysing one HD frame is about 85ms; 70 ms are needed for the fastHOG and 35 ms for blob detection. Due to the optimised parametrisation of the fastHOG the proposed approach's runtime is about 200 times faster than using the original CPU implementation. The feature point tracker's runtime for 1,500 feature points is 9ms. Compared to the CPU-implementation a speed-up factor of over 14 is achieved. Using two CUDA-capable GPUs the fastHOG, the blob detector and the feature point tracker can run in parallel at a runtime of 70ms. The remaining 15 ms are consumed by fusing the partial results and scoring.

5.5 Online Detection of Concepts and Actions

In a format-agnostic production context, the Virtual Director needs information about the presence of objects in the scene, of visual concepts in certain regions of the video (e.g., crowd) as well as specific actions involving people and objects in the scene (e.g., people dancing). This information can be used to assess the relevance of certain candidate regions. Detection of static concepts has been successfully applied for several years in order to automatically annotate global concepts, such as indoor/outdoor or day/night shots, scenery (e.g., landscape, cityscape) or salient visual objects in the scene (e.g., mountains). The research on such methods has been strongly driven by search and retrieval applications. Concept detection can be used to generate metadata from visual content, that is then indexed for supporting text-based search. Benchmarking initiatives such as TRECVID (Text Retrieval Conference Video Track), which include a concept detection task called Semantic Indexing (SIN) task (previously known as High Level Feature Extraction, HLFE) have fostered research in this area and helped to build standard evaluation methods and data sets (Smeaton et al., 2006).

Most of the approaches for concept and action detection make use of features extracted from key frames, that is, still image features. Only recently, research on concept detection in video increasingly considers dynamic concepts and actions, making better use of the temporal dimension of video. This is evident both from the literature as well as from the increasing inclusion of dynamic concepts in the TRECVID SIN task as well as the Multimedia Event Detection (MED) task introduced in 2010. Approaches for detecting dynamic concepts and actions jointly use features of video segments or include dynamic features (e.g., motion vectors). This can be done offline (collecting all features of a video segment and then performing detection) or online (performing detection in the live video stream). In a production environment, methods for online detection of concepts and actions are of course of particular interest.

The problem of online detection can thus be seen as the problem of classifying whether subsegments contain a concept or action in a video segment $\mathbf{S}$. The segment is represented by n samples s_i (e.g., frames, key frames sampled at regular or non-regular time intervals). Each of these samples is described by a feature vector $\mathbf{x}_i$, consisting of arbitrary features of this sample (or a temporal segment around this sample). In order to represent the video segment, the individual feature vectors are concatenated to form a feature vector $\mathbf{X} = (\vec{\mathbf{x}}_1, \dots, \vec{\mathbf{x}}_n)$ of the segment. This requires methods that are able to model concepts not only by samples from single key frames, but by the sequence of samples over the duration of a video segment. Clearly, not every segment has the same length and/or consists of the same number of samples, thus the lengths of the feature vectors of different segments will differ. It is thus necessary to be able to determine the similarity between such feature vectors with different lengths. In the online detection case, a stream of input samples s_i is encountered, and samples need to be processed in a time window, that is, a sequence of feature vectors $\mathbf{X} = (\vec{\mathbf{x}}_1, \dots, \vec{\mathbf{x}}_\tau)$. In the next step, the feature vectors from the next sliding window position $\mathbf{X}' = (\vec{\mathbf{x}}_{1+\delta}, \dots, \vec{\mathbf{x}}_{\tau+\delta})$ need to be processed.

Concept and action detection in video can be approached as a supervised machine learning problem by training a classifier on a set of annotated examples. Kernel methods, most notably Support Vector Machines (SVMs), have been widely applied to concept detection problems, also due to the availability of toolkits such as LibSVM (Chang and Lin, 2001). In the following, methods proposed for the detection concepts and actions in video are reviewed. Different types of kernels working on sequences of feature vectors have been proposed. These are discussed in a separate subsection, followed by a discussion about how these approaches can be adapted for the online case.

An early work that proposed the use of SVMs for the recognition of dynamic visual events has been described by Pittore et al. (1999). This approach simply normalises the lengths and value ranges of features to deal with segments of different length. Cao et al. (2004) propose a method for online event recognition based on SVMs. Recursive filtering is performed to obtain an image representing the motion characteristics in a short time window. This image is rescaled and represented as a feature vector. For each clip, the images resulting from recursive filtering of each frame are treated as feature vectors for this frame. Majority voting over the classification results of the frames of the test clip is performed. In the online case this is done for the frames in a sliding window covering the most recent set of frames, typically 5–20 (key) frames.

Other approaches use different types of kernels or use alternative types of classifiers. Fleischman et al. (2006) model events performed by humans as a tree, where the non-leaf nodes capture the temporal relations of the actions in the leaves or sub-trees. A tree kernel is used to classify the events. Relative margin support tensor machines have been recently proposed by Kotsia and Patras (2010) for spatio-temporal classification problems such as gait and action recognition. They are defined as extensions of SVMs for spatio-temporal data, however, they are restricted to data of the same temporal length and do not support flexible alignment. Zhang et al. (2012) propose a framework for event detection in video by defining se   ents of positive and negative examples and extract bags of features for these segments (thus not modelling a sequence). For classification, they use multi-instance SVMs trained on multiple samples of the event. Recently, an event classification framework based on rough set theory has been proposed by Shirahama et al. (2011), which addresses the issue of learning from a small number of samples.

Different approaches to sequence-based kernels are reviewed in Section 5.5.1. Section 5.5.2 discusses how sequence-based kernels can be adapted to the online detection case, and Section 5.5.3 reports the performance of the online kernels.

5.5.1   Sequence-based Kernels

Sequence-based kernels, that is, kernel functions that are able to determine the similarity of sequences of feature vectors, are one of the methods proposed for including temporal information in classification methods. They are commonly used in computational biology, where they are often restricted to discrete features. Recently, appropriate sequence-based kernels supporting a wide range of features have been proposed for applications in multimedia, such as concept or event detection. Kernel-based machine learning methods, most notably Support Vector Machines (SVMs), have been successfully applied to concept detection in video. Experiments have shown that concept classifiers using sequence-based kernels outperform those using kernels matching the individual feature vectors of the samples of a segment independently (Bailer, 2011a).

The general approach of sequence-based kernels is to define a kernel function on a sequence of feature vectors from two video segments (which may be regularly or irregularly sampled).

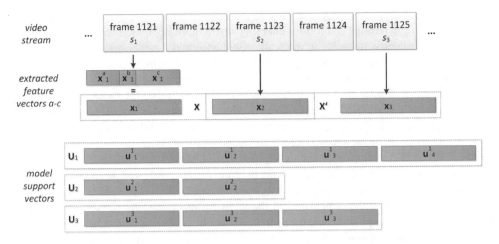

Figure 5.11   Schematic diagram of feature and model vectors of sequence-based kernels.

The term sequence denotes a possibly noncontiguous subsequence. Elements in the sequence represent the feature vectors of individual frames, and a base distance/similarity function (which can be a kernel itself) is applied to them. Then the kernel value for the two sequences is determined from the base distance similarity value, for example, by choosing some optimal alignment or a weighted combination of different alignments. The latter step includes many properties that discriminate between the different types of sequence-based kernels, such as thresholds for the base distance/similarity, or constraints on gaps in the alignment.

In the following, the most relevant of the sequence-based kernels proposed in the literature are presented. This discussion will enable a fuller understanding of Section 5.5.2 about the actions needed to make the kernels applicable online. In this section, specific notation is used. The set of support vectors of the model of a concept is denoted as $\mathcal{U} = \{U_1, \ldots, U_p\}$. Each U_k consists of feature vectors $U_k = (\vec{u}_1, \ldots, \vec{u}_n)$ of the samples in the segment, containing a sequence describing an instance of the concept. As it is intended to support arbitrary ground distances between the feature vectors of the input samples, a kernel for matching the feature vectors of elements of the feature sequences is used, denoted as $\kappa_f(\vec{x}_i, \vec{u}_j)$.

Figure 5.11 shows a schematic diagram of feature and model vectors of sequence-based kernels. From the input streams, three feature vectors x_i^a, x_i^b, x_i^c are extracted for the different types of features (e.g., global colour features, histogram of local texture features, motion histogram) from each of the key frames s_i (in this case, every second frame is used as key frame). These feature vectors are concatenated to vectors x_i for each of the key frames. Every sliding window position is represented by a sequence of feature vectors (in the example, X and X'). In a similar way, the support vectors of the model, that is, feature vectors of sequences of representative samples of the action, have been generated. In the example, the model contains three support vectors of sequences of different lengths. In order to match an input sliding window against the model, the sequence-based kernel is evaluated on each pair (X, U_j).

Some approaches design classifiers with an explicit temporal model of the action, such as a Hidden Markov Model (HMM). In Wu et al. (2003) a sequence alignment kernel for matching trajectories is proposed. Segments of trajectories are labelled to obtain both a symbol and a feature sequence. A pair-Hidden Markov Model (pair-HMM) is used to align the symbol sequences and to determine the joint probability distribution of the sequences. Gaps in the alignment are padded with zeros in the feature sequence and a radial basis function (RBF) kernel is applied to the feature sequence. The sequence alignment kernel is defined as the tensor product of the kernels evaluated on the symbol and feature sequences. Qi et al. (2008) propose a temporal kernel for concept classification. They train a HMM for each concept and define a distance between HMMs based on the Kullback-Leibler divergence, which is plugged into a Gaussian kernel. The approach outperforms classifiers using the same features without temporal information on 30 out of 39 of the concepts in LSCOM-lite, a subset of the LSCOM ontology of visual concepts described in Naphade et al. (2005).

Several approaches based on the idea of the pyramid match kernel have been proposed. The original pyramid match kernel proposed by Grauman and Darrel (2005, 2007) partitions the feature space in each of the dimensions of the input feature vector. Its efficiency advantage is based on avoiding explicit distance calculation, but only counting elements that end up in the same bin of the pyramid. This assumes that L_1 distances can be applied to the feature vectors. The vocabulary guided pyramid matching approach proposed in Grauman and Darrell (2006) addresses this problem, as it uses a clustering step to construct the pyramid, allowing the use of an arbitrary distance measure. The approach has been extended to spatio-temporal matching in Choi et al. (2008), using sets of clustered SIFT (Scale-invariant feature transform) and optical flow features as local descriptors. Their approach is similar to spatial pyramid matching proposed in Lazebnik et al. (2006), which applies pyramid matching only to the image space, (i.e., subdividing an image into a spatial pyramid, and counting features of the same type in each of the bins) but uses clustering in the feature space (i.e., the common bag of words approach).

In order not to constrain the choice of distances in the feature space, pyramid matching can be applied to the temporal domain only, in a similar way to that proposed for spatial (Lazebnik et al., 2006) or spatio-temporal (Choi et al., 2008) pyramid matching. As no clustering of the feature vectors is performed in advance, a threshold to determine whether two feature vectors match is defined.

Another temporal matching method based on the pyramid match kernel is described in Xu and Chang (2007, 2008). Temporally-constrained hierarchical agglomerative clustering is performed to build a structure of temporal segments. The similarity between segments is determined using the Earth Mover's Distance (EMD) and the pyramid match kernel is applied to the similarities on the different hierarchy levels. This approach explicitly assumes that the temporal order of the individual subclips is irrelevant (as is, for example, the case in news stories). The temporal order within the clips is aligned using linear programming.

Another group of approaches applies different types of distances from string matching, such as longest common subsequence (LCS) or the dynamic time warping (DTW) paradigm. Yeh and Cheng (2008) use the Levenshtein distance (Levenshtein, 1966) between sequences of clustered local descriptors for classification of still images. Recently, a kernel for event classification based on sequences of histograms of visual words has been proposed by Ballan et al. (2010). The authors consider different similarity measures between the histograms and use them instead of symbol equality in the distance (Needleman and Wunsch, 1970) and plug it into a Gaussian kernel. In Bailer (2011a) a kernel based on longest common subsequence (LCS) matching of sequences has been proposed. An arbitrary kernel can be plugged in to determine the similarity between two elements of the sequences, and the kernel value is determined as the normalised sum of the similarities along the backtracked longest common sequences. Similarity weighting can be achieved by performing backtracking of the longest sequence, summing the values of $\kappa_f(\cdot)$ of the matches and normalising. This requires backtracking of all sequences ending in the last element of either $\mathbf{X}$ or $\mathbf{U}$, i.e., $O(n^2)$ backtracking steps.

The dynamic time alignment kernel (DTAK) proposed by Shimodaira et al. (2001) is based on the dynamic time warping (DTW) approach for sequence alignment (Myers and Rabiner, 1981). DTW tries to align the samples of the sequences so that the temporal order is kept, but the distance (i.e., the sum of the distances of aligned elements) is globally minimised. Each sample of one sequence is aligned with one or more samples of the other sequence.

Cuturi et al. (2006) propose a kernel called Time Sequence Generalised Alignment (TSGA) that generalises the idea of dynamic time warping by using the soft-max of all alignments rather than choosing a specific alignment. It is assumed that the kernel between two sequence elements is a conditionally positive kernel of the form $\kappa_f = e^{-d(\cdot)}$, with $d(\cdot)$ being some distance function.

5.5.2   Kernels for Online Detection

All of the kernels described so far are designed to work on pairs of segments of input data, and thus cannot be directly applied to online concept detection. Note that the focus is on the detection (prediction) step here, but there is no online requirement for training. Lavee et al. (2009) surveyed a large amount of work on understanding video events. They note that SVM-based (support vector machine) approaches are straightforward to implement and have proved to be successful for other classification problems, but have difficulties in modelling the temporal dimension of events in video. This issue is being increasingly addressed, and approaches based on kernels performing, for example, sequence alignment or pyramid matching have been successfully applied. Most approaches targeting the online detection case rely on frame-based features or use statistical features of segments. Also, some of the proposed kernels limit the type of features to be used, for example, requiring them to be a discrete vocabulary. The application of sequence-based kernels to online detection is not straightforward for the following reasons. All kernels for sequences of feature vectors need some segmentation of the content (e.g., shots, subclips), which does not exist for a live stream. One cannot simply choose arbitrary segments in the input stream, as many of the kernels will not provide satisfactory results if the model and the input samples are not temporally well aligned. Many of the proposed sequence-based kernels determine some optimal alignment between the elements of the sequences to be matched. Thus, the value of the kernel function for a position of the sliding window cannot be easily derived from that of the previous position, but requires re-evaluation of the kernel, which is an extremely inefficient and impractical solution. Some sequence-based kernels require a specific base distance between the feature vectors of the individual samples. In order to be generally applicable, kernels that allow an arbitrary kernel function to be used to determine the similarity of feature vectors of two samples are of interest.

In order to efficiently evaluate the kernel function in the online detection case, the way the kernel is evaluated in the prediction step needs to be adapted. The training step is left unmodified for all the kernels. In the case of supervised learning the ground truth defines the segments to be used in training. For the sequence alignment kernels, such segments can even be obtained in a semi-supervised learning setting, as the boundaries of matching segments are a result of these algorithms.

Given the result for the samples of a time window $\mathbf{X} = (\vec{\mathbf{x}}_1, \ldots, \vec{\mathbf{x}}_\tau)$, the subsequent time window $\mathbf{X}' = (\vec{\mathbf{x}}_{1+\delta}, \ldots, \vec{\mathbf{x}}_{\tau+\delta})$ shall be evaluated. The naive approach of applying each of the kernels discussed above to a sliding window is to re-evaluate the kernel function independently for each temporal position. The size of the sliding window needs to be at most the number of elements of the longest support vector $|\mathbf{U}_{max}| = \max_{k=1}^{p} |\mathbf{U}_k|$. Kernels based on sequence alignment can handle sliding window sizes $< |\mathbf{U}_{max}|$, as they support partial matches. A lower boundary for the size of the sliding window is the shortest support vector $|\mathbf{U}_{min}| = \min_{k=1}^{p} |\mathbf{U}_k|$ of the model.

The different kernels impose different constraints on the sampling structure of training and test data. Different sampling rates are supported by all the kernels discussed in the previous section. Changes in the frame rate may require a change of the size of the sliding window, that is, when the sampling rate of the test data is higher than that of the training data, the sliding window duration for matching has to be adapted to ensure that the duration of the concept is adequately covered (e.g., a walk cycle, people moving towards each other and meeting). The kernels based on sequence alignment methods as well as EMD also support irregular sampling on one or both of the data sets, and are more robust when evaluating sequences that only partially represent the concept of interest.

The temporal offset δ between two positions of the sliding window can be in the range of $1 \leq \delta \leq |\mathbf{U}_{min}| - 1$ samples. The choice of this number has an impact on the latency of the detection as well as on the effort for re-evaluation in each step. If $\delta = 1$ (remember that one sample does not necessarily mean one frame, but typically represents a longer time span), only the last $n = |\mathbf{U}_k|$ samples need to be matched for the kth support vector, as all other possible matches are already covered by other window positions. This saves time compared to the segment-based variant for long segments, especially for rather short events, where the support vectors are shorter than the typical segment (e.g., shot).

To evaluate the kernel function for one position of the time window, $\kappa_f(\cdot)$ needs to be evaluated for each pair of elements $(\vec{\mathbf{x}}_i, \vec{\mathbf{u}}_j)$, which has a runtime complexity of $O(n^2 T_{\kappa_f})$ for each support vector, where T_{κ_f} is the time needed to evaluate $\kappa_f(\vec{\mathbf{x}}_i, \vec{\mathbf{u}}_j)$. Once this kernel matrix of κ_f is available, the respective sequence-based kernel can be applied.

It is possible to reduce the runtime by storing information from the kernel evaluation at the previous sliding window position. Note that data needs to be kept for each support vector of the current model. Obviously, when moving the sliding window by one element, only the distances between the new element $\vec{\mathbf{x}}_\tau$ and a support vector $\mathbf{U}^*$ need to be calculated. This reduces the runtime to $O(n T_{\kappa_f})$ per support vector, but requires storing the kernel matrix (memory complexity $O(n^2)$).

5.5.3   *Performance of Online Kernels*

The TRECVID 2007 HLFE data set is used for comparing segment-based and online versions of sequence kernels (Bailer, 2011b). Following the definition of the TRECVID HLFE task, the mean average precision (MAP) of up to 2,000 results (on shot granularity) per concept is used for evaluation. As a baseline, the results are compared with the RBF kernel on individual key frames, with the RBF kernel's γ parameter optimised by grid search. The results confirm those reported in the literature, (e.g., Bailer, 2011a), that is, all the kernels taking the sequence of feature vectors into account outperform the RBF kernel applied to individual feature vectors. The results for the different types of sequence kernels are quite similar. At least for the concepts of this data set, there is no significant difference in whether a kernel performs alignment of the sequences or not. The models used for the sliding window-based variants of the kernels have been trained using the segment-based variants. The results show that the sliding window variants perform slightly better than the segment-based ones, although the difference is not significant. The improvement in runtime performance when using the sliding window versions of the kernels has also been reported. The question of whether real-time performance is achieved depends on the number of classifiers that need to be applied (i.e., the number of concepts to be detected) and the complexity of each of these models. For all 20 concepts in the test set (e.g., walking/running, studio), prediction takes between 2.2–2.7× real time for the segment-based versions of the different kernels and between 1.5–1.9× real time for the sliding-window-based versions. The mean prediction time factor for the 20 models is between 0.11–0.13× real time in the segment-based case, and around 0.07× real time in the sliding-window case. For the most complex model (i.e., the one with the largest number of support vectors) it takes 0.55–0.65× real time in the segment-based case and around 0.46× real time in the sliding-window case. This means that with the current implementation real-time prediction on a single core is possible for 14 concepts (assuming models of average complexity).

5.6   Supporting Annotation for Automated Production

Advanced audio-visual content production is undergoing a paradigm shift beyond current programme production. In addition to manual production, examples of automatic or semi-automatic production are starting to appear.

In this section an approach for assisting automatic tools in the production process is described – an example of possible future user interface (UI) for semi-automatic production. The aim of the UI suggested here is to support professional production teams to work with ultra-high resolution panoramas by complementing the automatic tools and helping to define default view options. This includes adding annotations to the content highlighting interesting events and defining virtual cameras.

As these interactions have to be carried out in real-time in various settings, such an interface needs to adjust easily both in terms of the size of the production and the functionality needed, retaining the usability and efficiency of the user interfaces at the high level.

5.6.1   User Preferences and Functionality Definition

The user interface for semi-automatic production that is suggested here is meant to be used by a production team. The design is based on a set of user studies – production studies of professional galleries (Perry et al., 2009; Engström et al., 2010), workshops with BBC production teams and design workshops. BBC production teams were brought in on two workshop occasions, and their inputs have informed the design of an interface for semi-automatic production. Feedback with implications for the design and implementation of tools to assist the automatic production is summarised below.

- The use case of novel systems for advanced audio video content production, as a standalone production environment is feasible, primarily for small- to medium-scale productions. The operator would rely greatly on virtual cameras and tracking, and could manage a handful of moving ROIs.
- Mark-up of meaningful actions in the form of annotations, and the possibility for automated detection of for example, a goal kick, are seen as interesting tools.
- Selecting from a variety of camera angles is key in live TV. When competing with high-end television, novel systems must support both automatically generated ROIs and additional cameras, or risk removing essential storytelling elements that production teams rely on. Both types of images must be integrated and easily accessible. More details can be found in Sections 2.4.3 and 3.3.
- Touch screen controls are seen as interesting and even desirable for some production actions, such as overview and selection of ROIs. For time-critical actions such as cuts, physical and easily accessible hardware controls are preferred.
- In-system replay control would be an added value in low budget productions, but not necessarily for high-end productions where advanced alternatives are available.
- Any novel system for advanced audio-visual content production needs to integrate with an existing production gallery with respect to the current interactions and technologies used.

Bearing in mind the production team's preferences, and opportunities and challenges that the work with ultra-high resolution panoramas brings, a list of basic functionalities that the user interface for semi-automatic production should support has been assembled. The main focus is to support the core functionality of a production tool with panoramic and virtual cameras. A summarised list of functionalities for the semi-automatic production interface includes the following: displaying the panoramic and standard broadcast cameras' views; visualisation of view options automatically produced; monitoring and management of automatically produced view candidates; real-time annotation and indication of high-level events/actions (e.g., in the football domain goal, tackle); tracking support; preconfiguration; and the production of the main (i.e., linear) programme including the switching between different cameras. More details about these functionalities will be given when describing system components in Section 5.6.3.

5.6.2   Design Overview and Principles

Novel interfaces for semi-automatic production should be able to adjust to any size of production, and different environments. Consequently, the functionality required will depend on the type and the size of the broadcast event. A modular design approach, as suggested here, adjusts well to these needs. The basic principle is that the user interface consists of a set of workspaces, and a workspace consists of separate panels, each containing functionality relevant to a well defined (simple) task. Through the combining and positioning of these panels into workspaces, it is possible to adjust the interface to different needs. The benefits of this modular approach to the production interface design follow.

Adaptation A current state-of-the-art hardware video mixer contains over 1,000 buttons and other controls. But there are very few situations, if any, where the operator needs to have access to all of them at the same time. By dividing the functionality into separate panels it is possible to make the interface dynamic and adapt it to the users' current needs. Different panels can be displayed depending on production needs and production roles. The goal is to have only the functionality necessary for the task at hand.

Future functionality Another reason to have the functionality separated into panels is that future functionality can be added without having to redesign the entire interface, either by adding a new panel, or by modifying a current one, while panels unaffected by the updated functionality remain unchanged.

Integration with today's production galleries Integration with the existing technology is also made easier by having a modular design, for example, the functionality within a panel can be substituted by the functionality of a third party system, without affecting the rest of the system.

Support for external I/O units The concept of modularity enables support for external input and output units, for example, a vision mixer might prefer the tactile benefits of a physical hardware control unit such as a fader bar or the camera operator may prefer a pan-tilt interface (see Section 9.3.2, Figure 9.1).

Customisation The production needs will vary considerably between events of different sizes, types or importance. By having functionality divided into panels, the interface can be customised and functionality modified to suit the specific needs of the production at hand.

Scalability Production needs vary not only with different production types, but also with production size and scope. By having a modular system where functionality can be added, removed and adapted, the same system can be used for many production applications, and with different numbers of operators. Being entirely software based and using generic touch screens, the same set-up could be used for a wide variety of production situations. The user only needs to configure the production interface required for the particular application.

To include all functionalities as given in Section 5.6.1, the interface can be divided into components each covering a set of functionalities, where each component consists of a set of simple tasks. The components are: production, virtual camera, tracking, annotation, audio and replay control. The standard semi-automatic production interface comes with a set of predefined workspaces, in a way that each workspace represents one component, a logical group covering certain functionalities and is typically run by one operator. However, for audio and replay controls there will be no predefined workspace since for those separate stations are needed, integrated into the current production setup, and run by the sound engineer or replay operator.

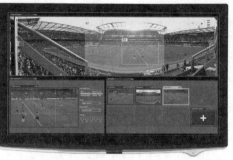

Figure 5.12 An example of a semi-automatic production gallery. Images reproduced by permission of The Football Association Premier League Ltd.

The user interface can be adapted to different needs and tasks by changing which panels are present in the workspace, and the position and size of them. Such new workspaces can be saved and used again together with the predefined workspaces. Since the core functionality within the panels is unaffected, the use of predefined workspaces or the creation of new ones is just a matter of selecting the preferred layout, and the number of available/needed operators.

In a typical setting, one workspace covers one screen. In the simplest case, there will be one operator working on one screen. The main workspace – the production control – will be open. In the case that some tasks not covered in that workspace are needed, the operator will switch to another workspace. However, typically several screens will be available each having one workspace opened. Further, in the case of two or more operators working on the production, each of them will be working on one or more screens.

An example of a setup for semi-automatic production is shown in Figure 5.12. In this case the production gallery consists of two screens. On the left side is the main one having the production control workspace active. The screen on the right shows the virtual camera control workspace. However, that workspace will be exchanged with other workspaces, like the tracking control workspace, during live broadcast as required.

5.6.3   Preconfiguration and Predefined Workspaces

The usage of the production interface can be roughly divided into preconfiguration carried out before the broadcast and the work during the live broadcast. The latter is achieved with the help of workspaces and panels. In this section, settings and the predefined workspaces with accompanying panels are described.

Preconfiguration

Preconfiguration is carried out at the start-up of the system. It enables the interface to be adjusted to the needs of the specific event and production team, that is, to decide on the workspaces required, as well as to define annotations on the semantic level, identify events or add view options preferences. Figure 5.13 shows an example of the pre-configuration page.

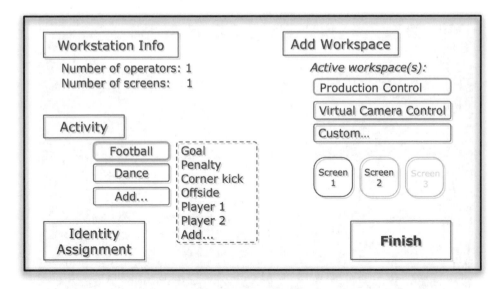

Figure 5.13   Pre-configuration workspace.

Production Control

The production control is the main component in the semi-automatic production interface. Functionalities available here are: displaying panoramic and standard broadcast cameras feeds, visualising virtual cameras, selection of video feeds for broadcast, video mixing the produced coverage of the event, support for the real-time annotation of high-level events and indication of salient points/regions. The production control operator monitors what is happening on the virtual cameras, and corrects them if needed (in the separate workspace). The workspace for the production control is comprised of five panels (shown in Figure 5.14):

Panoramic overview The large panoramic overview monitor displays the entire panoramic field of view (shown in Figure 5.15). The position and relative size of active virtual cameras and manned cameras are also shown. The virtual/manned cameras currently being broadcast and virtual cameras on the preview monitor are colour coded.

Gallery The gallery displays the currently available inputs that are to be used for vision mixing (shown in Figure 5.16), monitoring or direct broadcast. They could be manned cameras, virtual cameras (i.e., automatically produced view options), or any type of broadcastable material.

Programme/preview monitors The programme monitor to the left displays the output of the mixer, that is, what is being broadcast at the moment (e.g., a main programme). The preview monitor to the right displays the output of a selected camera (shown in Figure 5.17) in the gallery.

Figure 5.14 Production control workspace. Images reproduced by permission of The Football Association Premier League Ltd.

Figure 5.15 Panoramic overview panel. Images reproduced by permission of The Football Association Premier League Ltd.

Figure 5.16 Gallery panel.

Figure 5.17 Programme and preview panel.

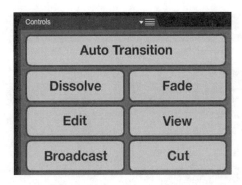

Figure 5.18   Action module panel.

Action module The action module displays the actions available on the input cameras (shown in Figure 5.18). It is possible to perform the following actions: dissolve, fade, auto transition, cut, edit, view and broadcast. *Dissolve*, *Fade* and *Auto Transition* are used to make a transition between the input currently being broadcast and the input currently being shown on the preview monitor. *Dissolve* mixes two sources, one fading in, one fading out, while *Fade* makes the source to fade to or from a solid colour, usually black. A *Cut* action instantly switches the camera currently being shown on the programme monitor with the one selected in the gallery. An *Edit* opens a selected virtual camera in the virtual camera control workspace where it can be reframed. This action is only available for virtual cameras. A *View* displays a selected camera on the preview monitor, while a *Broadcast* action sends a selected camera or any other input in the gallery to the broadcast output.

Basic annotation module The basic annotation module (shown in Figure 5.19) offers two ways of annotating high level events/actions: by identifying that the

Figure 5.19   Basic annotation module panel.

Figure 5.20 Virtual camera control workspace. Images reproduced by permission of The Football Association Premier League Ltd.

interesting event occurred, and by marking the start and end times of a particular event and connecting it to the one or more predefined labels, for example, in the case of a football event, a tackle or goal.

Virtual Camera Control

The virtual camera control shown in Figure 5.20 deals with monitoring virtual cameras, which includes controlling the automatically produced view options, proposing objects/regions as salient, and flagging virtual cameras (e.g., level of priority, quality content, not-broadcastable content). The virtual cameras in the virtual camera gallery are automatically available in the (sources) gallery for broadcasting (e.g., production workspace). The virtual camera control consists of: panoramic overview, virtual camera gallery and virtual camera info:

Panoramic overview As in the production control workspace.

Virtual camera gallery The virtual camera gallery displays each existing virtual camera as an icon with the output of the virtual camera displayed on it (Figure 5.21). It serves as an overview of the existing virtual cameras, as well as a way to select a certain virtual camera or to define a new one.

Virtual camera info The virtual camera information panel displays the selected virtual camera and information about it (Figure 5.22). Here virtual camera

Figure 5.21   Virtual camera gallery panel.

parameters can be changed/adjusted. Also, a priority level may be assigned to the virtual camera or it can be flagged as non-broadcastable content.

Tracking Control

The tracking control monitors the person/object detection and tracking, and helps (re)assigning identity to a person/object. Also, objects and people can be proposed here for tracking. An example interface is shown in Figure 5.23.

Figure 5.22   Virtual camera information panel.

Figure 5.23   Tracking control workspace example. Images reproduced by permission of The Football Association Premier League Ltd.

5.7   Conclusion

In this chapter an overview of the semi-automatic tools for annotating content in order to support the format-agnostic production process has been presented. The Virtual Director can benefit from richer metadata about the captured content particularly, if this metadata is not only available on a stream basis, but also for regions within panoramic video. Performing annotation on this level of granularity is a challenging task for both humans and machines, and thus needs a combination of automatic and manual methods. While automatic methods are efficient and reliable for extracting low- and mid-level features (such as detecting and tracking people), humans can add annotations on a higher semantic level and correct errors in the automatic processes.

It has been identified that the main challenges in content annotation for format-agnostic production are the real-time constraints in a live production scenario, dealing with ultra-high resolution media and the domain independence of the approaches. While there are established solutions for analysis architectures and metadata models, indexing and streaming metadata on a fine granularity while also keeping all metadata synchronised is still a challenging task, requiring the design and implementation of tailored components for metadata management.

The review in the chapter included methods for visual content analysis that are genre independent, like spatio-temporal saliency detection, focus on features such as people that are relevant across many content domains, or can be trained to a specific content domain, like supervised learning for concept and action detection. These methods share the common approach of extracting a set of low-level visual features from the video stream, in order to significantly reduce the amount of data that needs to be stored and processed. This initial feature extraction reduces the direct dependency of the runtime on the input resolution, but the real-time challenge remains. In the live production context, real-time does not only mean the ability to process all incoming data without accumulating a delay, but also to keep the latency of outputs under a certain threshold at all times. Otherwise, the information arrives too late for the Virtual Director to make editing decisions in time. Thus performance optimisation of these automatic algorithms is crucial for their practical value. This includes modification and optimisation of the algorithms themselves, as well as different parallelisation strategies, such as processing tiles of the panorama on different machines or using dedicated highly parallel hardware. For the latter, graphics processors (GPUs) are a very interesting technology. They can provide significant performance gains at affordable hardware costs. However, porting algorithms to a highly parallel architecture may need significant effort and this cannot be neglected.

Finally, the design of a specific user interface for monitoring, validating and possibly correcting the results obtained from the automatic content analysis tools as well as adding further annotations, has been discussed. In the manual part of the annotation process, the real-time requirement is the driving force. Any user interaction must be adapted to the content and must be as efficient as possible. In contrast to single purpose interfaces (e.g., for logging tennis) the user interface for a general purpose format-agnostic production system must be customisable to the requirements of different content genres.

Recent literature on automatic content analysis contains a number of approaches that are highly interesting in the context of format-agnostic production. This includes in particular methods for producing higher-level semantic information on events, interactions between people, or inclusion of structured external knowledge. However, many of these approaches are not yet applicable in this context due to their processing requirements. Thus further optimisation and performance improvements remain a challenge for providing richer metadata in a semi-automatic production process.

References

Alface, P.R., Macq, J.F., Lee, F. and Bailer, W. (2013) 'Adaptive Coding of High-Resolution Panoramic Video Using Visual Saliency'. *1st International Workshop on Interactive Content Consumption at EuroITV (2013)*, Como, Italy, 24 June.

Bailer, W. (2011a) 'A Feature Sequence Kernel for Video Concept Classification'. *17th Multimedia Modeling Conference*, Taipei, Taiwan, 5–7 January.

Bailer, W. (2011b) 'Sequence-Based Kernels for Online Concept Detection in Video'. *AIEMPro '11: 4th International Workshop on Automated Information Extraction in Media Production*, Scotsdale, AZ, 11 July, pp. 1–6.

Ballan, L., Bertini, M., Del Bimbo, A. and Serra, G. (2010) 'Video Event Classification Using String Kernels'. *Multimedia Tools and Applications* 48(1), 69–87.

Basex (2013) 'Basex. The XML Database', accessed 15 August 2013 at: http://basex.org/

Bishop, C.M. (2006) *Pattern Recognition and Machine Learning (Information Science and Statistics)*. New York: Springer.

Bosak, J. (1999) 'Shakespeare in XML', accessed 15 August 2013 at: http://www.ibiblio.org/xml/examples/shakespeare/

Calderara, S., Prati, A., Vezzani, R. and Cucchiara, R. (2005) 'Consistent Labeling for Multi-Camera Object Tracking', In F. Roli and S. Vitulano (Eds.) *Image Analysis and Processing ICIAP (2005)*. Berlin: Springer, pp. 1206–1214.

Canny, J. (1986) 'A Computational Approach to Edge Detection'. *IEEE Transactions on Pattern Analysis and Machine Intelligence* 8(6), 679–698.

Cao, D., Masoud, O., Boley, D. and Papanikolopoulos, N. (2004) 'Online Motion Classification Using Support Vector Machines', *International Conference on Robotics and Automation*, Barcelona, 18–22 April, pp. 2291–2296.

Chang, C.C. and Lin, C.J. (2001) 'LIBSVM: A Library for Support Vector Machines', accessed 15 August 2013 at: http://www.csie.ntu.edu.tw/~cjlin/papers/libsvm.pdf

Choi, J., Jeon, W.J. and Lee, S.C. (2008) 'Spatio-Temporal Pyramid Matching for Sports Videos'. *1st ACM International Conference on Multimedia Information Retrieval, Vancouver*, 26–31 October.

Chu, Y., Chia, L.T. and Bhowmick, S.S. (2004) 'Looking at Mapping, Indexing & Querying of MPEG-7 descriptors in RDBMS with SM3'. *2nd ACM international Workshop on Multimedia Databases*, New York: ACM, pp. 55–64.

Clark, J. and DeRose, S. (1999) 'XML Path Language (XPath)'. W3C Recommendation, 16 November.

Cuturi, M., Vert, J.P., Birkenes, O. and Matsui, T. (2006) 'A Kernel for Time Series Based on Global Alignments', accessed 15 August 2013 at: http://cbio.ensmp.fr/ jvert/publi/pdf/Cuturi2007Kernel.pdf

Dalal, N. and Triggs, B. (2005) 'Histograms of Oriented Gradients for Human Detection'. *IEEE Computer Society Conference on Computer Vision and Pattern Recognition, (2005)*, San Diego, 20–26 June, vol. 1, pp. 886–893.

DBLP (2013) 'The DBLP Computer Science Bibliography', accessed 15 August 2013 at: http://www.informatik.uni-trier.de/~ley/db/

DCMI (2003) *Information and Documentation—the Dublin Core Metadata Element Set*, ISO 15836. Dublin; Dublin Metacore Data Initiative

EBU (2012) EBU Core Metadata Set, (EBUCore) v1.4 EBU Tech 3293.

Engström, A., Juhlin, O., Perry, M. and Broth, M. (2010) 'Temporal Hybridity: Footage with Instant Replay in Real Time'. *SIGCHI Conference on Human Factors in Computing Systems*, Montreal, 22–27 April, pp. 1495–1504.

Excelon (2013) 'eXcelon Corporation homepage', accessed 15 August 2013 at: http://excelon-corporation.software.informer.com/

eXistdb (2013) 'Homepage', accessed 15 August 2013 at: http://exist-db.org/

Fassold, H., Rosner, J., Schallauer, P. and Bailer, W. (2009) 'Realtime KLT Feature Point Tracking for High Definition Video'. *GraVisMa – Computer Graphics, Vision and Mathematics for Scientific Computing*, Plzen, Czech Republic, 2–4 September.

Fischler, M.A. and Bolles, R.C. (1981) 'Random Sample Consensus: A Paradigm for Model Fitting with Applications to Image Analysis and Automated Cartography'. *Communications of the ACM* 24(6), 381–395.

Fleischman, M., Decamp, P. and Roy, D. (2006) 'Mining Temporal Patterns of Movement for Video Content Classification.' *8th ACM International Workshop on Multimedia Information Retrieval*, Santa Barbara, CA, 26–27 October, pp. 183–192.

Florescu, D. and Kossmann, D. (1999) 'A Performance Evaluation of Alternative Mapping Schemes for Storing XML Data in a Relational Database'. Technical report, INRIA, Rocquencourt, France.

Grauman, K. and Darrell, T. (2005) 'The Pyramid Match Kernel: Discriminative Classification with Sets of Image Features'. *IEEE Computer Conference on Computer Vision and Pattern Recognition*, San Diego, CA, 20–25 June.

Grauman, K. and Darrell, T. (2006) 'Approximate Correspondences in High Dimensions', *NIPS*, accessed 15 August 2013 at: http://books.nips.cc/papers/files/nips19/NIPS2006_0492.pdf

Grauman, K. and Darrell, T. (2007) 'The Pyramid Match Kernel: Efficient Learning with Sets of Features'. *Journal of Machine Learning Research* 8, 725–760.

Guo, C., Ma, Q. and Zhang, L. (2008) 'Spatio-Temporal Saliency Detection using Phase Spectrum of Quaternion Fourier Transform', *IEEE Conference on Computer Vision and Pattern Recognition*, Alaska, 24–26 June.

Hou, X. and Zhang, L. (2007) 'Saliency Detection: A Spectral Residual Approach'. *IEEE Conference on Computer Vision and Pattern Recognition*, Minneapolis, MN, 18–23 June, pp. 1–8.

Itti, L. and Baldi, P. (2005) 'Bayesian Surprise Attracts Human Attention', *19th Annual Conference on Neural Information Processing Systems*, Whistler, 5–6 December.

Itti, L. and Koch, C. (2001) 'Computational Modelling of Visual Attention'. *Nature Reviews Neuroscience* 2(3), 194–203.

Itti, L., Koch, C. and Niebur, E. (1998) 'A Model of Saliency-Based Visual Attention for Rapid Scene Analysis'. *IEEE Transactions on Pattern Analysis and Machine Intelligence* 20(11), 1254–1259.

Kaiser, R., Thaler, M., Kriechbaum, A., Fassold, H., Bailer, W. and Rosner, J. (2011) 'Real-Time Person Tracking in High-Resolution Panoramic Video for Automated Broadcast Production'. *Conference for Visual Media Production (CVMP)*, London, United Kingdom, 19–17 November, pp. 21–29.

Kalman, R.E. (1960) 'A New Approach to Linear Filtering and Prediction Problems'. *Transactions of the ASME – Journal of Basic Engineering* 82 (Series D), 35–45.

Kang, J., Cohen, I. and Medioni, G. (2003) 'Soccer Player Tracking Across Uncalibrated Camera Streams'. *Joint IEEE International Workshop on Visual Surveillance and Performance Evaluation of Tracking and Surveillance (VSPETS) In Conjunction with ICCV*, Nice, France, 11–12 October, pp. 172–179.

Khan, S., Javed, O., Rasheed, Z. and Shah, M. (2001) 'Human Tracking in Multiple Cameras', *Eighth IEEE International Conference on Computer Vision*, Vancouver, Canada, 9–12 July, pp. 331–336.

Kosch, H. (2004) *Distributed Multimedia Database Technologies Supported by MPEG-7 and MPEG-21*. Boca Raton, FL, USA CRC Press.

Kotsia, I. and Patras, I. (2010) 'Relative Margin Support Tensor Machines for Gait and Action Recognition', *ACM International Conference on Image and Video Retrieval*, New York: ACM, pp. 446–453.

Lavee, G., Rivlin, E. and Rudzsky, M. (2009) 'Understanding Video Events: A Survey of Methods for Automatic Interpretation of Semantic Occurrences in Video'. Technical Report CIS-(2009)-06, Technion – Israel Institute of Technology, Haifa, Israel.

Lazebnik, S., Schmid, C. and Ponce, J. (2006) 'Beyond Bags of Features: Spatial Pyramid Matching for Recognizing Natural Scene Categories'. *IEEE Conference on Computer Vision and Pattern Recognition*, Portland, OR, 23–28 June.

Levenshtein, V.I. (1966) 'Binary Codes Capable of Correcting Deletions, Insertions and Reversals'. *Soviet Physics Doklady* 10, 707.

Lucas, B.D. and Kanade, T. (1981) 'An Iterative Image Registration Technique with an Application to Stereo Vision'. *7th International Joint Conference on Artificial Intelligence*, Vancouver, August, pp. 674–679.

Mörzinger, R. and Thaler, M. (2010) 'Improving Person Detection in Videos by Automatic Scene Adaptation'. *Fifth International Conference on Computer Vision Theory and Applications*, vol. 2, Angers, France, May 17–21, pp. 333–338.

MPEG (2001) 'Information Technology – Multimedia Content Description Interface'. ISO/IEC 15938.

MPEG (2012) 'Information Technology – Multimedia Content Description Interface - Part 9: Profiles and Levels, Amendment 1: Extensions to Profiles and Levels'. ISO/IEC 15938-9:(2005)/Amd1:(2012).

Myers, C.S. and Rabiner, L.R. (1981) 'A Comparative Study of Several Dynamic Time-Warping Algorithms for Connected Word Recognition'. *The Bell System Technical Journal* 60(7), 1389–1409.

Naphade, M.R., Kennedy, L., Kender, J.R., Fu Chang, S., Smith, J.R., Over, P. and Hauptmann, A. (2005) 'A Light Scale Concept Ontology for Multimedia Understanding for TRECVID'. Technical Report RC23612 (W0505-104), IBM Research, USA.

Needleman, S.B. and Wunsch, C.D. (1970) 'A General Method Applicable to the Search for Similarities in the Amino Acid Sequence of Two Proteins'. *Journal of Molecular Biology* 48(3), 443–453.

NVIDIA (2009) 'Nvidia's Next Generation CUDA compute Architecture: Fermi', whitepaper, accessed 15 August 2013 from: http://www.nvidia.com/content/PDF/fermi_white_papers/NVIDIA_Fermi_Compute_Architecture_Whitepaper.pdf

NVIDIA (2013) 'CUDA Parallel Computing Platform', accessed 15 August 2013 at: http://www.nvidia.com/object/cuda_home_new.html

Perry, M., Juhlin, O., Esbjörnsson, M. and Engström, A. (2009) 'Lean Collaboration through Video Gestures: Co-ordinating the Production of Live Televised Sport'. *SIGCHI Conference on Human Factors in Computing Systems*, Boston, MA, 4–9 April, pp. 2279–2288.

Pittore, M., Basso, C. and Verri, A. (1999) 'Representing and Recognizing Visual Dynamic Events with Support Vector Machines'. *10th International Conference on Image Analysis and Processing*, Venice, Italy, 27–29 September. CIAP '99.

Prisacariu, V. and Reid, I. (2009) 'fastHOG – A Real-Time GPU Implementation of HOG'. Technical Report 2310/09, Department of Engineering Science, Oxford University.

Qi, G.J., Hua, X.S., Rui, Y., Tang, J., Mei, T., Wang, M. and Zhang, H.J. (2008) 'Correlative Multilabel Video Annotation with Temporal Kernels'. *ACM Transactions on Multimedia Computing, Communications and Applications* 5(1), 1–27.

Rubner, Y., Tomasi, C. and Guibas, L.J. (2000) 'The Earth Mover's Distance as a Metric for Image Retrieval'. *International Journal of Computer Vision* 40(2), 99–121.

Schmidt, A., Kersten, M. and Windhouwer, M. (2004) 'Efficient Relational Storage and Retrieval of XML Documents'. *3rd International Workshop on the Web and Databases*, Dallas, TX, 18–19 May.

Schmidt, A.R., Waas, F., Kersten, M.L., Florescu, D., Manolescu, I., Carey, M.J., et al. (2001) 'The XML Benchmark Project'. Technical report, CWI, Amsterdam.

Schöning, D.H. (2001) 'Tamino – A DBMS Designed for XML'. *17th International Conference on Data Engineering*, Washington, DC: *IEEE Computer Society*, p. 149.

Shi, J. and Tomasi, C. (1994) 'Good Features to Track'. *IEEE Conference on Computer Vision and Pattern Recognition* Seattle, WA, 21-23 June, pp. 593–600.

Shimodaira, H., Noma, K., Nakai, M. and Sagayama, S. (2001) 'Dynamic Time-Alignment Kernel in Support Vector Machine', accessed 15 August 2013 at: http://books.nips.cc/papers/files/nips14/AA20.pdf

Shirahama, K., Matsuoka, Y. and Uehara, K. (2011) 'Video Event Retrieval from a Small Number of Examples using Rough Set Theory'. *17th Multimedia Modeling Conference*, Taipei, Taiwan, 5–7 January, pp. 96–106.

Smeaton, A.F., Over, P. and Kraaij, W. (2006) 'Evaluation Campaigns and TRECVid'. *8th ACM International Workshop on Multimedia Information Retrieval*, Santa Barbara, CA, 23–27 October, pp. 321–330.

SMPTE (2010) 'Metadata Dictionary Registry of Metadata Element Descriptions SMPTE RP210.12', accessed 15 August 2013 at: http://www.smpte-ra.org/mdd/

Thaler, M. and Bailer, W. (2013) 'Real-time Person Detection and Tracking in Panoramic Video', *1st IEEE International Workshop on Computer Vision in Sports, Portland*, OR, 4 July.

Thaler, M. and Mörzinger, R. (2010) 'Automatic Inter-Image Homography Estimation from Person Detections'. *7th IEEE International Conference on Advanced Video and Signal Based Surveillance*, Boston, MA, 29 August–1 July, pp. 456 –461.

Tomasi, C. and Kanade, T. (1991) 'Detection and Tracking of Point Features'. Technical Report CMU-CS-91-132, Carnegie Mellon University, Pittsburgh, PA.

Tong, Y., Cheikh, F.A., Guraya, F.F.E., Konik, H. and Trémeau, A. (2011) 'A Spatiotemporal Saliency Model for Video Surveillance'. *Cognitive Computation* 3(1), 241–263.

Wang, J.R. and Parameswaran, N. (2004) 'Survey of Sports Video Analysis: Research Issues and Applications'. *Pan-Sydney Area Workshop on Visual Information Processing*, Sydney, Australia, pp. 87–90.

Wojek, C., Dorkò, G., Schulz, A. and Schiele, B. (2008) 'Sliding-Windows for Rapid Object Class Localization: A Parallel Technique'. *Proceedings of the 30th DAGM symposium on Pattern Recognition*. Berlin: Springer-Verlag, pp. 71–81.

Wu, G., Wu, Y., Jiao, L., Wang, Y.F. and Chang, E.Y. (2003) 'Multi-Camera Spatio-Temporal Fusion and Biased Sequence-Data Learning for Security Surveillance'. *ACM International Conference on Multimedia*, Berkeley, CA, 6–11 November.

X-Hive (2013) Homepage of EMC Developer Community, accessed 15 August 2013 at: http://www.emc.com/domains/x-hive/index.htm

XQL (2013) 'Infonyte QXL', accessed 27 August 2013 at: http://xmlbroker.de/prod_xql.html

Xu, D. and Chang, S.F. (2007) 'Visual Event Recognition in News Video Using Kernel Methods with Multi-Level Temporal Alignment'. *IEEE Conference on Computer Vision and Pattern Recognition*, Minneapolis, MN, 18–23 June.

Xu, D. and Chang, S.F. (2008) 'Video Event Recognition Using Kernel Methods with Multilevel Temporal Alignment'. *IEEE Transactions on Pattern Analysis and Machine Intelligence* 30(11), 1985–1997.

Yeh, M.C. and Cheng, K.T. (2008) 'A String Matching Approach for Visual Retrieval and Classification'. *1st ACM International Conference on Multimedia Information Retrieval*, Vancouver, Canada, 26–31 October, pp. 52–58.

Yilmaz, A., Javed, O. and Shah, M. (2006) 'Object Tracking: A Survey'. Computing Surveys 38(4), 13.1–13.45.

Yoshikawa, M. and Amagasa, T. (2001) 'XRel: A Path-Based Approach to Storage and Retrieval of XML Documents using Relational Databases'. *ACM Transactions on Internet Technology* 1(1), 110–141.

Zhang, T., Xu, C., Zhu, G., Liu, S. and Lu, H. (2012) 'A Generic Framework for Event Detection in Various Video Domains'. *IEEE Transactions on Multimedia* 14(4), 1206–1219.

6

Virtual Director

Rene Kaiser and Wolfgang Weiss
JOANNEUM RESEARCH, Graz, Austria

6.1 Introduction

Traditional broadcast productions typically utilise multiple cameras, yet broadcast only a single view of the scene they capture. Camera selection – decisions about when and how to cut – is performed by highly skilled human professionals, in cooperation with directors, camera operators and other members of the production team. The result is a single output stream where every viewer gets to see the same content. In contrast, a format-agnostic approach uses a scene capture model, and as automation comes into play, the result is less restricted. One implication is that users do not have to get the same view of the scene, and viewpoint selection could also be influenced dynamically by the users.

Automation of that view selection process is a fundamental research challenge for progress towards interactive TV. Its focus is to support a high degree of interactive content access and personalisation. Such software, usually referred to as a *Virtual Director*, has the potential to lead to more immersive user experiences, to unlock new content consumption paradigms, to increase the level of freedom for viewers in influencing what they get to see and to save production cost by generating content for a number of playout channels in parallel.

This chapter discusses how to further process content analysis results and manual annotations as described in detail in Chapter 5, for the sake of production automation. It gives insight into various aspects of Virtual Director technology and is organised as follows. The remainder of this section introduces the concept of a Virtual Director and puts existing implementations in literature into perspective. In Section 6.2 *Implementation Approaches*, potential technical approaches are discussed and the combination of event processing and rule engines is inspected. Next, Section 6.3 *Example Architecture and Workflow* presents an example of how to design a Virtual Director system. Selected subprocesses of a software framework executing Virtual Director behaviour are highlighted in Section 6.4 *Virtual Director Subprocesses*. The

Media Production, Delivery and Interaction for Platform Independent Systems: Format-Agnostic Media, First Edition. Edited by Oliver Schreer, Jean-François Macq, Omar Aziz Niamut, Javier Ruiz-Hidalgo, Ben Shirley, Georg Thallinger and Graham Thomas.

subsequent Section 6.5 *Behaviour Engineering: Production Grammar* explains the process of designing a Virtual Director's behaviour and takes a closer look at selected issues, focusing on visual output. A concrete existing implementation of a Virtual Director research prototype for a format-agnostic production system is presented in Section 6.6 *Virtual Director Example Prototype*. Conclusions, and considerations regarding limitations and evaluation are presented in Section 6.7.

6.1.1   What is a Virtual Director?

A format-agnostic approach for content creation moves from simple audio and video capture to scene capture, where a set of specific hardware, such as panoramic video cameras, allow exploitation of its full potential. To derive a set of individual streams for different viewers, a method to automatically extract streams matching the individual viewer's interest is required. A Virtual Director in the context of a format-agnostic production system is an intelligent software system that automatically frames and cuts between multiple camera views. Its purpose is to present the most relevant action in the scene in an aesthetically pleasing style, appropriate and specific to the production domain. Metaphorically, a Virtual Director replaces certain tasks of a human broadcast production crew. The extent to which such technology can be utilised in commercial productions is, in general, still subject of research.

Such a scalable automatic system may allow for a high degree of personalisation. Adaptation to the preferences of each individual user unlocks an enormous potential in added value, especially in the realm of a format-agnostic production system. Users may choose to bias camera selection to follow their favourite actors and actions, or influence audio presentation to, perhaps, focus on specific instruments in a concert broadcast. A Virtual Director may also offer a choice between multiple editing styles. The users' preferences may change during content consumption through any form of interaction, ultimately enabling novel interactive media consumption formats. Furthermore, automatic software can parallelise this decision-making process to a degree human production crews are not able to, for economic reasons.

Even without direct interaction, the system might learn about the users' preferences, for example, through some form of relevance feedback, and is able to keep a rich user profile. A Virtual Director system can also adapt its behaviour to the concrete playout device properties. Beyond taking the loudspeaker setup and aspect ratio of the screen into account, the level of detail of shots, transition speeds and patterns in sequences of shots need to be adapted to a concrete playout system. The shots used to broadcast a basketball match, for example, will look significantly different when consumed on a mobile phone in contrast to a large projection screen, so alternative shot selections are appropriate.

The research problem of automating camera selection and framing for a certain production domain in high quality is multifaceted. The area is far from being solved in the sense of universally applicable generic frameworks. With today's state-of-the-art technology, the scope of Virtual Director technology is still limited. With an increase of available internet bandwidth and omnipresent mobile devices, and live audio-visual services beginning to move beyond traditional consumption formats, one can expect to see such technology applied in various production domains reasonably soon. Directing and programming a Virtual Director's behaviour requires significant resources for professional applications, but customers' expectations for live broadcast may be met sooner in smaller, local markets. However, such programmed behaviour may always lack the creative brilliance and intuition of human professionals. In research, the term *quality of experience* (QoE) (see Callet, Möller and Perkis, 2012) serves as a key paradigm in the assessment of Virtual Director technology.

A Virtual Director may keep only a short-term memory of its input and decisions for performance reasons. It holds features in a limited time window and quickly decides what to do, depending on current information. To achieve visual aesthetic balance with respect to the camera selection, however, it also has to consider its recent decisions. And, due to the delay of reacting to its sensors, a certain level of algorithmic prediction is desired. In total, these considerations illustrate the complexity of such solutions.

The Virtual Director of a format-agnostic production system needs to be a distributed component, as its intelligence is required to take effect at various points in the production workflow between the production site and the terminal. On a high level, the realisation of a Virtual Director system can be divided into two aspects that can be separated: implementation of a software framework and implementation of a Virtual Director behaviour. The production behaviour of a Virtual Director can, in turn, be divided into pragmatic behaviour (frame what is most important) and aesthetic behaviour (follow cinematographic principles).

In order to take reasonable decisions, a Virtual Director engine needs information about what is currently happening in the scene and which camera streams are capturing that action. It therefore depends on its sensors, which send information about the action in the scene towards it, and has to interpret its metadata as uncertain information. A Virtual Director implementation is not necessarily a fully automated process. A production team could keep informing (by manual annotation) and steering it via an interface at the production site, for example, by annotating fouls in a basketball match that could not be detected automatically, or by adjusting the framing of a static shot. The production professionals are not taking decisions on a single AV stream output, but injecting metadata that can be used for the parallelised personalisation process to serve the individual needs of many. A Virtual Director typically decides how to present both audio and video content to the users, however, the majority of this chapter will focus on visual aspects.

6.1.2   Features Enabled by a Virtual Director

A format-agnostic system supports a set of features that allow the viewer to go beyond traditional broadcast viewing and to achieve a higher degree of personalisation, through individual static parameters or interaction. This necessitates automation in view selection, and the concept of a Virtual Director in such a context can be extended in many facets. Viewers are given increased freedom to interact with systems in order to individually select what they want to see – directly through some form of interaction, via more abstract menus, or even implicitly as the system learns about their preferences without targeted interaction. At one extreme, a viewer would lean back and not interact with the system at all, watching the default coverage as suggested by the Virtual Director. At the other, continuous interaction may be used for free view navigation within high-resolution content.

A schematic example of concurrent virtual camera crops from a high-resolution panoramic video stream is visualised in Figure 6.1. It can be assumed that most users prefer a position between the previously mentioned extremes. This may be supported by dedicated devices with sophisticated interaction modes. More details of this are discussed in Chapter 8.

Figure 6.1   Different virtual cameras can be positioned within a high-quality panoramic video stream, covering different aspects of the scene and the actions within, with shots that differ in type and aspect ratio.

A Virtual Director enables a user to have an individual viewpoint of the scene. In an example production of a track-and-field athletics meeting, viewers could choose to closely follow an object such as a certain athlete, certain groups such as people from a certain country, or any individual combination. The Virtual Director's intelligent behaviour might ensure that despite these settings, actions of a high priority are overruling specific user preferences to the benefit of the viewer experience. As an example, in a situation where a sprint race (very high priority event) is about to start, the system might cut away from a selected close-up view of an unrelated athlete warming up apart from the main action, to make sure the viewer does not miss the occurrence of the more important action.

Based on this basic concept, a multitude of value-added features could be envisaged, depending on the user's needs of each production domain. To enable such ideas, specific low-level sensors, interaction modes and production behaviour are needed. One example of such an advanced feature is viewer-specific replay generation.

6.1.3   Definitions

The following brief definitions are relevant for the rest of the chapter. Note that individual terms might be used differently in other chapters, or related literature.

> **Production:** We refer to a *production* as the concrete broadcast of an event, for example, a whole basketball match or a rock concert. A Virtual Director might process it both live and later as recorded footage. Note that the term *broadcast* is used in this chapter even when referring to a production system with a back-channel or personalised content selection.

> **Production genre, production domain:** A *production genre* or *production domain* is the type or the domain of a production. The behaviour for a beach volleyball match will be different from a basketball match, and therefore they represent different production domains. However, to a certain extent, implemented behaviour should be re-usable for other live productions as well, if configured appropriately.

Scene: In the context of this chapter, we use the term *scene* to refer to a spatial area that is covered by audio and video sensors of the production over time. That means a scene is regarded in a spatio-temporal sense, for example, the court and part of the stands of a basketball match, for the duration of the production. Note that the term is used differently in other disciplines, in some denoting a temporally limited content segment containing coherent actions.

Shot, virtual camera: A *shot* or *virtual camera*, in this context, is a rectangular area selected as a view of the current scene. It could be a crop within a panoramic video stream or the stream from a broadcast camera. Note that in related literature the term may be used differently, for example, to denote a temporal segment bounded by shot boundaries such as cuts or transitions.

Production grammar, screen grammar, production logic: A Virtual Director's scene understanding and decision-making behaviour with respect to framing and cutting, for example, is referred to as *production grammar*, *screen grammar* or *production logic*. Examples are the balancing of shot types used, the definition of concrete shot sizes, panning speeds and camera smoothing behaviour, cutting effects and transition properties. Most importantly, the behaviour defines the reaction to production domain specific events that occur. Ideally, all production behaviour is encapsulated in an exchangeable *domain model*.

Production script: Production scripts are messages that are compiled and sent by a Virtual Director implementation to other components in a format-agnostic broadcast system, or to other Virtual Director instances along the production workflow. They include instructions and metadata about decisions that are further processed, for example, by renderers.

Sensors: In the realm of this chapter, the term *sensors* is used to refer to other components in a production system that inform the Virtual Director about actions in the scene, for example, audio-visual content analysis components. Note that in other chapters, the same term is used to refer to microphones and cameras.

6.1.4 Requirements for Virtual Director Technology

The following presents several key requirements that need to be considered when implementing a Virtual Director system.

Real-time decision making: In times of considerable social media presence, live broadcast services cannot afford large delays, or their content could be regarded old news, especially in the sports domain where the final result is of most interest and shared via the Internet quickly. Since a Virtual Director typically does not access multimedia content directly, but only its metadata, technical approaches exist that are fast and scalable enough for most productions (details in Section 6.2). It might be that the sum of sensor delays, decision-making processes and instruction of subsequent components in the workflow adds up to less than the inherent delay in content preparation and transmission itself, leading to interesting effects that might be perceived by users as intelligent prediction, that is, a cut to an action just before it starts to happen. A Virtual Director is part of the production chain, therefore, it is in turn highly dependent on the speed of the sensors informing it. The overall synchronising of processes in a format-agnostic backend system is not trivial, since many subprocesses need to be managed, and the individual delay of some components tends to vary over time.

Parallel decision making: Besides the requirement to take fast decisions, a Virtual Director system may also have to serve many users in parallel. This implies an advantage if many subprocesses are designed to operate independently of individual user preferences, that is, for all users, while at least some have to realise the user-specific decisions. Delays may play a role here as well: if several different renderer systems are instructed in parallel, and they differ in overall delay, one of the components has to buffer instructions for synchronising.

Re-usable production grammar: Every broadcast production domain is different. Regarding the production grammar, which is the concrete behaviour of the Virtual Director, this uniqueness is true for almost every aspect. However, ideally, the behaviour should be configurable, extendable and exchangeable. While any production domain consists of very specific aspects, some basic parts of it might be re-usable, especially behaviour that defines the style of camera framing and panning. It should therefore be formalised in a way such that it can be re-used or re-configured for another domain. Re-use may be supported by additional authoring tools.

Re-usable software framework: In order to enable production grammar re-use, the other part of a Virtual Director system, the software framework executing the behaviour, should be a generic system that could also execute a different set of production behaviours. Ideally, it would not have to be adapted beyond simply switching to a model of a different production genre. Such a framework contains generic interfaces to: low-level audio and video analysis sensors; manual annotation interfaces; and other format-agnostic workflow components such as renderers and content transmission optimisers. Using standard data and metadata representations like MPEG-7 (MPEG, 2001) for interfaces is especially advantageous here.

Pragmatics of decision making: A Virtual Director's scene understanding intelligence should be good enough to identify the most important and interesting actions, as relevant in a certain production genre. The system should also be able to work out the action's location in the scene, so that appropriate physical cameras or virtual camera crops from high-resolution panoramas can be selected. Missing very important actions strongly affects the quality of experience and should thus be avoided.

Aesthetics of decision making: Beyond capturing what is most interesting, a Virtual Director's behaviour also has to define how to frame and cut the action as video, and how to steer the audio feed. The requirement is to take decisions that are visually pleasing enough to be accepted by its users. While a piece of software might never outperform a dedicated team of human professionals in terms of storytelling skills and intuition, the software may have certain advantages over humans, for example, in speed of reaction or decision-making consistency. It should be expected that at least a certain number of cinematic principles are followed. A wide range of literature on such principles esixts (see, for example, Arijon, 1976 and Brown, 2002).

Homomorphism of production behaviour: In the realm of declarative approaches (Section 6.2.1), it is desired that the representation language (executed by the machine) supports statements in the same form as, or at least resembling what is expressed by experts in knowledge elicitation processes. The advantage implied is that production behaviour can be modified much more easily, and knowledge representation skills may not be required from production experts. Section 6.2 will discuss this in more detail.

Keeping state: A Virtual Director must be able to keep track of its state and past decisions, since not only information about the present influences decision making, but also the recent and long-term past. For example, it might aim to keep a certain balance of different shot types used in a production.

Temporal reasoning and pattern matching: When processing considerable quantities of multimodal low-level information streams (see examples in Chapter 5), efficient algorithms are needed for concept detection. Many pattern matching frameworks are good candidates for the task, and many event-processing engines implement a set of temporal operators that are useful to define such patterns. For an overview on event-processing technology, refer to (Etzion, 2010).

Prediction capability: An advanced feature, yet quickly required in professional productions, is the ability to go beyond a posteriori *scene understanding* by predicting future actions. A purely reactive system behaviour may not satisfy user expectations. For example, when framing dedicated cameras to cover a successful shot in a basketball match, it is not feasible for the Virtual Director to wait for a specific basket detector to trigger. It might react earlier instead and cut to this likely event, accepting odd behaviour as a certain probability. Predictive production behaviour may be realised, for example, by partial recognition of patterns on the live event stream.

6.1.5   *Existing Implementations and Research Activities*

This section highlights several recent research activities related to the concept of the Virtual Director. A number of related automatic view selection approaches using recorded content have been proposed: the FILM system (Amerson and Kime, 2001); the interactive storytelling system proposed by (Kim et al., 2011); various video clip remixing concepts, for example (Kaiser, Hausenblas and Umgeher, 2009); or the narrative structure language (NSL) (Ursu et al., 2011) developed within the NM2 project (NM2, 2007). However, here we will focus on the more specific research problem of real-time Virtual Directors in live broadcast systems, or enabling technology that could be utilised in such a context. The challenge is significantly different due to the real-time decision making requirement, the lack of comprehensive high-level annotations and the lack of thoroughly pre-authored story frames in the sense of a scripted motion picture.

Overall, there are a number of research activities investigating *beyond HD*, which would be well suited for combining with Virtual Director technology to create a range of different shots from a single ultra-high definition image. NHK's Super Hi-Vision (NHK, 2013) system, for example, offers an even larger resolution than the OmniCam (cf. Schreer, Kauff and Sikora, 2013), all-be-it as a flat (2D) image. NHK targets several features such as programme customisation, recommendation and social TV services. However, it is not only the higher resolution but the wide range of potential new features that are noteworthy. A high degree of production automation is not only interesting for economic reasons but also because it enables a range of features for the benefit of the viewer, parallelising personalisation capabilities and more.

The following discussion of selected research activities gives a review of the state-of-the-art, technical approaches and application domains.

Virtual Directors for Event Capture and Broadcast

The APIDIS (Apidis, 2013) project (Chen and De Vleeschouwer, 2011a,b; Chen, Delannay and De Vleeschouwer, 2011; Chen et al. 2011) aimed to automatically produce personalised multimedia content in the field of team sport. Personalisation in this context means that the user's preferences such as the preferred team or player, the user's profile, history and device capabilities are considered for camera selection. The system builds on computer vision sensors to automate the production of video summarisations. A multiview analysis is implemented that detects and tracks players automatically. It monitors the scoreboard to recognise the main actions and the status of the game. To produce semantically meaningful and visually appealing content, the system balances between the following three principles: First completeness, which refers to the integrity of selecting a view and storytelling in summarisation. Second, fineness: which refers to the level of detail provided by the chosen shots. Spatially it prefers close views and temporally it implies redundant storytelling. Finally, smoothness: denoting the graceful change of virtual camera viewpoints and temporally appropriate switching between cameras.

The generic architecture for creating personalised offline video summaries consists of several consecutive processing steps. The first step segments the videos based on game states, identifies salient objects, and detects highlighted events. Then, for each segment, the production strategy is planned by determining the temporal boundaries, determining the view type of each shot and inserting necessary replays. The last step is camera planning where a viewpoint of a certain camera is selected by an optimisation function that considers completeness and fineness. The smoothness is applied afterwards with a two-layer Markov chain. Fulfilling the user preferences is modelled as a resource allocation problem that evaluates all optimal combinations of clips by their benefits and costs.

Another algorithm for automated video production from multiple cameras is presented in (Daniyal and Cavallaro, 2011). An object and frame-level feature ranking implementation is described. The problem of selecting cameras is modelled as a decision process. Therefore, the approach estimates object visibility scores and employs an optimal control policy for maximising the visibility over time, and minimising the number of camera switches. The implementation consists of a Partially Observable Markov Decision Process where the video content such as object size, location or scene activity is a Markov Process and the camera scheduling process is based on recursively estimating and updating the belief state. An additional reward model allows the approach to control the number of camera switches. The proposed method was evaluated on a basketball match monitored by five cameras with partially overlapping fields of view. The outcome was compared to a manually generated edit (regarded as ground truth) with an overlap of 95 percent. The result of the subjective test is that 26 out of 31 people considered the automatically generated video to be as good as the manually generated video.

The European research project, 'LIVE: Live Staging of Media Events' (LIVE, 2009) (Jiang et al., 2013) allowed the real-time involvement of users in live and interactive TV productions while also changing the role of the traditional live TV director. In this project, a director becomes a video conductor that communicates with the audience in real-time, and therefore has to direct the actors on stage, the orchestra in the pit, and respond to the mood of the audience, to ensure the highest possible quality of entertainment. Collecting feedback from TV viewers enables modelling and tracking their preferences and allows the production of personalised content for the target viewer groups. Viewers have the possibility of giving explicit feedback by voting, for example, choosing from a fixed set of answers. Implicit feedback is automatically collected by tracking channel switches and consumption durations of channels. Based on user feedback and other data, the following factors are analysed: the number of viewers per channel, the preferences of the channel audience, an analysis of the user groups on an observed channel, the trends of viewers, and voting statistics. The analysed data is an input for the production team who can then adapt the content. The system developed for live interactive TV production consists of the following four main parts: (1) A consumer's Internet Protocol television (IPTV) application to display notifications to a specific group and allow the users to send feedback to the director. (2) The real-time notification and feedback channel transfers the viewer's responses and channel switching information to the TV production team as well as the notifications to the IPTV application. (3) The services for real-time analysis of the audience, including their preferences and responses, are called feedback collection and analysis services. (4) The adjustments and the results of the analysis for the production team are displayed in the feedback application. The live intelligent TV production use case aims to develop tools to produce a multichannel live TV broadcast more efficiently. These include managing media assets, generating metadata on media items, conducting videos and providing intelligent decision support. One production team is then able to track and synchronise individual content items on up to five output channels in parallel. Therefore, a knowledge-based middleware is used to support and manage the production process in preparing and staging media events. The knowledge model is enriched with low-level annotations extracted by audio-visual analysis and annotations generated manually by users. Then, a recommender system processes the knowledge model for on-the-fly content selection from the TV archives during live production for personalisation of TV channels, according to the preferences of the target group. A field trial was conducted at the Olympic Games Beijing 2008 with 489 users. The interactive multichannel TV format was positively rated by the consumers. One director stated that the new real-time communication and feedback options became very important for him during the trail (Jiang et al., 2013, p. 11).

In another research activity, the My eDirector 2012 project (Patrikakis et al., 2011a,b) aimed to create context-aware and personalised media in real-time streaming environments for large-scale broadcasting applications. This allows end users to direct their own coverage of large athletic events and to create their own personal Virtual Director. Raw video content is enriched with annotations generated by scene analysis, person tracking and other sensor data. This annotated media stream is used to make camera decisions based on the user's profile.

Virtual Directors for Human Communication

Live broadcast is not the only application for camera selection automation. Virtual Director technology has also been applied for mediated video communication.

The Virtual Director in the TA2 (TA2, 2012) and Vconect (Vconect, 2013) systems (Falelakis et al., 2012; Kaiser et al., 2012) aims at automatic camera selection. The *Orchestration Engine* selects between a range of streams in order to support group-to-group communication in scenarios of social video conferencing. Low-level event streams from audio-visual content analysis are processed based on a rule set that defines how to abstract from it for the detection of higher-level events (*Semantic Lifting*). Both this and the decision making process are realised using the JBoss Drools event processing engine.

The AMI and AMIDA projects (AMI, 2013) developed a browser for recorded business meetings and to enable remote business meetings. A typical meeting room in this project consists of a table, a whiteboard and a projector screen. In one room there are four people where each person is captured by a fixed close-up camera and a separate microphone. Additionally, there are three overview cameras (left, right and centre) and two microphone arrays for far-field recordings. Al-Hames et al., (2007) developed a Virtual Director that permanently evaluates and selects the best suitable camera either for the remote location or for recorded content to create a summary of the meeting. The authors formulate this task as a pattern recognition problem and therefore they apply machine learning techniques using Hidden Markov Models and the Viterbi algorithm (Viterbi, 1967). The first layer of the Hidden Markov Model is fed with a global motion feature that identifies the major characteristics of the motion, a skin blob feature representing the activities of the participants' hand and head movements, and acoustic features including Mel frequency cepstral coefficients (cf. Al-Hames et al., (2007) and the energy. The second layer models individual actions such as standing up, sitting down, nodding or shaking the head. Finally, the Viterbi algorithm is used to segment the video stream into a sequence of camera switches. The machine learning algorithms have been trained based on previously recorded and annotated content. The recorded meetings were annotated by different people in terms of editing ground truth. The authors note that there is a rather low agreement of the annotations between the annotators, and therefore they were not consistent enough to be regarded as ground truth. They then decided to use only two sets of annotations to ensure a consistent training set. In an experiment the desired output was compared to the actual result of the system, with the result that the system selects the wrong camera (frame error rate) in 27 percent of the cases.

Virtual Directors for Free Viewpoint Video

Another application domain for Virtual Director technology is 3D video. While there seems to be little specific innovation potential in standard stereoscopic 3D productions, Virtual Directors for free viewpoint video (FVV) are becoming a hot topic as capturing and rendering technology becomes more mature. The consumers' advantage using free viewpoint video is that they may freely choose their viewpoint within a scene in 3D space, though possibly restricted spatially, or of limited quality from certain angles. However, users may not necessarily want to consume 3D FVV content following a *lean forward* paradigm for extended periods where they have to actively interact to be able to follow the scene from a suitable viewpoint. Interaction might be combined with a more *lean back* experience where a Virtual Director selects the most interesting viewpoints automatically, cutting and transitioning in 3D space in an aesthetically enjoyable manner.

This research field is moving fast, and many research advancements have not been made accessible to the public because of commercial interests. A good overview is given by Smolic, (2009, 2011). A spherical approach for a 3D virtual video camera is presented in Lipski et al., (2010). Concrete examples from the sports domain are presented in Inamoto and Saito (2005) and Hilton, Kilner and Starck (2006). Free viewpoint video is also discussed in Schreer, Kauff and Sikora, (2005).

To identify viewpoints that are currently most relevant, a high-level understanding of what is currently happening in the scene (actions, objects, etc.) might not only be achieved through content analysis and manual annotation, but also by interpreting active or implicit viewpoint recommendations. These recommendations can be extracted through real-time analysis of other users' interaction and viewpoint paths. In this domain, Virtual Director technology can build on technology from computer-generated 3D content such as games. Interaction design is especially crucial in this application domain to allow users to benefit from Virtual Director technology to the full extent.

6.2   Implementation Approaches

The following briefly discusses technical approaches for implementing key subprocesses of a Virtual Director system. The choice is crucial, since it implies a lot of technical advantages and disadvantages, or may pose limits in runtime performance and scalability.

As illustrated by research examples in the previous section, there is considerable variety in complexity and in the extent to which proposed systems can be generalised. Perhaps the clearest distinction is the ability to operate *live* within real-time constraints, as this is clearly more challenging than automatically producing an offline edit of an event with similar production logic. It should not be underestimated that the real-time requirement has strong implications on the design of such a system, and that it rules out applying some established video post-production processes for recorded content. On the one hand, existing approaches may simply not be fast enough, on the other, there is much less scope for manual content annotation and decision intervention in a real-time context.

The remainder of this section provides information about software implementation aspects, choices to take, options and candidate technologies. The following section *Technical Approaches* discusses pros and cons of approaches such as Description Logics or Machine Learning. It presents neither a complete list of options nor a comprehensive pro/contra comparison of alternate approaches: only main issues are highlighted. The subsequent section *Rule engines and event processing technology* provides more detail and discusses advantages of combining rule engines and event processing technology for implementing a Virtual Director.

6.2.1 Behaviour Implementation Approaches

The following gives information about a number of technical approaches for structuring production behaviour. Some approaches are declarative, which means that knowledge bases are separate from reasoning procedures, and are expressed in representations comprehensible at least to the technical people editing it. A particular behaviour is defined by expressing it as a set of statements. Changing the behaviour requires amending only those statements, not the software engine executing it.

Surveys of methods for automatic interpretation and computational understanding of video events in general are presented in Lavee, Rivlin and Rudzsky (2009) and Jiang et al. (2013). Methods of abstracting video data and how to model events are investigated. For event modelling, the authors discuss a multitude of approaches, such as finite-state machines, Bayesian Networks, Hidden Markov Models, or constraint satisfaction algorithms.

Description Logics

Description Logics (DL) (Baader et al., 2010) are a family of knowledge representation languages widely used for modelling ontology models – see Staab and Studer, 2009; and Hitzler et al., 2012 for details. Practically more relevant, the Resource Description Framework (RDF) and the Web Ontology Language (OWL) data formats are an integral part of Semantic Web technology. Description Logics are decidable fragments of first order logics and allow representation of the knowledge of an application domain in a formal and precise way by using so-called *concepts*, *individuals* and related *properties*. An example for such *triples* are static shots that are configured for a certain production, including all their properties such as position and size within a panoramic video. Furthermore, DL allows formal reasoning over facts and inference of new knowledge by combining existing knowledge. These features make the use of ontologies interesting for modelling the behaviour of Virtual Directors. DL is very well suited to static modelling of concepts and their relations. In the context of Virtual Director behaviour, ontologies may be used for:

- Representing a production domain, with all of its concepts and relations.
- Formalising the production behaviour, referencing to the concepts of the production domain.
- Structuring and keeping the Virtual Director's input and decisions at runtime.
- Knowledge-base for runtime reasoning that takes decisions based on the inputs the Virtual Director receives, for example, through OWL-DL reasoning, or rules, such as Jena rule inference (Jena, 2013).

Despite continuous progress in DL reasoning research, however, it has been reported that performance and scalability issues remain for the practical application of such runtime reasoning for Virtual Director systems (Kaiser, Torres and Höffernig, 2010). An ever-increasing knowledge base may lead to a system slowing down over time. To optimise, the reasoner would have to continuously check which parts of the knowledge base are still relevant for future decisions. This proves to be a practical challenge in the context of this approach and in the context of real-time requirements. Besides that, the runtime performance of OWL-DL reasoning or rule execution in this context seems to be slower compared to other approaches. Hence, we suggest using DL only for static behaviour models, that is, the Virtual Director's domain model, not for runtime decision making.

Machine Learning

Using Machine Learning techniques allows the behaviour of a Virtual Director to evolve based on empirical data. The idea of *learning* production grammar based on metadata of manual edits is very appealing, as it could save on declarative behaviour engineering effort. It appears that a generic software framework could be implemented that executes behaviour that has been learned before. But, without going into detail on concrete Machine Learning algorithms, there are a number of issues with such an approach.

A crucial problem is that creating useful test data for learning the behaviour might be expensive or even practically impossible. Some productions, such as covering a basketball game, will be very similar each time, allowing behaviour learned from one game to be applied to another. However, this is not the case for productions in general. A specific question in that context is who is defining the behaviour. Here, this process is less explicitly driven through human intention than with most other approaches. When utilising the decisions of not only one but multiple human directors as ground truth, the question how to merge different decision sets appears. As pointed out by Al-Hames et al. (2007), selecting cameras and cutting videos can be done in many different ways that are all reasonable, and no objectively correct behaviour can be defined.

Another issue is the tweaking of behaviour once it is learned, as fine-tuning typically requires new training data. It is also not clear what to change in the ground truth in order to achieve the desired effect. Changing ground truth data might require several human professionals to repeat their work over and over. Tweaking the abstract models themselves as an alternative might be practically impossible. This is a research challenge on its own, as such changes are not really deterministic and hard to foresee. Tweaking and fine-tuning, however, needs to be fast so that adapted production grammar can be tested over and over.

Another potential problem is that behaviour models might have to be re-learned whenever any detail in the production setup changes, for example, cameras are re-positioned or added. Because of these and further limitations, we do not recommend a Machine Learning approach for the complete decision making of a Virtual Director implementation, however, it might be suited for subprocesses, depending on the concrete production. To conclude, more research is needed in that domain and new approaches need to be developed.

Event Processing

A Virtual Director in a live broadcast production environment continuously receives data from the sensors that it processes in order to take its decisions. These data can be interpreted as events. Over time they can be regarded as event streams with common basic properties such as timestamps. A comprehensive overview of this research field has recently been compiled by Etzion and Niblett (2010) and Cugola and Margara (2012). A term often used in that domain is Complex Event Processing (CEP). The term *complex events* stems from the detection of predefined meaningful patterns, that is, complex events consist of patterns of simpler ones. Stable event processing solutions are available, many of which come with a specific query language, and offer built-in capabilities such as querying for (spatio-)temporal event patterns Anicic et al. (2012). Several language styles have been investigated, for example, stream-oriented, imperative and rule-oriented languages.

Rule Engines

A Virtual Director has to react to incoming events with a certain action, based on the information just received and its own state. An obvious approach to that problem is to define the desired behaviour in the form of rules. Rule-based systems are used to store and manipulate knowledge to interpret information in a predefined way. Event-condition-action (ECA) rules are well-suited because they are typically in line with human thinking regarding technical behaviour. Human broadcast professionals are usually able to express desired behaviour in natural language rules resembling ECA format, which can be transformed into a formal language (homomorphism). Rule-based systems may even deal with uncertain information (Wasserkrug, Gal and Etzion, 2012a; Wasserkrug et al., 2012b).

Stream Reasoning

An interesting approach researchers recently proposed is Stream Reasoning (Valle et al., 2009; Barbieri et al., 2010). It combines the ideas of reasoning over Description Logics ontologies and event stream processing. Stream reasoning might be applied as the basis for a higher-level decision-making process that requires complex and real-time reasoning over noisy data streams and rich background knowledge. Frameworks are based on Machine Learning algorithms and existing reasoning mechanisms extended for continuous event processing. Currently, this promising approach is subject of research and a lot of challenges remain. There are only a few experimental implementations that can be used (see Hoeksema and Kotoulas, 2011; Stuckenschmidt et al., 2010; Anicic et al., 2011). One example is Sparkwave, a library for continuous pattern matching over RDF data streams. Sparkwave uses and extends the Rete algorithm (Forgy, 1982) with additional features to meet the requirements of data stream processing (Komazec, Cerri and Fensel, 2012).

Behavioural Programming

Behavioural Programming (BProg) (Harel et al., 2011; Harel, Marron and Weiss, 2012) is a relatively novel way of programming reactive systems. BProg facilitates the process of specifying intended behaviour by breaking up this task into the specification (called *play in*) of simple example scenarios. Some BProg environments offer useful tools to detect unexpected interactions between the scenarios, which can be blocked by adding additional scenarios. Several implementations exist, however, this is one more example of a promising research development where it remains to be seen how far it can eventually be applied to solve larger practical challenges, such as implementing the complex behaviour of a Virtual Director.

Cognitive Architectures

A completely different approach, *cognitive architectures*, allows intelligent systems to be built that provide a broad range of capabilities similar to those of humans. Cognitive architectures emerged from the field of artificial intelligence research. These systems focus on creating structures that support learning, encoding (in the sense of knowledge representation), and using knowledge to perform tasks in dynamic environments. Cognitive architectures have fixed processes and memories to acquire, represent and process knowledge about the current task enabling moment-to-moment reasoning, problem solving and goal-oriented behaviour. Knowledge is made available through different types of memories such as short-term memory or long-term memory. A declarative long-term memory could be encoded using production rules. A cognitive architecture might also provide a declarative learning module using activities in short-term memory to add facts back into this memory. The short-term memory also uses a procedural long-term memory, which has knowledge about what to do and when to do it. Based on the available knowledge, an *agent* can select actions to achieve its goal (Duch, Onentaryo and Pasquier, 2008; Laird, 2012).

Cognitive architectures could be used to implement the production grammar of a Virtual Director to allow it to adapt dynamically to a specific user group in a specific domain. This would increase its flexibility and re-usability. However, this approach also raises concerns, such as: could Virtual Director behaviour based on temporal dependencies be implemented with reasonable effort, and could runtime execution fulfil the real-time requirements that are crucial in a live broadcast system? Some of these concerns are yet to be addressed by research efforts.

6.2.2   Combining Rule Engines and Event Processing Technology

Event processing has been identified as a powerful enabler and feasible solution to implement a Virtual Director as it can be integrated well with the advantages of rule engines. Recent research advances of event processing systems, which are based on the idea of processing events that include logic for filtering, transforming or detecting patterns in events, created a lot of powerful and scalable frameworks. Furthermore, their combination with rule-based systems enable substantial expressivity in this domain.

Not all rules that are part of a Virtual Director's behaviour are general; many tend to depend on a domain, a production or a style. It is difficult to regard rules in isolation because a Virtual Director's behaviour results from their dynamic interplay. Most rule engines are fast, allowing temporal patterns to be defined over time windows. The continuous queries within CEP modules are proven to be very fast, and most query languages allow the definition of powerful triggers on (spatio-)temporal patterns. One key reason for their favourable performance are efficient optimisation algorithms. One remaining research question is how to effectively model larger corpora of rules (e.g., via dynamic layering), and, foremost, how to resolve decisions based on contradicting/competing principles – more on that will be discussed in Section 6.5.5.

The combination of event processing and rules engines has been exploited before in the realm of video event understanding, that is, detection of relevant high-level actions appearing in a scene that are not directly observable by audio or video sensors themselves.

A key reason that makes the decision-making process of a Virtual Director challenging is that it can almost never be assumed that the information at the engine's disposal is perfect or complete. It should be processed by algorithms dedicated to reasoning with uncertain information, or interpreted in a way such that expected error probabilities are taken care of by design. Sometimes, confidence values from sensors are available and can be accessed. In the very simplest case, thresholds are used to dismiss useless events. Beyond that, several frameworks facilitate clever reasoning with uncertain information by design.

A range of commercial event processing engines is available, some of them as Open Source. The distinction between CEP engines and rule engines is sometimes not clear. The following will introduce a selected set of candidate frameworks.

Drools

JBoss Drools (Drools, 2013) has grown into a unified integrated platform for rules, workflows and event processing. Drools is designed for executing forward chaining and backward chaining rules. It also provides event processing capabilities and therefore implements interval-based time event semantics (cf. Allen, 1983) to allow temporal reasoning. Timers and calendars are supported for time-based tasks and sliding windows can be defined by temporal duration or amount of events. Drools provides comprehensive functionalities for forward chaining rules and integrates event processing capabilities neatly. Drools uses an extended version of the Rete algorithm (Forgy, 1982), called ReteOO, for efficient pattern matching. Its rule language is first order complete. In Drools, facts (data) and rules (logic) are separated. Rules are separated from the application code as well, which makes it possible to re-use rules across applications.

Esper

Esper (Esper, 2013) is an open source library for Complex Event Processing and event stream processing and uses a stream-oriented language, called Esper event processing language (EPL), which is an extension of SQL. It is best suited for real-time event driven applications, such as business process management and automation, stock trading, network and application monitoring and sensor network applications. The library is available for the Java and .NET platforms. The Java version can be integrated into any Java application. The language supports both time-based and length-based event windows, causality patterns, aggregates, filters and joins; however, it has limited expressivity in comparison to some other frameworks.

ETALIS

ETALIS (Anicic et al., 2010; Anicic, 2012; Etalis, 2013) is an expressive logics-based language for specifying and combining complex events. This language enables efficient run-time event recognition and supports deductive reasoning. ETALIS uses a logics-based approach for event processing. The declarative (logic) rules approach has several advantages, for example, that the language is expressive enough to represent complex event patterns and the logics-based event model allows for reasoning over events. ETALIS implements event-driven backward chaining rules, is implemented in Prolog, and runs on several Prolog systems. Interfaces for external programming languages such as Java depend on the connectivity of the Prolog implementation used. The ETALIS system provides two languages: the ETALIS Language for Events (ELE) and Event Processing SPARQL (EP-SPARQL).

Event Calculus and ProbLog

Event Calculus (Artikis et al., 2010) is a many-sorted, first-order predicate calculus for representing and reasoning about events and their effects. Various dialects of Event Calculus exist but it is typically expressed as a logic program, for example, Prolog. ProbLog (ProbLog, 2013) is a probabilistic extension of the logic programming language Prolog, which allows the use of probabilistic facts. Skarlatidis et al., (2013) describe an approach for making Crisp Event Calculus compatible to ProbLog and therefore be able to process events that contain probabilities.

Prova

Prova (Prova, 2013) is a Java-based rule engine. The roots of Prova lie in logic programming and its approach can be compared to Prolog, but Prova provides additional semantics and more features, for example, rule-based workflows and rule-based event processing. Additional operators are available, for example, for detection of out-of-order events. A fundamental aspect of the system and its design is reactive messaging. It is used to organise distributed Prova engines into a network of communication agents. Messages can be sent both synchronously and asynchronously. The library further supports rule interchange and dynamic access to external data sources and Java APIs.

To conclude, for the implementation of Virtual Director technology, we recommend utilising engines or frameworks that allow the benefits of event processing engines and rule-based systems to be exploited. They facilitate implementing multimodal fusion over parallel event streams. Query languages support pattern matching with temporal operators well. Perhaps most importantly, these approaches are fast and scalable.

6.3   Example Architecture and Workflow

Previous sections have described the purpose of Virtual Director technology and technical approaches that might be exploited to implement such a software system and the behaviour it executes. To obtain a complete picture and a good idea of how to dissect the task of designing and implementing a Virtual Director, this section presents typical subcomponents and subprocesses.

Naturally, since a Virtual Director in a format-agnostic production environment it could be connected to any component as input and output, and since it could be used in any conceivable production domain, the following example is neither complete nor the only possible solution.

6.3.1   Workflow of the Production System

The workflow of a format-agnostic production system comprises all processes between the content sensors and the playout devices. A generic example is illustrated in Figure 6.2,

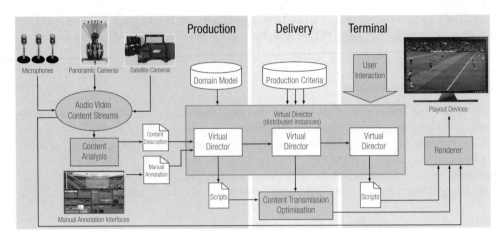

Figure 6.2 An example workflow of a Virtual Director for format-agnostic live broadcast production in the context of related components. The workflow is divided into three segments: production, delivery and terminal.

depicting relevant processes and components only. We chose to dissect the workflow into the following three steps along the production chain:

- **Production:** At the capture venue, the Virtual Director has interfaces to its sensors, such as content analysis and manual annotation interfaces. One side aspect is that content transmission optimisation may be steered by the Virtual Director through metadata defining the currently most important regions.
- **Delivery:** One or more Virtual Director instances may sit along the network chain, influencing camera selection by taking various aspects into account, for example, excluding parts of the content because of copyright issues in a certain delivery network or country, or ruling out potential privacy issues such as closeup zooms on the event's audience – compare with example factors illustrated in Figure 6.3 as arrows.
- **Terminal:** Virtual Director intelligence may also work close to the playout system, given that terminal devices support that. At the end of the workflow chain, direct input through user interaction can be supported.

The workflow of a Virtual Director does not have to be a strict chain. In some cases, information has to be sent back to the production end as well, for example, to enrich the user profiles stored at the production site.

6.3.2   Workflow of the Virtual Director

Focusing on the Virtual Director component only, from a high-level point of view, we suggest dividing the typical workflow into three main (sequential) subprocesses: (i) scene understanding, (ii) identification and preparation of shot candidates and (iii) decision making for shot selection and cutting behaviour, the latter for each viewing device individually.

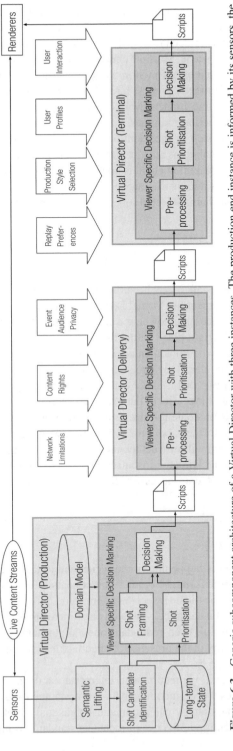

Figure 6.3 Generic subcomponent architecture of a Virtual Director with three instances. The production end instance is informed by its sensors, the terminal end instance is instructing the renderers. Parallelizable decision making processes that operate for each user independently are highlighted by darker boxes. The term *sensors* denotes audio-visual content analysis modules that inform the Virtual Director.

Without understanding what is happening in the scene, it would be impossible to take reasonable decisions on shots and camera movements. Therefore, the first step in the proposed workflow is to achieve an abstract understanding of which domain-specific actions are currently happening in certain areas of the scene. Multiple channels may inform the system about the real world, for example, automatic feature extraction modules and manual annotation interfaces. The Virtual Director will subsequently process a real-time stream to bridge the *semantic gap* between low-level annotations and higher-level concepts (cf. Hare et al., 2006). The latter are defined as part of the production grammar, the rules and principles that define how the action is framed. The second step is the identification of shots and their preparation for subsequent use, a process that maintains a list of suitable shot candidates based on the current situation. The last part in this processing chain is taking the actual decisions, that is, selecting the actual shots for each viewer and instructing the renderers or subsequent Virtual Director instances along the production chain. Selected subprocesses will be discussed in more detail in Section 6.4.

6.3.3 Distributed Nature

While in small setups a single instance may be sufficient, in larger format-agnostic production systems it makes sense to implement the Virtual Director as a distributed component, in order to take full advantage of the processing power at different stages of the production chain, and in order to take decisions at the right place within the workflow.

Any number of instances may be implemented, yet from an engineering point of view, it is recommended to separate them clearly by decision aspects. The first instance in the chain is typically connected to the sensors, that is, other components informing it with metadata about the scene. Every instance may have a local data store to keep its long-term state. Multiple instances may communicate with content transmission optimisation components, in order to instruct them, and also to be informed about the current bandwidth restrictions.

The final instance of the chain will instruct the renderers. The messages emitted to instruct renderers or subsequent Virtual Director instances are called *production scripts*. Besides instructions, scripts contain metadata necessary for the subsequent components to inform further processing.

6.3.4 Sources of Knowledge

To be able to *understand* what is happening in the scene, and to take high-quality decisions, a Virtual Director needs to be informed by other components, often referred to as sensors. Without any information from the outside, such a system could only take random decisions. A continuous stream of information has to be generated or extracted in real-time and transmitted to it. Any sensor might be useful, even redundant ones might help to reduce noise and to improve pattern matching quality in the presence of uncertain information, especially through multimodal fusion.

As elaborated in Section 6.2.2, we recommend interpretation of incoming information as events. This further guarantees that all incoming data is timestamped and the event processing engine's internal optimisation strategy can automatically dismiss events no longer needed. For data representation of such input, the use of metadata standards is highly recommended, most notably the MPEG-7 (MPEG, 2001) format. The following gives a brief overview of potential sources:

Content analysis: Since a Virtual Director decides on the presentation of audio and video content, it is logical to apply content analysis techniques, as described in Chapter 5, for standard or panoramic cameras with fixed position, and moving cameras that are operated by humans or steered by the Virtual Director system itself. The same holds for audio, where processing is often simpler, computationally cheaper and quicker. Fixed, wearable or virtual microphones (cf. Del Galdo et al., 2011) of any kind may be exploited.

Manual annotation: Because of the aforementioned deficiencies of automatic content analysis algorithms (see Chapter 5), professional applications typically require in addition manual annotation interfaces that are operated by professionals. Manual annotation is expensive, even for professional broadcast productions, and has deficiencies as well, for example, with respect to delays in comparison to a direct vision mixer. Such interfaces may both visualise the internal state of a Virtual Director and send information and commands towards it. The most obvious option is to manually annotate high-level actions that are not directly observable by automatic sensors. Another example task would be identity assignment for people that are automatically tracked, for example, the most interesting people in a concert or sports match, to ultimately enable the users to follow their favourite characters closely. It could further serve as a tool for validating and correcting the results of content analysis. An interface may also serve for selection of, or re-prioritisation between shot options provided by the Virtual Director system.

User profiling and interaction: Another obvious influence on Virtual Director decisions are metadata related to the users themselves. Creating and managing rich user profiles requires the ability to identify users across content consumption sessions. Profiles may also contain general preferences. In addition, dynamic interaction during content consumption can be exploited. One option is to automatically learn user preferences and domain-specific interests. Even without users consciously telling the system what they prefer to see, their interaction may be interpreted as relevance feedback. For all of these possibilities, privacy concerns have to be considered and respected in interaction design.

The Web: Sources outside the format-agnostic production system may also be considered. User-related information might be accessed from online resources in order to exploit them to the benefit of the users. Analysis of topical user interests, of relationships and of metadata about the event that is broadcast, might be employed. Event metadata could also be used to inform the users about the content.

As an advanced example, a geographically disparate community is watching a live broadcast event. The Virtual Director of that broadcast system might be granted access to their social media profiles to find out about their common interests. Such information might be used to turn distributed live content consumption into a shared experience (cf. the approach of Ursu et al. (2011) and Stevens et al. (2012). If information from online sources is available in structured

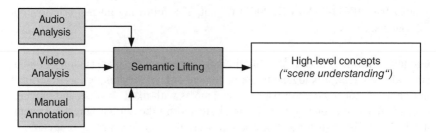

Figure 6.4   Schematic illustration of the Semantic Lifting component's purpose.

form such as RDF/OWL (cf. Hitzler et al., 2012), it is easier to process. In case the Virtual Director's domain model is represented in the same form, they can be linked efficiently.

6.4   Virtual Director Subprocesses

In this section, typical subprocesses of a Virtual Director system are elaborated on in detail. This discussion assumes a certain production setup and gives general advice on implementation issues, while Section 6.6 will present a concrete Virtual Director implementation. As stated before, the following is only one possibility of how to design such an engine and how to separate concerns into modular processes. The proposed architecture and interplay of processes is depicted in Figure 6.3, in a setup with three instances along the format-agnostic production chain. The following refers to an imaginary basketball production as a worked example, and partially assumes a rule-based event processing approach to be implemented, as described above. To simplify, it focuses on the directing of visual output and neglects automatic steering of audio playout.

6.4.1   Semantic Lifting

Semantic Lifting (Figure 6.4) refers to the process of achieving computational *scene under-standing*. It deals with the problem that semantically-abstract events are typically not directly observable by content analysis components, and that the incoming information about the scene is on a lower semantic level than the production rules for camera selection decisions. In order to bridge this semantic gap, this component aims to achieve an understanding of what is currently happening. From a technical point of view, it aims to derive domain-specific higher-level concepts from low-level sensor input streams. This can be implemented by, for example, detecting certain spatio-temporal constellations of low-level events in the streams as triggers. It emits a range of higher-level events to inform subsequent components. These events ideally correspond directly to domain concepts used in rules within subsequent processing steps.

> **Input:** A real-time low-level event stream, potentially a very high number of events received per frame via different automatic and human input sensors. Examples: bounding boxes of persons tracked, position of ball tracked, audio events of domain-specific actions, game clock, manually annotated game actions.

Output: Semantic concepts (i.e., events) of a higher abstraction level. These events describe the action in the scene on a certain semantic level, such that their occurrence can directly trigger subsequent processing components. These concepts are part of the domain model and the production grammar. Examples: Foul action, successful shot.

Frequency: Input events occur whenever they are detected or annotated, while some are received in fixed intervals, for example, every certain number of frames. High-level concepts are emitted whenever detected on the event stream.

Approach: Semantic Lifting is the first step in the processing chain, therefore it must be able to handle a high volume of input events. Incoming low-level events need to be received through a software interface and be prepared to be added to the internal input event stream. Multimodal fusion might be realised implicitly, and the issue of receiving out-of-order events from multiple sources might have to be addressed. This step might include several pre-processing steps such as coordinate transformation, filtering, or enrichment with metadata such as timestamps and confidence values. Inherently, the input to the Semantic Lifting step has to be regarded as uncertain information, and has to be processed appropriately.

The Semantic Lifting process itself may continuously execute (spatio-)temporal queries on the event stream to detect relevant high-level concepts. An artificial raster over the (panoramic) videos might be used to facilitate such spatial queries, for example, to locate the area with the highest person density. Since temporal and spatial patterns are essential for detection, language support for such relations within event processing query languages is a significant advantage. An example is to detect the global direction of the movement of people over a certain number of frames. Further, the observation of trends, fuzzy classification, or reaction to sudden changes over time may be used to trigger the detection of semantic concepts.

6.4.2   Shot Candidate Identification

Shot Candidate Identification (Figure 6.5) creates and maintains a list of useful shot candidates, that is, a list of physical and virtual cameras from panoramic and other cameras. A virtual camera is a crop of the omni-directional panorama that can be static or moving along a path. The results are generic options that allow subsequent components to take personalised camera decisions, choosing to cut between available options as appropriate. Shot Candidate Identification builds on the higher-level understanding achieved by Semantic Lifting to decide which views to select as candidates, while also keeping its options balanced with respect to the diversity of shot types. Some candidates can be directly derived from the input concept stream,

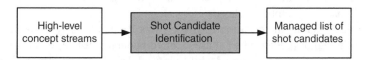

Figure 6.5   Shot Candidate Identification continuously creates and manages a list of global virtual camera candidates, which may subsequently be used for user-specific decisions.

some need to be detected by specific algorithms. Candidates might be added or removed at every frame.

Input: Shot Candidate Identification is informed by a stream of high-level concepts that correspond to abstract notions of events that happen in the scene. Events consist of several kinds of metadata, for example, the type of event, timestamps, confidence values, reference to sensors by which the event was created, or linked low-level events that led to its detection. High-level concepts ideally correspond to semantic concepts used in the domain model. Examples: Free throw attempt, or successful layup shot in a basketball match.

Output: The list of shot candidates is managed such that an appropriate number of options is available at all times. The list should never be empty, and may include several default views as configured for the production. The ideal length of the list certainly depends on the production setup and user interaction capabilities, however, a minimum amount of options is always needed to enable personalisation in shot selection. Candidates in the output list are defined by a location within a panoramic or normal camera, including a set of metadata. Their default size can be derived from the domain model, given a standard size is defined for every high-level action there. The concrete shot framing and size is however defined by subsequent components. Example: New virtual camera with a close-up shot on the free-throw shooter added to the list of options.

Frequency: In order to react fast to the action in the scene, shot candidates need to be updated every frame.

Approach: Shot Candidate Identification aims to determine suitable candidates based on the high-level event information as provided by Semantic Lifting. The output are not final decisions, but options that subsequent components use to take decisions for each individual user, based on a range of factors prioritising different types of shots and the event/action concepts assigned to them.

Using a similar approach to Semantic Lifting, the component observes a real-time event stream to detect useful concepts. Detection triggers the addition or removal of candidates. A number of factors influence this decision, for example, the priority level of new candidates compared to existing options, avoiding options with large spatial overlap, keeping a balance over shot types, making sure a minimum number of shots is available.

Besides dynamically-detected shot options, static ones might be configured and available throughout the production, for example, fixed shots of generally interesting areas of the scene. The current list of shot candidates might also be made accessible to the production team through a dedicated user interface, such that useless candidates can be removed in real-time. One option for shot candidate visualisation that is especially useful for high-resolution panoramic cameras is to use semitransparent colour overlays.

Another critical issue is the duration for which shot candidates stay available, since this component itself cannot predict when subsequent components decide to start or stop using them. When a shot candidate is dismissed while it is used for at least one user, decision-making behaviour in subsequent components has to define how to react to this situation.

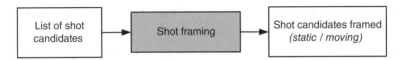

Figure 6.6   Shot framing prepares both static and moving shots.

6.4.3   Shot Framing

Since the shots used by a format-agnostic Virtual Director are typically virtual cameras, that is, crops from a high-resolution video stream, these shots need to be positioned automatically. The aim of Shot framing (see Figure 6.6) is to optimally frame shot candidates, i.e. to define the bounding boxes for static and moving virtual cameras in a way such that the action is covered well, respecting well-known cinematic principles of film production. The shot size and aspect ratio depend on the viewing device.

A Virtual Director's Shot Framing module might work for each playout device in parallel, as Figure 6.3 suggests, or it might be a general process preparing shots independently of the specific viewing devices they are used for. In the latter case, the aim of the Shot framing component is to define only the outer bounding box of a shot, and the specific size and aspect ratio are decided later by a Virtual Director instance that is aware of the specific playout device's properties.

When framing moving objects with virtual cameras within a large panoramic video stream, smooth camera pans are required to capture their movement in a visually-pleasing manner. The camera movement smoothing logic has to take the panorama boundaries into account. One option for that is to use a *spring model* for smoothing out minor movements and avoiding sudden stops, taking the object type, direction and speed of movement into account. As an example, a horizontally moving athlete could be positioned side of the image centre so that more of the running direction area is seen (*looking room*). Further, the bounding box size should depend on the distance of the object to the camera, that is, more distant objects are covered by smaller boxes so that they appear larger.

Another advanced option for shot framing for non-panoramic cameras is to steer the physical camera viewpoint itself. Pan-tilt-zoom (PTZ) cameras could be used, which can be steered remotely. One convenience of such cameras is that specific intelligence like automatic person tracking could be implemented on the camera itself. Another important advantage in comparison to panoramic video cameras is that the focal length is not fixed but can be changed dynamically, allowing full resolution to be maintained for tight zooms, and also allowing limited depth-of-field effects to make the subject appear sharper than the background. As a result of the complexity in logic required and inherent delay issues, however, automated control of physical cameras is difficult to implement in a real-time production environment.

> **Input:** The Shot Framing process receives a list of shot candidates, which are linked to metadata that allows their size and position to be optimally defined within the camera from which they are cropped. Note that not all shot candidates might have to be framed, since also manually-steered broadcast cameras can be used by a Virtual Director or the full shot size might be used for certain renderers. Example: A basketball player has been identified to dribble straight to the basket

from half court and a new virtual camera within a panoramic video stream was created to capture this action.

Output: The output of the Shot Framing process is an updated list of shot candidates where their position within physical camera streams is defined. Framing respects well-known cinematic principles, for example, the well-known *rule of thirds* (*golden section*) (see Bowen and Thompson, 2013). Example: A basketball player mentioned above is framed such that the person is fully covered throughout the action, and a certain amount of space is left at the top and bottom. Capturing the athlete from the side of the court, running to the right, the shot is positioned such that he is left of the centre until arriving at the basket for the layup attempt. The shot follows the player smoothly.

Frequency: Framing for moving virtual cameras needs to be updated every frame.

Approach: Iterating over the current list of shot candidates, new and moving ones need to be (re-)framed, taking into account the type of action they cover, and parameters of the shot framing rules as mentioned above. Bounding boxes are framed in parallel for each viewer, ideally taking the display's size and aspect ratio into account. For each shot candidate, completely different shot framings might be created, since, as a practical example, the maximum appropriate camera panning speed also depends on the display size in relation to the distance to the viewer, that is, a fast pan on a small hand-held screen might be fine whereas the same panning speed might be visually disturbing in a large cinema projection setup.

Virtual camera smoothing for close-up shots of moving people and objects needs to react to changes of the subject's speed and direction. The moving virtual camera also needs to stop smoothly at the boundary of the physical camera stream. The shot size may need to change as well if the distance of the covered objects or people to the camera changes. According to the proposed architecture in Figure 6.3, Shot Framing can be a parallel process to Shot Prioritisation.

6.4.4   Shot Prioritisation

Shot Prioritisation (Figure 6.7) might be a distributed process along multiple Virtual Director instances in the production chain. For example, the first instance can assign standard priority values to each shot candidate, according to the action it covers. Each action – ideally corresponding directly to one of the semantic concepts relevant to the specific production domain – should be configured with a default priority.

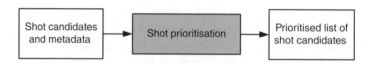

Figure 6.7   In an iterative step along the production chain, shots are (re-)prioritised based on various inputs to the decision-making process and to filter out un-usable shot candidates.

Each subsequent instance might update that priority value, respecting the preferences of the user and production grammar. Each instance in the production chain might be responsible for re-prioritisation based on one concrete aspect, for example, introducing a bias in shot selection for one of two competing basketball teams.

Input: All available shot candidates, independent of users, with underlying metadata. Example: A close-up of a certain athlete which can be of both high and low priority to a given viewer, depending on individual preferences.

Output: An updated list of shot candidates with metadata and updated shot priorities per user, that is, per playout device. Example: A user chooses to follow her favourite player closely to learn about the player's unique skill-set. All shots framing that player are assigned very high priority, especially close-ups. The priority is influenced by the ratio between the shot size and the size of the player's bounding box that is provided by the person tracking component (see Chapter 5).

Frequency: Prioritisation has to be calculated for every playout device independently. As appropriate, priorities might be updated every frame or whenever significant metadata changes occur.

Approach: Default priorities per shot type should be configured in the domain model, depending on the action that is framed. Any priority scale might be used. Any number of re-prioritisation iterations might be implemented along the production chain. This component operates for each user in parallel and should be able to take individual preferences into account, which is a key step in enabling content consumption personalisation.

For each Virtual Director instance in the chain, the priorities are re-calculated according to the instance's purpose, while keeping previous decision factors for future instances. Depending on the production setup and domain, any aspect might have an influence. Some options are illustrated in Figure 6.3. For certain viewers in a given delivery network, shots might be filtered, for example, because they show inappropriate content such as tight zooms onto the audience, or because content rights do not permit that shot to be shown.

6.4.5   Decision Making

Decision making (Figure 6.8) is the final process within a particular Virtual Director instance, as suggested in the architecture proposal of this section. It prepares decisions and sends information down the production chain. The last instance in the production chain is responsible for sending final instructions to the renderer. Information packets sent along the chain

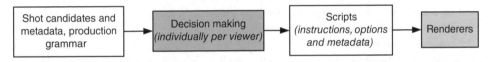

Figure 6.8   Decisions are taken at the end of the production chain, to inform the renderers.

of Virtual Director instances and to other components are referred to here as (production) *scripts*. Intermediate Virtual Director instances might require a separate subcomponent for pre-processing scripts from previous instances (see Figure 6.3). The same component might be responsible for pre-processing user interaction or updates to the user profiles.

The core task of a Virtual Director is to automatically take real-time decisions: when to cut, how to cut, and to which shot candidate to cut. Since the concept of a format-agnostic approach is to produce a number of audio-visual content streams in parallel, this process naturally has to work in parallel for each viewer (decisions per playout device).

Decision making is typically the most demanding task in the process, since automatic execution of cinematic rules is still very much a research topic and commercial state-of-the-art solutions are generally limited. This subcomponent exploits inputs prepared by other components to create a real-time edit, therefore it depends on their quality as well. A lot of factors are involved that might affect the audio-visual *quality of experience* strongly. Besides the quality of decisions, however, the speed in decision making is of utmost importance, since in a real-time format-agnostic broadcast system, decisions have to be taken fast and with constant low delay. Aspects and options for this have been discussed in Section 6.2.

Input: Shot candidates accompanied by comprehensive metadata on the one hand, and the production behaviour on the other, the latter typically part of the domain model. Example: A list of shot candidates, one of which is currently shown to a particular user.

Output: For each user associated with a certain device, decisions on live content presentation are taken to be sent towards the corresponding renderer. Production scripts contain decisions, options, metadata, and concrete instructions for renderers and playout devices. Example: Triggered by the occurrence of an event with very high priority in relation to other available options, an instruction to cut to a close-up on a certain athlete is issued for a particular viewer.

Frequency: Scripts that are sent to subsequent Virtual Director instances (cf. Figure 6.3) are typically sent every frame. Instructions to renderers are sent whenever necessary, that is, a new command like a change in viewpoint angle is issued.

Approach: The decision-making step, which is executing the production grammar, processes incoming information together with the production behaviour that defines how the Virtual Director decides. Any technical approach might be implemented, for example, a rule-based approach as suggested in Section 6.2.

The Decision Making component decides not only to which shots to cut, but also when to do so. The behaviour can be triggered by the occurrence of high-level events, or state changes that are specific to each viewer. It might also be triggered by cinematic rules, for example, based on a timer to make sure that no shot lasts longer than a certain number of seconds, as defined for this shot type in the domain model.

By analysing user profiles and by interpreting user interaction throughout the production, personalisation of content selection can be achieved in whichever form the production grammar defines.

6.5   Behaviour Engineering: Production Grammar

This section presents how a Virtual Director's production behaviour can be defined as a set of rules/principles, in order to be executed by a rule engine, ideally in combination with an event processing approach as suggested in Section 6.2.2. Part of this section assumes such an approach has been chosen. Its aim is to describe how capturing production principles from human experts can be conducted. It presents the cinematic techniques and shot types a Virtual Director component might implement in detail. It also highlights the possibility of steering audio playout, and presents selected issues in production grammar implementation.

Production grammar, a Virtual Director's behaviour definition, is not trivial to develop, since it involves automating many aspects that a human production team usually specialises in. As mentioned before, one of the key advantages of the Virtual Director approach is that it can help to realise many novel and advanced personalisation features. In a format-agnostic approach where virtual cameras are framed within high-resolution panoramic cameras, this is especially attractive. See Figure 6.9 for a simple illustration of simultaneous shot candidates.

6.5.1   Production Knowledge Elicitation Process

It is generally a good idea to get first-hand input from production professionals. This was the approach taken in the design of the Virtual Director in Kaiser, Weiss and Kienast (2012), where, through observations and interviews with production staff, a set of rules has been captured. Universally valid cinematic principles can be found in literature. Section 6.5.2 will discuss some examples. However, when crafting the production grammar for a certain production, it has to be evaluated in detail which rules apply and which not, which work together well and which do not lead to a pleasing result in a certain context.

If available, observing existing broadcast productions is a good way to start. By understanding manual framing and cutting decisions, their aesthetic balance and intended storytelling

Figure 6.9   Shots defined as virtual cameras on a high-resolution panoramic video stream. This example illustrates one of the benefits of using panoramic cameras for a Virtual Director, where users might watch different parts of the scene. Reproduced by permission of the Berlin Philharmonic Orchestra.

function, such production grammar could be automated step by step. However, technical people implementing a Virtual Director system typically do not have enough understanding of production grammar to derive the desired behaviour themselves. When engineers speak to broadcast directors, a problem known as the *knowledge acquisition bottleneck* (cf. Wagner, 2006) needs to be kept in mind. The vocabulary of knowledge engineers (technology experts) might be very different from the domain experts' language. By making all people involved aware of this issue in advance, by assisting the knowledge capture process with tools and by introducing feedback cycles, this issue can be minimised (cf. the approach of Gordon and Harel, 2011).

Parts of the production grammar potentially are directly related to the interaction design of the overall format-agnostic system. Since Virtual Director behaviour is partially triggered by user interaction, the choice of methods and abstraction levels, through which users interact with the system, has to be carefully designed as well. An important aspect in this process is to involve users at an early stage, to assess their expectations, understand their needs and learn how they would use the system.

Regarding the formal representation of production grammar, ideally in a textual format that both humans and machines can read, a widespread interoperable standard unfortunately does not exist. More on this is discussed in Section 6.7.

6.5.2   Cinematic Techniques

Cinematic techniques (see Arijon, 1976; Brown, 2002; Ronfard, 2012; Bowen and Thompson, 2013) refer to methods and common conventions in video, film and TV productions. The term *cinematographic* is sometimes used synonymously, as varying definitions exist. These techniques cover creative methods that can be used by film makers to communicate meaning, to entertain and to evoke a particular emotional or psychological response by the audience. This includes for example lighting, using depth of field, focus, camera position, camera movement, framing, special effects, shot cutting effects. Subsequently, we discuss a selection of cinematic techniques that a Virtual Director can automatically execute.

Shot Types by Size

A shot (Bowen and Thompson, 2013, p. 8–24) can be of any duration and is defined by the distance between the camera and the subject (in relation to the subject size). It defines the ratio of the size of the visible part of the object to the total area in the shot. Shot types by size may be interpreted differently in different setups. For example, a camera operator would use a different focal length for a wide shot when showing a landscape than for showing a wide shot of a theatre stage. An overview of the shot taxonomy suggested in the following is depicted in Figure 6.10, with examples from a soccer match.

Automatic shot classification has also been investigated in research, see for example the shot taxonomy classification approach in Wang and Cheong, (2009). For additional information, please refer to Section 2.4.3.

> An **extreme long shot** (see Figure 6.10a) is taken from distance where the view is so far from the subject that they are not visible. It is often used as an 'establishing shot' to show the audience where the action is taking place and shows, for example, the landscape and the surroundings of the scene.

Figure 6.10 Examples for shot types capturing people, taken from the same event, using two different cameras: (a) extreme long shot; (b) wide shot; (c) medium-long shot; (d) medium shot; (e) close-up.

In a **wide shot** (Figure 6.10b) the full height of the subject is shown and the shot also shows a lot of the surrounding area. In this example of a soccer game, the wide shot shows several players.

A **medium-long shot** (Figure 6.10c), also known as a full shot, is between a wide shot and a medium shot. The example covers two or more players.

A **medium shot** (Figure 6.10d) shows some part of the subject in more detail for example, a human body from the knees/waist up. It is often used for dialogue scenes.

The **close-up** (Figure 6.10e) shot is used to focus on certain features or parts of the subject which take up most of the frame while showing very little background. It is a very intimate shot and highlights the importance of details, possibly taking the viewer into the mind of a character.

An **extreme close-up** can be used to show only small parts of the subject, for example, the mouth or eyes. This shot needs extra care to produce, and is used rarely, for very dramatic effects.

Shot Types by Angle

The angle of a shot (Bowen and Thompson, 2013, p. 45–61) refers to the direction and height from which the camera records the scene. It gives emotional information to the audience. The more extreme the angle is, the more symbolic and 'heavily-loaded' the shot might be.

The **bird's eye view** shows a scene from straight above. It is an unnatural and strange point of view but is very useful in sports and documentaries.

A **high angle** also shows the subject from above but is not as extreme as the bird's eye view. The subject nevertheless appears smaller and less significant.

The **eye level** shot is most common view and natural shot where the camera shows the subject as humans typically see it.

The **low angle** shows the subject from below and tends to make them appear more powerful.

Slanted or **canted angle**, also known as **Dutch tilt**, is when the camera is not placed horizontally to floor level. It suggests imbalance or instability and creates an emotional effect.

Camera Movements

A story can be told by a series of cuts from one shot to another or to move (Media College, 2013; Wilson, 2012) the camera (Bowen and Thompson, 2013, 165–181). The basic methods of moving a camera are:

A **pan** is a movement that scans a scene horizontally. To achieve a stable image, the camera might be mounted on a tripod.

The vertical movement of a camera is called **tilt** where the camera points up or down from a stationary location.

In a **dolly shot** the camera is mounted on a moving vehicle and moves alongside the action.

For **hand held** shots, the camera is usually held on the shoulder, or it is mounted on a steadicam to stabilise it. This kind of equipment gives the camera operators lots of freedom in moving the camera.

Crane shots allow the camera to move up, down, left, right, swooping on the action or moving diagonally out of it.

A **zoom** is technically not a camera movement but produces an effect similar to moving the camera closer to the subject by changing the focal length.

6.5.3   Audio Scripting

Even though a Virtual Director is a true multimedia application, as it deals with both audio and video content, a strong bias in research towards the visual aspect can be observed. This chapter is focuses on automating camera framing and cutting first and foremost, however, this also implies that a lot of research is yet to be conducted regarding the audio aspects. While it is quite common to use results from audio analysis to inform a Virtual Director, few examples are known that intelligently steer audio playout in innovative ways, a process often referred to as *audio scripting*. Notable exceptions can be found, for example, in the related domain of group video conferencing.

What makes this situation so interesting is that there is a lot of value in high-quality audio not only in video communication, but in live event broadcast in general. The reasons are not entirely clear. One explanation attempt would be to look at the differences in audio playout systems in domestic environments. Their capabilities differ from each other more strongly than is the case for video, where almost any 2D video stream can be rendered by consumer devices such as a computer screen, projector or TV set.

As a simple first step in a format-agnostic scene capture approach, the viewers' quality of experience could be enhanced by intelligently matching the audio feed with the visual viewpoint. In an example of sports broadcast, if there is an audience shot visible after a successful score, the audio should correspond and loud cheers should be heard. Thinking of personalisation, viewers may want to hear the fans of their favourite team more than those of the opposing, potentially leading to a more immersive viewing experience.

As a general principle, the visibility of objects should correspond to their audibility. Objects that are currently not visible may not be audible at all or at a reduced level, depending on the distance and other objects between them and the viewpoint. Audio close-ups might become problematic though, as they may interfere with the privacy of a player-coach discussion, or may cover inappropriate language of players and so forth.

6.5.4   Domain Model

According to the approach recommended in this chapter, a Virtual Director's behaviour and configuration is defined in the so-called domain model. We recommend encapsulating all production/domain-specific information, including the production grammar, in a single exchangeable model. For storing and exchanging this information, we suggest exploiting the advantages of ontologies. Since no widely accepted standard exists, a flexible data model can facilitate information exchange between components.

A practical question that needs to be solved when implementing a distributed Virtual Director is how to propagate changes in the configuration or model, for example, through direct user interaction, to all relevant instances along the production chain. In a complex production setup, information flow up the production chain is almost unavoidable, even though it appears to be a good design decision to avoid it as much as possible because of delay implications.

Apart from the production grammar itself, the following paragraphs briefly discuss some suggested key concepts that a Virtual Director could define in its domain model.

Shot types: A shot-type model defines all shot types used by a Virtual Director, including metadata such as the basic priority, basic bounding box size around enclosed high-level concepts such as a person. These definitions are essential for a number of subprocesses, according to the approach proposed in Section 6.4. It most notably concerns Shot Candidate Identification, which decides when shot candidates should be created or changed to cover a certain action. The shot type model might also define relations between shots, such that for example the type of a shot can change over time. It further might define fading effects used when cutting from one shot to another.

Static shots: It appears to be useful to define a set of static shot candidates for each production. They can be framed very well manually, and can always be kept available as fall-back views, for example, a close-up shot of the goal area in a soccer production. The concrete shot size and aspect ratio depends on the configuration of the viewer's playout device.

Dynamic shots: Besides static shots, the behaviour of moving shots can be pre-defined in the domain model. For panning and zooming movements, a set of parameters such as speed, size and movement smoothing behaviour should be defined. Dynamic shots are triggered by the occurrence of certain high-level concepts in a certain context. Their concrete behaviour might be adapted through external information sources, for example, the panning path might be informed by a person tracker. An example would be a pan of two seconds across the local basketball match audience that is shown after an action of type *slam dunk* has been detected, that is, an action with very high priority took place.

Content information: Basic information about the cameras and microphones (resolution, aspect ratio, etc.) is needed in a Virtual Director's content selection processes. Typically, the information can be considered fixed for the duration of a broadcast, apart from the position and orientation of movable microphones and cameras.

6.5.5   Limitations in Rule-based Behaviour Engineering

Assuming a rule-based approach, the following discusses selected key issues that Virtual Director behaviour engineers might encounter when designing a complex production grammar rule-set. Further aspects are reflected in the Section 6.7.

A Virtual Director's rules need to define both the pragmatic scope of how to capture domain-dependent actions, and the cinematographic scope of how to do that in a visually aesthetic manner. However, a key issue is that, in general, production rules are not independent. The engineering effort necessary to structure their interplay and to balance their effects and side-effects is considerably high. Competing and contradicting principles need to be resolved so that

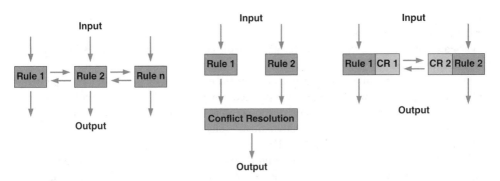

Figure 6.11 When several rules are not independent, some form conflict resolution needs to be implemented. This figure illustrates the problem and two options for its resolution: (a) Multiple rules fire in parallel and interfere with each other (left part of the diagram). (b) One option is to add separate rules for conflict resolution. This might lead to a certain hierarchy of rules in the rule-set. One of the disadvantages of that solution is that this layering is typically quite static (middle part of the diagram). (c) Another option is to include logic for resolving the dependencies in the actual code of the rules themselves, which might turn out to make it difficult to maintain the code engineering-wise (right part of the diagram).

the desired decisions are made, which is especially challenging in this context. As Figure 6.11 illustrates, resolving contradicting rules is a serious implementation issue.

A simple example for contradicting rules would be a situation where a cut to a certain virtual camera happened three frames ago. If at that time a high-priority action appears, the question is how to react. While one rule would define that a cut to a virtual camera covering that action appropriately shall be issued immediately before the moment is lost, another rule could make sure cuts do not happen too frequently, for example, not within two seconds. In that case, the behaviour engineer might introduce a static hierarchy that makes sure the two-second rule is never violated, even though important actions might be missed.

To get useful output from the event processing components, it is necessary to process events in the order they occur (Etzion and Niblett, 2010). However, since the Production Scripting Engine (PSE) is informed by several distributed audio-visual analysis components, it has to assume that detected events arrive out-of-order. This is because every machine has its own clock that is used to assign a timestamp to generated events and the clock might be out of sync in comparison with the other machines. Furthermore, the detection of an event may be delayed due to thread scheduling of the operating system, and the transmission of data over the network also takes varying times. JBoss Drools, for example, is able to process events retrospectively, which means that when it receives a delayed event, it may process it in the correct temporal order. However, it relies on timestamps in order to be aware of the real-life sequence of events. Technical approaches to address the problem have to be carefully chosen in the context of a Virtual Director's real-time requirements.

6.6 Virtual Director: Example Prototype

To give more insight into practical design and implementation issues, this section discusses a number of selected aspects of an example Virtual Director implementation, the PSE, presented in Kaiser et al. (2012). This Java software component is a research prototype that has been built as part of the FascinatE (FascinatE, 2013) research project and is embedded in this format-agnostic production system.

The FascinatE project investigates a format-agnostic approach (Niamut et al., 2013) and aims to create an end-to-end system for interactive TV services. It allows users to navigate in a high-resolution video panorama, showing live or recorded content, with matching accompanying audio. The output is adapted to the viewing device, covering anything from a mobile handset to an immersive panoramic display with surround sound, delivering a personalised experience. For lean-back consumption, the PSE automatically selects camera views.

Alongside regular broadcast cameras, FascinatE uses a high-resolution panoramic camera with a 180° horizontal field of view, the OmniCam (Schreer et al., 2013). This camera is key to a scene-capture approach and allows the PSE to frame virtual cameras as static or moving crops of that panoramic video stream. The PSE has been built as a software framework, and production grammars have been developed for three example production domains.

As a continuous process, the PSE fuses the sensor information streams it receives and produces several live content streams for different playout devices in parallel, enabling new forms of content production and consumption beyond straightforward personalisation. In contrast to classic TV, different viewers may choose to watch different parts of the scene, framed in their style of choice.

The PSE is based on an event processing engine, an approach chosen mainly as a result of performance reasons based on previous experiences, as it is required to react to real-time input streams and needs to take decisions with very little delay. A number of its subcomponents take a rule-based approach, that is, most of its logic is executed by a rule-engine, which is used as a library/framework. The PSE's behaviour is defined by a set of domain-dependent production principles, implemented in a format specific to the rule engine. These principles define how the PSE is automatically framing virtual cameras within the omni-directional panorama, how camera movements are smoothed, when cuts and transitions to other cameras are issued, and so on.

As part of its workflow, it fuses different sources of information as the basis for taking decisions. The PSE is informed by several other software components. It is tightly integrated with a user interface for the professional production team, the *Editor UI* toolset. They allow creating live annotations for concepts not covered by audio-visual content analysis, creating virtual cameras that can be both fixed and moving in a smoothed manner.

Regarding automatic audio-visual content analysis, independent of the production domain, a person-tracking module as described in Kaiser et al. (2011) informs the PSE about the location of people in the scene. Another content analysis component is the extraction of a visual saliency measure for regions and the detection of scenario-specific events. Implementation details for those components can be found in Chapter 5.

This Virtual Director implementation is designed to support replay functionality by communicating with interfacing components to trigger recorded content transmission, and to retrieve metadata stored in a semantic datastore, the *SLMS*. The PSE does not keep a long-term memory itself for performance reasons, in the sense that it investigates a temporally limited time window to decide quickly what to do. It mostly depends on information regarding the present. To overcome the limitation of losing past information, it communicates with its semantic datastore in both notification and query form.

In a nutshell, the PSE's production grammar implements a set of pragmatic rules (capture what is important) and aesthetic rules (when and how to cut between cameras). In that realm it implements a set of cinematographic principles and takes decisions on camera selection within real-time constraints. The prototype system only handles virtual cameras that are crops from the high-resolution OmniCam panoramic video stream, however, it could also access content streams from other cameras.

The remainder of this section discusses selected technical implementation aspects.

6.6.1   Architecture and Software Framework

The Production Scripting Engine is a research prototype that has been implemented following the approach and recommendations discussed earlier in this chapter. Its architecture in a simple single-instance setup is abstractly depicted in Figure 6.12.

The underlying aim behind this work was to create a generic software framework that could be quickly adapted to different production genres and complexities. Therefore, a strong design principle was to allow re-configuration of the system without changing the framework itself, that is, to enable new instances to be added along the production chain, to enable new content analysis modules to be plugged in seamlessly, and to allow the domain model, comprising the production grammar for pre-processing and decision making, to be exchanged easily.

Since a format-agnostic production system takes customised decisions for many users in parallel, the number of different decisions and hence different video streams tends to explode. A Virtual Director's collaboration with a content transmission optimisation component is therefore crucial. One option to address that issue is to group decision-making paths as much as reasonably possible, that is, to group along delivery networks and content right configurations, playout device types, and user-specific preferences.

The PSE prototype comprises a number of utility components, some of which are explained in the following. For communicating with other FascinatE components such as content analysis,

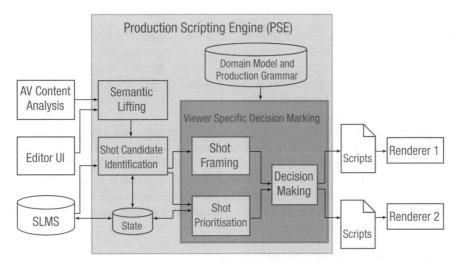

Figure 6.12   Simplified architecture of the Production Scripting Engine. The approach fits the suggestion illustrated in Figure 6.3.

the Internet Communications Engine (ICE) (ZeroC, 2013) is used. It is a highly efficient middleware and supports a variety of programming languages such as Java and C++.

With the exception of the production scripts, most information that is transmitted between components is based on the MPEG-7 standard (MPEG, 2001). Choosing MPEG-7 for that purpose implies certain disadvantages, as the overhead in comparison to the payload is quite high, and many developers consider it impractical. However, other metadata standards have specific disadvantages as well. Overall, MPEG-7's benefits for exchanging metadata in a research-heavy domain stand out, and therefore we recommend using it to design Virtual Director interfaces.

To bind the MPEG-7 XML document to an object oriented hierarchical class tree, the PSE uses Apache XMLBeans to create a binding based on an XML Schema. To be specific, it uses the MPEG-7 Audio-Visual Description Profile (AVDP) (see MPEG, 2012). The main functionality of AVDP is to describe the results of automatic content analysis with low-, mid- and high-level features for audio and video content. Typical application scenarios would be action detection, face tracking or speech recognition.

Another useful utility component in a scene-capture approach is coordinate transformation. Between the content from multiple cameras and microphones, and their corresponding content analysis components, different coordinate systems are used in the FascinatE system. Content analysis for the OmniCam, for example, processes the six panorama tiles independently before fusing results into information about the whole panorama. In contrast, the system's renderer regards the panorama as six independent projections into the cylindrical space, and the remote viewpoint selection interface accepts vectors that are basically offsets to the centres of the tiles. In this context, the PSE uses a coordinate system over the OmniCam panorama with the top left corner as reference point. A coordinate transformation component maps the different coordinate systems to a single one that is covering the complete video scene. This enables it to apply spatial reasoning over different information sources.

6.6.2 Production Scripts

Another key design decision to make when implementing a Virtual Director is the data format of the decisions it emits. The PSE's output is referred to as *production scripts*. Their core content are options of (virtual) camera views for further decision making, and ultimately decisions that instruct the renderers. Scripts are quite rich in metadata, such that the receivers can reason with the events that led to a certain decision. Scripts are sent quite frequently, at almost every frame, containing updates for each user.

Information is sent from PSE instances to subsequent instances along the production chain (further refining of decision through re-prioritisation of options) and to the renderer as instructions. It would also make sense to send scripts to the content transmission optimisation component, to assist the preparation and optimisation of audio-visual content streams. Receivers of scripts parse those parts that are relevant to them, but usually do not send information back to the sender in a direct response. The PSE's scripts are encoded as XML snippets. The prototype PSE scripts are designed to contain most notably:

- Configuration and metadata on cameras, users;
- Decisions on shot candidates to be selected for a certain playout device type, group of viewers or individual viewer;

- Shot candidates for subsequent selection, including priorities, metadata, high-level concepts contained;
- Additional metadata relevant to the decision making process;
- Renderer-specific instructions.

6.6.3 Behaviour Implementation

Using a rule-based approach, the production grammar typically consists of rules that can be regarded in isolation. However, it is important to understand that a Virtual Director's behaviour is a result of the interplay of all rules.

Rules defining visually aesthetic panning behaviour, for example, cannot simply be added as a separate rule on top of an existing rule-set, but have to be interwoven to a certain degree to achieve a well-thought-out balance overall. Behaviour engineering typically requires an iterative process where rules have to be added/modified and thresholds have to be adjusted. Decision making depends highly on the quality and semantics of its content analysis and manual annotation sensors.

The PSE prototype's rule-base consists of both general and production-specific rules, however, all rules have to be configured to a certain degree to fine-tune the system's behaviour. To what extent these issues can be avoided by more clever approaches remains to be investigated in future research activities. A concrete example for domain-independent behaviour is illustrated in Figure 6.13.

The PSE's behaviour is defined as a set of production rules, which a rule engine can execute. The PSE's implementation uses JBoss Drools, which can be seen as both an event processing engine and a rule engine. Its advantageous processing performance in the context of real-time

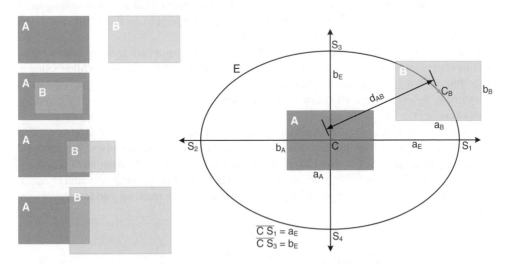

Figure 6.13 Illustration of the PSE's logic that decides to use either a hard cut or a transition when switching from shot candidate *A* to shot candidate *B*. Depending on the shot's relative size, distance, moving speed and direction, and spacial relation (one rectangle comprises the other fully or partially), the decision is taken.

requirements has been discussed earlier in this chapter. To create moving virtual cameras, the capabilities of the Java Timing Framework (Timing Framework, 2013) are exploited to achieve more natural animations. Furthermore, to smooth camera movements, a combination of de Boor's algorithm for spline curves in B-splines and a damped spring algorithm are used (de Boor, 2001).

The PSE's behaviour consists of several sub-aspects, of which the most important are:

- Deciding where to place virtual cameras as shot candidates in the panoramic video. They might be static or moving as pan, tilt and zoom, have a certain static or changing size (type of shot, e.g., close-up), and are moving at a certain speed. The latter properties might depend on preferences of individual viewers.
- Deciding when to drop shot candidates since the action they cover is no longer relevant or covered by duplicate virtual cameras.
- For each viewer (group) individually, deciding at which point in time to cut from one virtual camera to another.
- Deciding how to perform those cuts: depending on the location and content of the two virtual cameras, either a hard cut or a transition is decided. An example for the PSE's logic for that is illustrated in Figure 6.13. Several types of fades might be used, as far as they are supported by the renderer.
- If in addition to a panoramic video one or more broadcast cameras are available, deciding when to use those sources, which offer greater level of detail.

In order to define the PSE's intended behaviour, the engineers observed TV broadcasts and interviewed professionals. The production rules were initially captured in natural language before they were translated to JBoss Drools rules. Naturally, rules can only be triggered by an event that is directly observable or can be derived through Semantic Lifting. An exchangeable domain model defines these higher-level events, and also the primitives, that is, the low-level events that are automatically extracted or manually annotated.

Since virtual cameras are used heavily in a format-agnostic approach as crops within high-resolution panoramic video, the quality of automatic shot framing is crucial. While the logic of the PSE's *Shot Framing* sub-component is kept quite simple, it could be improved easily, for example, by executing further cinematic principles for camera animation, or by avoiding people being cut in half at the edge of the shot.

In general, the PSE's decision-making approach can be seen as a proof of concept for the approach suggested in this chapter. Even though this research prototype has obvious limitations, from a technical approach it is fast and scalable enough, the behaviour engineering process proved to be sufficiently efficient, and the decision quality that such a Virtual Director implementation can achieve appeared to be very promising, at least for certain production domains.

6.6.4   Production Grammar Example

The following discusses technical details of an actual Production Scripting Engine behaviour example taken from the implementation of this Virtual Director's decision-making process. The rules illustrate how to implement the cinematic constraint to stay no shorter than two seconds on a shot, and how to handle event triggers continuously in that case to define camera cutting behaviour.

The rules in the following example use the Drools rule language. Drools is a rule engine for executing forward and backward chaining rules and it includes event processing capabilities allowing temporal reasoning. Rules represent the logic of an application and facts and events are the data to process. Technically, facts and events are objects in the working memory. Facts, for example, can be used to save a state of the application. Events are transient, and they have temporal relationships that facts do not have.

The subsequent rules use the forward chaining principle and therefore they consist of a *when* part and a *then* part. Observing the continuous event stream, when all conditions in the *when* part can be evaluated to *true* then the *then* part will be executed. See also the Drools documentation (Drools, 2013) for further explanation of the rule language.

In the example, rules are triggered by inserting event objects into an event stream named *CameraDecisionStream*, which is performed by software components in the earlier workflow. The output of the rules presented in the following on the other hand are *cut-to-camera* events that obviously trigger further action.

By default, the system is ready to cut to a new shot. This state is expressed by a *FreeTo-CutFact*, which implies that an *ActiveCameraEvent* can be issued. This is implemented by the rules in the Listings 6.1 and 6.2. The first rule covers shots of the type *FIXED*, *MOVING* and *ZOOMING*. These shot types have individual expiration durations configured in the domain model. Apart from that case, the rule in Listing 6.2 handles predefined *STATIC* virtual cameras. This type of virtual camera does not have an expiration duration configured in the PSE implementation. It therefore is statically assigned a minimum duration to block new camera cutting decisions. This is implemented by retracting the *FreeToCutFact* and inserting a *Free-ToCutLetGoEvent*. The new *FreeToCutLetGoEvent* causes firing of the rule in Listing 6.3 after two seconds which inserts a new *FreeToCutFact* in turn.

Listing 6.1   Pass through rule 1

```
rule "pass through proposed cut - 1"
 when
  $ace : ActiveCameraEvent() from
   entry-point "CameraDecisionStream"
  not(ActiveCameraEvent(this after $ace,
   this != $ace) from
   entry-point "CameraDecisionStream")
  CameraFact(id == $ace.id,
   virtualCameraType == VirtualCameraType.FIXED ||
   virtualCameraType == VirtualCameraType.MOVING ||
   virtualCameraType == VirtualCameraType.ZOOMING)
  $f2cf : FreeToCutFact()
 then
  // Pass through the active camera and remove the
  // FreeToCutFact.
 end
```

Listing 6.2   Pass through rule 2

```
rule "pass through proposed cut - 2"
 when
  $ace : ActiveCameraEvent() from
   entry-point "CameraDecisionStream"
  not(ActiveCameraEvent(this after $ace,
   this != $ace) from
   entry-point "CameraDecisionStream")
  CameraFact(id == $ace.id,
   virtualCameraType == VirtualCameraType.STATIC)
  $f2cf : FreeToCutFact()
 then
  // Pass through the active camera, remove the FreeToCutFact
  // and insert a FreeToCutLetGoEvent. This behaviour is
  // intended for static virtual cameras which do not have an
  // expiration time.
end
```

Listing 6.3   Reset free to cut rule

```
rule "reset free to cut after 2 seconds"
 when
  $ftclge : FreeToCutLetGoEvent() from
   entry-point "CameraDecisionStream"
  not(FreeToCutLetGoEvent(this after[0, 2000ms] $ftclge,
   this != $ftclge) from
   entry-point "CameraDecisionStream")
 then
  // Insert a new FreeToCutFact.
end
```

The case when the system is not ready to cut, which means that there is no *FreeToCutFact* at the given time, is covered by the rule in Listing 6.4, which creates a new *NextCameraFact* and retracts the corresponding *ActiveCameraEvent*. This creates a list of *NextCameraFacts* in the working memory as a backlog to cut in the future when the system is *free-to-cut* again. The process of picking up a stored notification is done by the rule in Listing 6.5. It automatically fires when a *FreeToCutFact* gets inserted into the knowledge base and when there are one or more *NextCameraFacts*. The rule selects the next virtual camera with the highest priority where the expiration value, in an interval between 0 and 1, is bigger than 0.5.

Listing 6.4   Wait until free to cut rule

```
rule "wait until free to cut"
 when
  $ace : ActiveCameraEvent() from
   entry-point "CameraDecisionStream"
  not(ActiveCameraEvent(this after $ace,
   this != $ace) from
```

```
  entry-point "CameraDecisionStream")
 not(FreeToCutFact())
then
 // Create a new NextCameraFact and remove the
 // ActiveCameraEvent.
end
```

Listing 6.5 Delayed cut rule

```
rule "delayed cut"
 auto-focus true
 when
  $ftcf : FreeToCutFact()
  $acf  : ActiveCameraFact()
  $ncfs : ArrayList(size > 0) from
   collect ( NextCameraFact() )
 then
  // Select the best suitable camera based on its priority
  // value and expiration time. Remove all NextCameraFacts.
end
```

6.7 Conclusion

This section summarises the chapter's insights on the the state-of-the-art of Virtual Director research, as well as the related practical challenges, and technical approaches to software frameworks and behaviour engineering.

6.7.1 Summary

This chapter presented the concept of Virtual Director technology for automating view selection in the context of a format-agnostic approach for live broadcast. Typical features and requirements were discussed. An overview on related research activities was given, covering multiple application areas and very different approaches to the challenge of implementing such a system. A number of technical approaches for behaviour engineering have also been discussed, and the recommendation to utilise event processing capabilities in combination with rule-based production grammar was explained. A generic architecture and information workflow for Virtual Director software systems was suggested, and typical subprocesses were investigated from a practical point of view. Another section highlighted the task of engineering a Virtual Director's behaviour: its production grammar. Finally, detailed aspects of a Virtual Director prototype have been elaborated.

6.7.2 Limitations

One core challenge when designing and developing a Virtual Director is to process an event stream under real-time conditions and to take decisions based on a complex, predefined and exchangeable set of rules (production grammar). Several delay factors have to be respected and may affect the experience. Beyond the pragmatic aim to show the most relevant actions, production grammar needs to ensure that aesthetically pleasing camera selection behaviour is applied. Automation in that regard is a multifaceted research challenge and far from being considered solved.

It has to be kept in mind that the information available to the Virtual Director, which corresponds to the level of understanding about the current situation, is rather limited compared to what human operators are able to perceive. In that sense, limiting factors are both the quality and the number of different automatic content analysis components available. In any case, content analysis sensors cause a certain amount of delay, often on the critical path of the workflow. To make up for that to a certain degree, semi-automatic approaches should be considered where full automation can be avoided, for the sake of overall quality. Producers are in need of capable tools for interacting with a Virtual Director during the live production.

In cases with potentially many interesting actions taking place in parallel, the Virtual Director might be required to automatically overrule user decisions to make sure highly important events are not missed. To take such decisions fast enough, this may in turn require basic prediction functionality that can also lead to improper behaviour at times.

Another limitation is the elaborate process of production grammar engineering, where new approaches, design patterns and authoring tools need to be developed to assist the engineers. The issue with interdependencies between rules, that is, conflicting and contradicting principles, has previously been mentioned. New methods are needed to test the interplay of rules and their effect on the system, and ultimately on the user experience. To help reduce the effort required for production grammar engineering, effective rule re-use is another current limitation that needs to be investigated further.

6.7.3   Testing and Evaluation

Evaluation of a Virtual Director's performance is inherently difficult. There is no objectively right sequence of decisions, since there is an infinite number of appropriate ways of how to broadcast an event. Some decisions can be regarded as objectively wrong in isolation, however, in a sequence of decisions, they might be acceptable given the temporal context of other options and decisions.

The extent to which a Virtual Director system meets its requirements can only be answered in the context of the concrete production's quality demands. Users may take time to get used to novel production formats based on special cameras and format-agnostic approaches. Using panoramic cameras, for instance, has cinematic advantages but also comes with limitations. Users are used to a certain cinematic language. Regarding evaluation, the extreme example of using a single viewpoint only during a whole production would expose limitations that impede direct comparison to professional TV.

The central issue with evaluation is that there is no ground truth, no single objective camera selection that can be considered ideal and can be compared against. Another question is how to compare research results. Since standardised definitions and measures are lacking, results are difficult to compare with each other. Comparing on the level of individual decisions is also problematic because of the decision's temporal dependencies. In practical terms, setting up qualitative evaluation experiments is an elaborate and expensive process where a lot of effort is necessary to limit outside factors.

Evaluation of a Virtual Director can be performed on several levels. What is appropriate strongly depends on the production, the required quality of the automatic behaviour, the production system capabilities, the interaction possibilities, and so on. Examples for evaluation are:

- The performance and scalability of individual subcomponents.
- Compare the automatically produced content with content produced by professionals.
- Compare alternative technical approaches or alternative production grammar sets regarding certain indicators (delays, precision/recall, etc.).
- Evaluate the user experience with real test users as subjective evaluations.

One key requirement for Virtual Director systems is to take decisions in real-time. Therefore, an evaluation process has to take delays of individual components and of the whole workflow into account, at least of those processes on the critical path regarding delay. On the system level, it has to be ensured that delays do not cause components to lose sync. Further, it has to be ensured that delays are not unnecessarily propagated through the workflow or have detrimental effect on the overall user experience when it can be avoided.

On the software level, iterative unit-testing of components helps to improve quality and helps to assess the Virtual Director's performance regarding both the correctness and speed of execution. If not possible any other way, recorded or synthetical dummy data can be used. Stress tests (e.g., processing an extremely high number of low-level input events) can be used to test the stability and scalability of the Virtual Director. Regarding underlying event processing frameworks, making general performance statements over different implementations is difficult due to different requirements of applications and outside influences on performance metrics. In a rule-based environment, the performance metrics are mainly influenced by the number of events per time unit, the number of attributes of an event, the number of rules in the rule-base, the complexity of the rules, the current number of entities in the working memory, and so on. To facilitate the testing process, additional tools for monitoring the state of the system and for visualising inputs and decisions are recommended.

Important performance indicators of a Virtual Director regarding real-time requirements are the average latency and the maximum latency of decision-making components. However, we cannot straightforwardly measure the latency of the whole processing chain. The reasons are (i) temporal operators in the rules and (ii) rules that deliberately wait for other decisions to be taken first, like, for example, the 2-second rule mentioned in the previous section.

6.7.4 Research Roadmap

Implementing Virtual Director technology is a multifaceted research challenge where many questions remain. A Virtual Director can not look into the future, but also cannot delay decisions too long, otherwise important actions will be missed. Beyond a rule-based approach where production grammar reacts to events in the scene with a certain delay, intelligent algorithms for prediction in that realm need to be found. Other modalities besides video need further consideration. An increase in research efforts regarding audio scripting can be identified in the literature.

Virtual Director systems, as discussed in this chapter, are part of a format-agnostic production system and interact with other components of that system. A key success factor of such systems is the interplay and collaboration between content selection and intelligent content transmission, a field with considerable future potential. To support the producers, better approaches for semi-automatic decision making and tools are required.

While the approach suggested in this section is viable for certain productions as individual technical components become more mature, the theoretic and practical limits of this approach remain to be reflected further. It might turn out that the suggested approach does not transfer well to very large rule-sets. There is little literature on design patterns for structuring complex sets of production grammar, and formalisms to represent Virtual Director behaviour have yet to be defined and standardised. Another open research question in that regard is to understand how rule re-use can be encouraged and simplified. What is contributing to good solutions is a high degree of configurability. Another dimension is to carefully split domain-specific rules from generic rules, the latter group being more likely to be re-used in other production domains.

Apart from the core system, user interaction design has strong effects on the user experience. Some users, but not all, prefer to interact frequently with Virtual Director systems. Different interaction devices and methods exist and emerge from related research fields, for example second screen interaction.

In assessing the state-of-the-art of Virtual Director technology, it has to be understood that automated personalisation still comes with quality limitations. One key factor when comparing format-agnostic production systems to traditional productions is that different cameras might be used, and these differences lead to different viewing experiences. A direct comparison is complicated since exterior factors come into play, that is, users are used to a certain production style.

Interviews with experienced broadcast directors revealed that producing with a panoramic camera as the main source is completely new to them and the production team. There is currently little knowledge and experience on how to use it to the benefit of the users' experiences (cf. Zoric et al., 2013). A drawback is that a crop of a panoramic view is not the same as a real zoom, meaning there is no variation of the angle of view or the depth of field. Panoramic cameras typically have a fixed position and the director may be restricted to a single point of view for certain objects in the scene. This means that more knowledge is required on how to use panoramic cameras in broadcast productions.

A human production team including the director takes decisions on the basis of a large number of rules they execute consciously, based on their knowledge, intuition, and a certain aesthetic sense. Humans may adapt their behaviour ad-hoc to assess the current situation, and might also anticipate what is going to happen next, based on experience. Apart from human creativity and adaptability, the challenge of automating a high number of static production rules is a research task that comprises many challenges. Many cinematographic principles are well documented in literature, but mostly in plain natural language. It is not straightforward to model them in a way that a machine can automatically execute them, and to make their effect dependent on a number of parameters at runtime, such as different cutting styles.

Ideally, the knowledge and sense of human directors, camera operators and other members of the broadcast team would be downloaded to a database for automatic execution – which is of course not possible as such. However, visual production grammar, sets of cinematographic principles, to a certain extent can be captured and modelled. The elicitation of desired production behaviour is an elaborate process, however. We recommend to capture desired behaviour in the form of ECA (event–condition–action) rules because it is an intuitive format for video production experts to explain cinematic principles. Further, such rules can be processed by machines. However, the research community for Virtual Director technology is currently lacking standards and design patterns on how to model and formally represent complex behaviour.

While sensors typically emit low-level information bits, directors think in more abstract concepts. Limited scene understanding can be achieved by abstracting the input to a higher semantic level. Virtual Directors aim to bridge the semantic gap by detecting certain situations automatically, utilising semantic inference, spatio-temporal pattern matching algorithms, and so on. In a real-time stream of events, higher-level concepts relevant for production grammar have to be detected via continuous queries on the recent time window of the event stream. Event processing frameworks are well-suited for that task since they make the behaviour developer's task easier. More research is however needed on how to deal with fuzzy, uncertain knowledge.

6.7.5   Concluding Thoughts

This chapter discussed the concept and architecture of Virtual Director systems that support live event broadcast production and are able to frame shots within a high-resolution panoramic video stream. Virtual Directors are a core technology of format-agnostic broadcast systems since their decisions have direct influence on the user experience. The key advantage of automation in view selection is that it can help to produce personalised content for a large number of users in parallel, even though compromises regarding its quality may have to accepted. Viewer interaction can be taken into account both before and during content consumption.

Besides the quality of decision making, the real-time requirement is perhaps the most important. Virtual Director systems typically have to process a large amount of real-time events that describe the scene. Regarding implementation frameworks, we conclude that approaches that effectively use a small sliding window, for keeping only the data necessary for future decision in its memory, are favourable. Hence, the rule-based event processing approach is recommended.

Delay management is a key success factor for automatic broadcast systems. Beyond individual components, the critical delay path of the overall production system needs to be designed carefully. At times, processing metadata and taking automatic decisions is faster than content preparation and transmission anyway, which leaves some scope for intelligent processing and may even lead to interesting effects that could be perceived as intelligent prediction by users. For example, cuts to a certain action may appear to be issued even before the action took place as an implication of overall system delays under certain circumstances.

The Production Scripting Engine presented in Section 6.6 is an actual research prototype based on the approach suggested in this chapter. It validated the approach in the sense that it proved to be suitable to manually define and automatically execute production behaviour for several genres. On a technical level, the overall delay of the processing chain was sufficiently small to fulfil real-time and scalability requirements. Defining the set of production rules and their interplay requires engineering effort, and even though replicating the creative brilliance of experienced directors seems to be nearly impossible, one can be confident that a basis is now in place that will allow future research activities to extend the concept, and to see format-agnostic productions based on it being realised. It can be concluded that Virtual Director technology is in its early stages, but already, its enormous potential towards interactive live content access becomes apparent.

References

Al-Hames, M., Hörnler, B., Müller, R., Schenk, J. and Rigoll, G. (2007) 'Automatic Multi-Modal Meeting Camera Selection for Video-Conferences and Meeting Browsers'. *Proceedings of the IEEE International Conference on Multimedia & Expo (ICME'07)*, pp. 2074–2077.

Allen, J.F. (1983) 'Maintaining Knowledge about Temporal Intervals'. *Communications of the Association for Computer Machinery* 26, pp. 832–843.

Amerson, D. and Kime, S. (2001) 'Real-time cinematic camera control for interactive narratives'. Working Notes of AAAI Spring Symposium on Artificial Intelligence and Interactive Entertainment, Stanford University, CA, 26–28 March.

AMI (2013) AMI Consortium home page, accessed 19 August 2013 at: http://www.amiproject.org/

Anicic, D. (2012) *Event Processing and Stream Reasoning with ETALIS: From Concept to Implementation*. Saarbrücken: Südwestdeutscher Verlag für Hochschulschriften.

Anicic, D., Fodor, P., Rudolph, S. and Stojanovic, N. (2011) 'EP-SPARQL: A Unified Language for Event Processing and Stream Reasoning'. *Proceedings of the 20th International Conference on World Wide Web (WWW'11)*, pp. 635–644, Hyderabad, India, 28 March–1 April.

Anicic, D., Fodor, P., Rudolph, S., Stühmer, R., Stojanovic, N. and Studer, R. (2010) 'A Rule-Based Language For Complex Event Processing and Reasoning'. *Proceedings of the 4th International Conference on Web Reasoning and Rule Systems (RR'10)*, Berlin: Springer-Verlag, pp. 42–57.

Anicic, D., Rudolph, S., Fodor, P. and Stojanovic, N. (2012) 'Real-Time Complex Event Recognition and Reasoning – A Logic Programming Approach. *Applied Artificial Intelligence* 26(1–2), 6–57.

Apidis (2013) Autonomous Production of Images based on Distributed and Intelligent Sensing, homepage. Accessed 19 August 2013 at: http://www.apidis.org/

Arijon, D. (1976) *Grammar of the Film Language*. Amsterdam: Focal Press.

Artikis, A., Paliouras, G., Portet, F. and Skarlatidis, A. (2010) 'Logic-Based Representation, Reasoning and Machine Learning for Event Recognition'. *Proceedings of the Fourth ACM International Conference on Distributed Event-Based Systems (DEBS'10)*, New York: ACM, pp. 282–293.

Baader, F., Calvanese, D., McGuinness, D.L., Nardi, D. and Patel-Schneider, P.F. (2010) *The Description Logic Handbook: Theory, Implementation and Applications*, 2nd edn. New York: Cambridge University Press.

Barbieri, D.F., Braga, D., Ceri, S., Valle, E.D., Huang, Y., Tresp, V., Rettinger, A. and Wermser H. (2010) 'Deductive and Inductive Stream Reasoning for Semantic Social Media Analytics'. *IEEE Intelligent Systems* 25(6), 32–41.

Bowen, C.J. and Thompson, R. (2013) *Grammar of the Shot*, 3rd edn. Abingdon: Taylor & Francis.

Brown, B. (2002) *Cinematography: Theory and Practice: Imagemaking for Cinematographers, Directors and Videographers*. Boston, MA: Focal Press.

Callet, P.L., Möller, S. and Perkis, A. (2012) 'Qualinet White Paper on Definitions of Quality of Experience'. European Network on Quality of Experience in Multimedia Systems and Services (COST Action IC 1003), Dagstuhl, Germany.

Chen, F., Delannay, D. and De Vleeschouwer, C. (2011) 'An Autonomous Framework to Produce and Distribute Personalized Team-Sport Video Summaries: A Basketball Case Study'. *IEEE Transactions on Multimedia* 13(6), 1381–1394.

Chen, F., Delannay, D., De Vleeschouwer, C. and Parisot, P. (2011) 'Multi-Sensored Vision for Autonomous Production of Personalized Video Summary'. *Lecture Notes of the Institute for Computer Sciences, Social Informatics and Telecommunications Engineering* 60, pp. 113–122.

Chen, F. and De Vleeschouwer, C.D. (2011a) 'Automatic Summarization of Broadcasted Soccer Videos with Adaptive Fastforwarding'. *Proceedings of the IEEE International Conference on Multimedia and Expo (ICME'12)*, Barcelona, 11–15 July.

Chen, F. and De Vleeschouwer, C.D. (2011b) 'Formulating Team-Sport Video Summarization as a Resource Allocation Problem'. *IEEE Transactions on Circuits and Systems for Video Technology* 21(2), 193–205.

Media College (2013) 'Video tutorial', accessed 19 August 2103 at: http://www.mediacollege.com/video/

Cugola, G. and Margara, A. (2012) 'Processing flows of information: From data stream to complex event processing'. *ACM Computing Survey* 44(3), 15:1–15:62.

Daniyal, F. and Cavallaro, A. (2011) Multi-camera scheduling for video production In Proceedings of the 9th European Conference on Visual Media Production, pp. 11–20.

de Boor, C. (2001) *A Practical Guide to Splines number Bd.27 in Applied Mathematical Sciences*. Berlin: Springer.

Del Galdo, G., Thiergart, O., Weller, T. and Habets, E.A.P. (2011) 'Generating Virtual Microphone Signals using Geometrical Information Gathered by Distributed Arrays 2011'. Joint Workshop on Hands-free Speech Communication and Microphone Arrays (HSCMA'11), Edinburgh, UK, pp. 185–190.

Drools (2013) 'Drools – The Business Logic integration Platform', accessed 19 August 2013 at: http://www.jboss.org/drools/

Duch, W., Oentaryo, R.J. and Pasquier, M. (2008) 'Cognitive Architectures: Where Do We Go From Here?' Proceedings of the 2008 Conference on Artificial General Intelligence: *Proceedings of the First AGI Conference*. Amsterdam: IOS Press, pp. 122–136.

Esper (2013) 'Event Stream Intelligence', accessed 19 August at: http://esper.codehaus.org/

Etalis (2013) 'Event-driven Transaction Logic Inference System', accessed 19 August at: https://code.google.com/p/etalis/

Etzion, O. (2010) Temporal perspectives in event processing Principles and Applications of Distributed Event-Based Systems IGI Global, pp. 75–89.

Etzion, O. and Niblett, P. (2010) Event Processing in Action. Manning Publications Co.

Falelakis, M., Groen, M., Frantzis, M., Kaiser, R. and Ursu, M. (2012) 'Automatic Orchestration of Video Streams to Enhance Group Communication'. *Proceedings of the 2012 ACM MM International Workshop on Socially-Aware Multimedia*, Nara, Japan, 29 October–2 November, pp. 25–30.

FascinatE (2013) Fascinate Project Homepage, accessed 19 August 2013 at: http://www.fascinate-project.eu/

Forgy, C.L. (1982) 'Rete: A Fast Algorithm for the Many Pattern/Many Object Pattern Match Problem'. *Artificial Intelligence* 19(1), 17–37.

Gordon, M. and Harel, D. (2011) 'Show-and-Tell Play-in: Combining Natural Language with User Interaction for Specifying Behavior'. *Proceedings of the IADIS Interfaces and Human Computer Interaction Conference (IHCI)*, 22–24 July, pp. 360–364.

Hare, J.S., Sinclair, P.A.S., Lewis, P.H., Martinez, K., Enser, P.G. and Sandom, C.J. (2006) 'Bridging the Semantic Gap in Multimedia Information Retrieval: Top-down and Bottom-up Approaches Workshop on Mastering the Gap: From Information Extraction to Semantic Representation', 3rd European Semantic Web Conference, Budva, Montenegro, 11–14 June.

Harel, D., Marron, A. and Weiss, G. (2012) 'Behavioral Programming'. *Communications of the ACM* 55(7), 90–100.

Harel, D., Marron, A., Wiener, G. and Weiss, G. (2011) 'Behavioral Programming, Decentralized Control, and Multiple Time Scales'. *Proceedings of the SPLASH'11 Workshops*, Portland, OG, 22–27 October, pp. 171–182.

Hilton, A., Kilner, J. and Starck, (2006) 'A Comparative Study of Free Viewpoint Video Techniques for Sports Events'. 3rd European Conference on Visual Media Production CVMP (2006) Part of the 2nd Multimedia Conference, 2006, London, 29–30 November, pp. 87–96.

Hitzler, P., Krötzsch, M., Parsia, B., Patel-Schneider, P.F. and Rudolph, S. (2012) 'OWL 2 Web Ontology Language: Primer' (Second Edition) W3C Recommendation, accessed 19 August at: http://www.w3.org/TR/owl2-primer/

Hoeksema, J. and Kotoulas, S. (2011) 'High-Performance Distributed Stream Reasoning Using s4'. First International Workshop on Ordering and Reasoning (OrdRing2011), ISWC'11, Bonn, 23–24 October.

Inamoto, N. and Saito, H. (2005) Free viewpoint video synthesis and presentation from multiple sporting videos *Proceedings of the IEEE International Conference on Multimedia and Expo (ICME'05)*, pp. 322–325.

Jena (2013) 'Reasoners And Rule Engines: Jena Inference Support', accessed 19 August 2013 at: http://jena.apache.org/documentation/inference/

Jiang, Y.G., Bhattacharya, S., Chang, S.F. and Shah, M. (2013) 'High-Level Event Recognition in Unconstrained Videos'. *International Journal of Multimedia Information Retrieval* 2(2), 73–101.

Jiang, J., Kohler, J., MacWilliams, C., Zaletelj, J., Guntner, G., Horstmann, H., et al. (2011) 'Live: An Integrated Production and Feedback System for Intelligent and Interactive TV Broadcasting'. *IEEE Transactions on Broadcasting* 57(3), 646–661.

Kaiser, R., Hausenblas, M. and Umgeher, M. (2009) 'Metadata-Driven Interactive Web Video Assembly'. *Multimedia Tools and Applications* 41(3), 437–467.

Kaiser, R., Thaler, M., Kriechbaum, A., Fassold, H., Bailer, W. and Rosner, J. (2011) 'Real-Time Person Tracking in High-Resolution Panoramic Video for Automated Broadcast Production'. *Proceedings of the 2011 Conference for Visual Media Production (CVMP'11)*, Washington, DC: *IEEE Computer Society*, pp. 21–29.

Kaiser, R., Torres, P. and Höffernig, M. (2010) 'The Interaction Ontology: Low-Level Cue Processing in Real-Time Group Conversations'. 2nd ACM International Workshop on Events in Multimedia (EiMM10) held in conjunction with ACM Multimedia 2010, Firenze, Italy, October 25–29.

Kaiser, R., Weiss, W., Falelakis, M., Michalakopoulos, S. and Ursu, M. (2012) 'A Rule-Based Virtual Director Enhancing Group Communication', *Proceedings of the IEEE International Conference on Multimedia & Expo*, Melbourne, 9–13 July, pp. 187–192.

Kaiser, R., Weiss, W. and Kienast, G. (2012) 'The FascinatE Production Scripting Engine Advances in Multimedia Modeling', *Lecture Notes in Computer Science*, vol. 7131, 682–692.

Kim, S., Moon, S., Han, S. and Chang, J. (2011) Programming the Story: Interactive Storytelling System. *Informatica* 35(2), 221–229.

Komazec, S., Cerri, D. and Fensel, D. (2012) 'Sparkwave: Continuous Schema-Enhanced Pattern Matching over RDF Data Streams'. *Proceedings of the 6th ACM International Conference on Distributed Event-Based Systems*, Berlin, 16–20 July, pp. 58–68.

Laird, J.E. (2012) *The Soar Cognitive Architecture*. Cambridge, MA: MIT Press.

Lavee, G., Rivlin, E. and Rudzsky, M. (2009) Understanding video events: a survey of methods for automatic interpretation of semantic occurrences in video. *IEEE Transactions on Systems, Man and Cybernetics Part C: Applications and Reviews* 39(5), 489–504.

Lipski, C., Linz, C., Berger, K., Sellent, A. and Magnor, M. (2010) 'Virtual: Image-Based through Space and Time'. *Computer Graphics Forum* 29(8), 2555–2568.

LIVE I (2009) LIVE Staging of Media Events homepage, accessed 19 August 2013 at: http://www.ist-live.org/

MPEG (2001) Information Technology – Multimedia Content Description Interface ISO/IEC 15938.

MPEG (2012) Information Technology – Multimedia Content Description Interface – Part 9: Profiles and Levels, Amendment 1: Extensions to Profiles and Levels ISO/IEC 15938-9:2005/Amd1:2012.

NHK (2013) Homepage (in Japanese), accessed 19 August 2013 at: http://www.nhk.or.jp/

Niamut, O.A., Kaiser, R., Kienast, G., Kochale, A., Spille, J., Schreer, O. et al. (2013) 'Towards a Format-Agnostic Approach for Production, Delivery and Rendering of Immersive Media'. *Proceedings of the 4th Multimedia Systems Conference*, Oslo, Norway, 27 February–1 March.

NM2 (2007) 'New Millenium, New Media', homepage, accessed 19 August at: http://www.ist-nm2.org/

Patrikakis, C, Papaoulakis, N, Papageorgiou, P, Pnevmatikakis, A, Chippendale, P, Nunes, M, et al. (2011a) 'Personalized Coverage of Large Athletic Events'. *IEEE MultiMedia* 18(4), 18–29.

Patrikakis, C., Papaoulakis, N., Stefanoudaki, C., Voulodimos, A. and Sardis, E. (2011b) 'Handling Multiple Channel Video Data for Personalized Multimedia Services: A Case Study on Soccer Games Viewing'. *2011 IEEE International Conference on Pervasive Computing and Communications Workshops*, Seattle, 21–25 March, pp. 561–566.

ProbLog (2013) 'Introduction', accessed 19 August 2013 from: http://dtai.cs.kuleuven.be/problog/

Prova (2013) 'Prova Rule Language', accessed 19 August 2013 from: http://prova.ws/

Ronfard, R. (2012) 'A Review of Film Editing Techniques for Digital Games'. Workshop on Intelligent Cinematography and Editing, Raleigh, NC, 28–29 May.

Schreer, O., Feldmann, I., Weissig, C., Kauff, P. and Schäfer, R. (2013) 'Ultra High-Resolution Panoramic Imaging for Format-Agnostic Video Production'. *Proceedings of the IEEE* 101(1), 99–114.

Schreer, O., Kauff, P. and Sikora, T. (2005) *3D Videocommunication: Algorithms, Concepts and Real-Time Systems in Human Centred Communication*. New York: Wiley.

Skarlatidis, A., Artikis, A., Filippou, J. and Paliouras, G. (2013) 'A Probabilistic Logic Programming Event Calculus'. *Theory and Practice of Logic Programming (TPLP)*.

Smolic, A. (2009) 'An Overview of 3D Video and Free Viewpoint Video'. *Lecture Notes in Computer Science* 5702, 1–8.

Smolic, A. (2011) '3D Video and Free Viewpoint Video – From Capture to Display'. *Pattern Recognition* 44(9), 1958–1968.

Staab, S. and Studer, R. (2009) *Handbook on Ontologies*, 2nd edn. Berlin: Springer.

Stevens, T., Cesar, P., Kaiser, R., Farber, N., Torres, P., Stenton, P. and Ursu, M. (2012) 'Video Communication for Networked Communities: Challenges and Opportunities'. 16th International Conference on Intelligence in Next Generation Networks (ICIN), Berlin, 8–11 October, pp. 148–155.

Stuckenschmidt, H., Ceri, S., Della Valle, E. and van Harmelen, F. (2010) 'Towards Expressive stream reasoning'. *Proceedings of the Seminar on Semantic Aspects of Sensor Networks*, Dagstuhl, Germany, 24–29 January.

TA2 (2012) 'Together Anywhere, Anytime', Homepage, accessed 19 August 2013 at: http://ta2-project.eu/

Timing Framework (2013) Homepage, accessed 19 August 2013 at: https://java.net/projects/timingframework/

Ursu, M.F., Thomas, M., Kegel, I., Williams, D., Tuomola, M., Lindstedt, I., et al. (2008) Interactive TV narratives: Opportunities, progress, and challenges. *ACM Transactions on Multimedia Computing, Communications and Applications* 4(4), 25:1–25:39.

Ursu, M.F., Torres, P., Zsombori, V., Franztis, M. and Kaiser, R. (2011) 'Socialising through Orchestrated Video Communication'. *Proceedings of the 19th ACM international Conference on Multimedia*, Scottsdale, AZ, 28 November–1 December, pp. 981–984.

Valle, E.D., Ceri, S., van Harmelen, F. and Fensel, D. (2009) 'It's a Streaming World! Reasoning upon Rapidly Changing Information'. *IEEE Intelligent Systems* 24(6), 83–89.

Vconect (2013) Homepage, accessed 19 August 2013 at: http://www.vconect-project.eu/

Viterbi, A. (1967) 'Error Bounds for Convolutional Codes and an Asymptotically Optimum Decoding Algorithm'. *IEEE Transactions on Information Theory* 13(2), 260–269.

Wagner, C. (2006) 'Breaking the Knowledge Acquisition Bottleneck through Conversational Knowledge Management'. *Information Resources Management Journal* 19(1), 70–83.

Wang, H.L. and Cheong, L.F. (2009) 'Taxonomy of Directing Semantics for Film Shot Classification'. *IEEE Transactions on Circuits and Systems for Video Technology* 19(10), 1529–1542.

Wasserkrug, S., Gal, A. and Etzion, O. (2012a) 'A Model for Reasoning with Uncertain Rules in Event Composition Systems', accessed 19 August 2013 at: http://arxiv.org/ftp/arxiv/papers/1207/1207.1427.pdf

Wasserkrug, S., Gal, A., Etzion, O. and Turchin, Y. (2012b) 'Efficient Processing of Uncertain Events in Rule-Based Systems'. *IEEE Transactions on Knowledge and Data Engineering* 24(1), 45–58.

Wilson, K. (2013 'Camera Angles', accessed 19 August 2013 at: http://www.mediaknowall.com/camangles.html

ZeroC (2013) 'The Internet Communications Engine', accessed 19 August at: http://www.zeroc.com/ice.html

Zoric, G., Barkhuus, L., Engstrom, A. and Önnevall, E. (2013) 'Panoramic Video: Design Challenges and Implications for Content Interaction'. EuroITV, 11th European Interactive Conference, Como, Italy, 24–26 June.

7

Scalable Delivery of Navigable and Ultra-High Resolution Video

Jean-François Macq[1], Patrice Rondão Alface[1], Ray van Brandenburg[2],
Omar Aziz Niamut[2], Martin Prins[2] and Nico Verzijp[1]

[1]*Alcatel-Lucent Bell Labs, Antwerp, Belgium,*
[2]*TNO, Delft, The Netherlands*

7.1 Introduction

In a review of the evolution of television (TV) the number of technological breakthroughs that have arisen since the 1950s is notable. There has been progress on many fronts, such as the move from black-and-white to colour, the booming diversity of content offered and an ever-increasing picture quality. The latter has itself been regularly improved by disruptive changes: the transitions from analogue to digital formats, from standard to high definition and more recently the advent of stereoscopic 3D content.

Despite an impressive list of technological improvements, which started many decades ago, it can be argued that the TV experience itself has barely changed. TV offers, almost by definition, a set of fully formatted pieces of media, defining how the content will be delivered to end users in terms of viewing experience (camera shot selection, post-processing effects, colour grading, and so on), content duration and sequencing.

However, in recent years, we have witnessed significant changes in the way people consume media in particular in their digital form. Among them one of the clearest trends is the increasing level of interactivity and of personalised consumption. People are now used to browsing web pages in order to immediately access the content they want, but progressively this interactivity is also impacting more advanced forms of media. In some novel imaging and digital mapping applications – think of Google Maps and Streetview (Google Maps, 2013) – the end user now has the ability to freely navigate large panoramic or omni-directional images with

Media Production, Delivery and Interaction for Platform Independent Systems: Format-Agnostic Media, First Edition. Edited by Oliver Schreer, Jean-François Macq, Omar Aziz Niamut, Javier Ruiz-Hidalgo, Ben Shirley, Georg Thallinger and Graham Thomas.
© 2014 John Wiley & Sons, Ltd. Published 2014 by John Wiley & Sons, Ltd.

panning and zooming commands. Bringing the same flexibility to video content is the logical next step and has the potential to revolutionise TV and more generally digital entertainment services.

The digital environment of end users is also evolving as, among home devices, high definition (HD) displays and surround sound systems are becoming commonplace. With 'Ultra HD' screens (4 K and beyond) being developed, this evolution to larger display sizes, finer resolutions and more immersive content reproduction is clearly not finished. This requires audio-visual content to be captured, produced and delivered at a matching and always higher fidelity.

These trends have resulted in various responses from the TV industry. Content is often produced at different fidelity levels, for instance with standard definition (SD) and HD (and soon 4 K) offerings of the same TV channel. Reframed versions are sometimes created for viewing on small mobile displays. Personalisation of the TV experience is now realised with the aid of the so-called 'second screens' where a separate application can be run in parallel to the linear TV program, with varying degrees of synchronisation. The delivery channel for the second screen usually relies on the internet and is then fully decoupled from the main screen broadcast channel. With the advent of 'Smart' and 'Connected' TV, these companion applications can also be run on the TV set itself. These solutions are usually referred to as Hybrid Broadcast Broadband TV (HBB TV) (Merkel, 2011).

From a high-level perspective, it can be seen that these approaches are actually contributing to a multiplication of formats, being deployed over several media delivery chains. Relatively modest interactivity or personalisation features are offered and are, most of the time, confined to the second screen only.

In contrast, the format-agnostic paradigm that we advocate in this book has the ambition to change the way content is produced at the source, so that the highest level of quality and flexibility is maintained as long as possible in the delivery chain, leaving more possibilities to customise the content as function of the end user's preferences and of the devices used for audio-visual rendering.

This chapter starts, in Section 7.2, with a review of different aspects related to *format-agnostic delivery*. The concept is first treated in a broad sense, with examples of mechanisms that allow postponing the actual content formatting to the end of the delivery chain.

In the remainder of the chapter the focus is on the capability to acquire and deliver audio-visual content at ultra-high quality and with extended field-of-view. The acquisition technologies for layered, panoramic or omni-directional content described in Chapter 3 enable the production of immersive content, which the end users may have the freedom to navigate. As the decisions on what part of the content will be presented on the device display are pushed to the end of the delivery chain, a key problem faced for delivery is the high data rate that has to be ingested in the delivery system. Another challenge is the high level of interactivity that the system has to cope with to offer fine-grained navigation of the content. Therefore, building a system to efficiently deliver navigable and ultra-high resolution video requires revisiting some fundamental aspects of the way video data is represented, coded and transported. In Section 7.3 we address the central problem of high data rate for panoramic video and explain how to efficiently reduce the content data rate, while still allowing direct access to independent subsets of the content. Section 7.4 shows how techniques addressed in Section 7.3 can be modelled and lead to optimisation mechanisms with respect to compression and delivery. Next, Section 7.5 covers transport of interactive and immersive content. We look at two types of evolution in transport technologies and detail to what extent they can be adapted for immersive and interactive video delivery. Section 7.6 summarises the problems and solutions discussed in this chapter and concludes with a view on the future of interactive media delivery.

7.2 Delivery of Format-Agnostic Content: Key Concepts and State-of-the-Art

In recent years many developments have addressed the generic objective of delivering audio-visual content based on a single representation made available at the source, and where the network gets the ability to adapt the content on an end user basis. This section gives a brief overview of the techniques that have been developed as of today to support such format-agnostic content delivery.

In that respect, the notion of 'format-agnostic content' can take different forms depending on the kind of adaptation required to 'format' the content at the user end and on the type of functionalities this enables. A first set of examples includes the possibility for the network to adapt the video representation in terms of scale, bitrate and fidelity. Other examples require the ability to have a personalised play-out of content, either from user interaction or from an automated semantic analysis. To support this, the system has to continuously adapt the content temporally and spatially. Finally, this section looks at the capability for end users to truly navigate into the video, allowing them to view the video with personalised zooming and panning controls, and at some of the solutions that can be deployed in today's delivery systems.

7.2.1 Delivery Agnostic to Content Formats, Device Capabilities and Network Bandwidth

As introduced in Chapter 2, a plethora of formats have been coexisting for digital representation of audio-visual content. Often, even a single audio-visual item (e.g., a TV channel or a video clip) is offered in different formats over a given delivery platform. The reasons are usually twofold.

First, the content format may be set to suit the capability of the end user device. A typical example in TV broadcast is the availability of the same channel both in SD and HD. Although most current TV productions follows an HD workflow, the SD format is still largely offered to support older generations of TV sets or Set-Top Boxes (STBs). Content can also be made available at sub-SD resolution, for instance, to be consumed on handheld devices or to be used as a thumbnail video content. Picture-in-Picture (PiP) is a classical application where a thumbnail video is displayed in inset windows overlaid on the main TV program. More recently, an analogous format issue arose with the introduction of stereoscopic 3D video content and the need for backward compatibility with 2D displays.

A second factor is the actual capacity of the delivery chain in terms of data rate. Despite a continuous growth of the bandwidth offered in residential networks, many service providers have to cope with the differences in terms of the bandwidth that is available to deploy video services for each of their customers. In the case of managed Internet Protocol (IP) networks, the service provider often has to adapt the format of video streams (in terms of resolution and hence bitrate) in order to guarantee access to the service to a majority of its customers.

The quantification of this access is often called 'service reach'. In the case of TV delivery over networks with a mix of network access technologies, such as fibre and Digital Subscriber Line (DSL) over copper, the service provider usually has to cope with differences in reach depending on the bandwidth required for the service (Vanhastel, 2008). The coexistence of SD and HD versions of the same TV channels provides operators a solution to deploy a TV service with different reach: an SD service available to most customers (low bitrate requirement, broad service reach) and an HD service for a restricted set of customers (high bitrate requirements, narrow service reach). Similarly, in the case of unmanaged or 'over-the-top' delivery networks, the Content Delivery Network (CDN) provider aims to adapt the format of video streams (in terms of resolution and hence bitrate) in order to offer the largest service reach to its customer, typically content providers.

The Cost of Multiple Formats

Regardless of why different formats for the same content are made available, this coexistence incurs a cost at the server side. That is, more bandwidth has to be provisioned to output multiple formats of the same TV channel, or more storage capacity for on-demand content. To mitigate that risk, an elegant solution is to aggregate the multiple formats in a unique stream. Doing so offers the opportunity to decrease the amount of redundant information across the different formats. Several developments in video compression technologies have addressed this requirement and are generally known as Scalable Video Coding (SVC). In particular, an instance of SVC known as spatial scalability supports the encoding of a single video stream, from it is possible to extract sub-streams representing the source content at different resolutions (Segall and Sullivan, 2007). Using this functionality, combinations of HD, SD and sub-SD versions of a video stream can efficiently be compressed, even when targeting displays with different aspect ratios (Lambert et al., 2009).

In a similar manner, stereoscopic 3D video can be efficiently streamed, using a joint compression of the left-eye and right-eye views. Techniques known as Multiview Video Coding (MVC) also support the extraction of a single (2D) view from the compressed bitstream (He et al., 2007). This can be used by legacy video decoders without support for 3D. Therefore, backward compatibility can be ensured without offering explicitly two versions of the same channel (a 2D one and 3D one).

Dynamic Format Adaptation

The examples given above have in common that a single format is statically assigned to a given receiver. However, there exist many situations, in regard to delivery aspects, where one would want a more dynamic way to adapt the format. A typical example concerns video streaming over the internet. In this case, the service provider seldom has control over the end-to-end bandwidth between his servers and the connected devices at the other side of the network. It may indeed have to face significant variations of that bandwidth over time, depending on the traffic load in each of the network segments traversed. The situation is even worse where the properties of the physical channels are changing over time, as with wireless technologies. Again, the SVC technology offers a solution as it supports finer-grained adaptation of the bitrate (and hence video fidelity) to match the channel capacity (Schierl, Stockhammer and Wiegand, 2007). In the case of video streaming to mobile devices, the use of SVC in conjunction with packet scheduling at the physical layer leads to an optimised usage of time-varying wireless channel (Changuel et al., 2010).

Despite the technical performances of the SVC-based solutions to address the adaptation problem, the aforementioned techniques have not been widely adopted in video delivery plat-forms. One of the hurdles faced is that most of these schemes assume some form of cross-layer optimisation (e.g., at physical and transport layer). This is in contrast with currently deployed architecture where the video delivery mechanism is mostly controlled at the application layer. We will review and give more details on this approach in Section 7.5.

7.2.2   Delivery Agnostic to Video Timing – the Temporal Interactivity Case

Other research tracks in the literature have investigated the possibility of adapting the content format in the temporal dimension, with the objective of personalising the content at the semantic level. Typical examples consist of the ability to deliver a given source of content in different adapted versions, each with a different duration, or even to personalise the sequence in which video fragments are played out.

Personalised Summarisation

Personalised summarisation consists of editing the video fragments in the temporal dimension. The process results in a summary video output that can be significantly shorter than the source content and may also take the personal preferences of the user into account. In Chen and De Vleeschouwer (2011), the problem formulation starts with an input video divided in predefined temporal segments, each characterised by duration and benefit value (for a given class of users). Given a user-defined length for the output video, the summarisation problem aims to maximise the total benefits of the selected segments and is thus defined as a classical allocation problem under limited resources (that can be solved by Lagrangian optimisation). Note that the paper also proposes models of the segment benefits as a function of realistic annotation of the video footages (relevance, emotional level, etc). A significant part of this work was conducted in the APIDIS project, which aimed at autonomous systems for producing and delivering video summaries for controlled scenarios, such as sport events recording (APIDIS, 2008). As described in Chapter 6 (in Section 6.1.5), this work also looked at combining summarisation and automated camera view selection. Examples of optimisation models can be found in Chen and De Vleeschauwer (2010).

Delivery Framework for Personalised Play-out

The Celtic RUBENS project introduced a delivery framework dedicated to personalised video play-out (RUBENS, 2009). The approach relied on the temporal segmentation of audio-visual sources. A segment is defined as a self-supporting atom of media that can be efficiently handled or even cached by the network segments. One of the possible uses of segment-based delivery is the transmission of video at various levels of Quality of Experience (QoE), when for instance each video segment is made available at multiple bitrate levels (see Section 7.2.1). However,

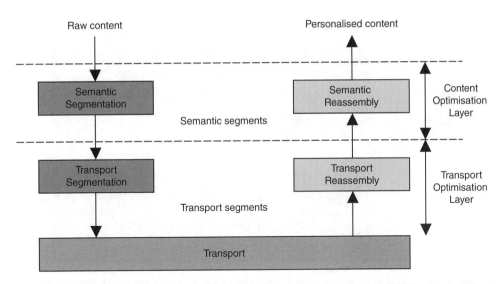

Figure 7.1 Semantic and transport segmentation layers proposed in RUBENS. Adapted from Huysegems, De Vleeschauwer and De Schepper (2011).

other use cases pertain to semantic personalisation of the content, assuming a semantically-consistent segmentation. The two levels of segmentation can be seen as two optimisation layers that allow the raw input content to be efficiently represented and transported for personalised and interactive video services, as depicted in Figure 7.1 (Huysegems, De Vleeschauwer and De Schepper, 2011).

A typical use case enabled by this framework is the possibility to create personalised video summaries. From a given set of segments, if all possible summaries were prepared before their transmission as distinct monolithic video clips, this would lead to a huge waste of bandwidth and intermediate storage, since the redundancies between the numerous summaries could not be exploited. This is the reason that lies behind the concept of 'late assembly' in this framework. The content is transported in segmented form as far as possible on the transport path and the segments are eventually reassembled based on the user preferences and interaction commands. It is essential to guarantee that this segment re-assembly is always performed in a meaningful way for the end user, either to ensure a consistent storyline or to respect the creative vision of a director. This is the rationale for introducing the model of a playmap, as illustrated in Figure 7.2. Concretely the playmap describes the possible sequences of segments, formatted as an XML file. The transition between segments can only happen for certain events triggered either explicitly by a user command, such as 'Channel Change', 'Next', 'Expand' and so on (dashed arrows in the figure), or implicitly when a segment ends (solid arrows).

The knowledge contained in the playmap is also useful for optimising the delivery of the segments, in particular in the context of a CDN, where video segments are transported towards the end points through multiple tiers of caching nodes. Knowing the possible segment transitions significantly improves the efficiency of the algorithms for segment caching and transmission scheduling (De Schepper et al., 2012).

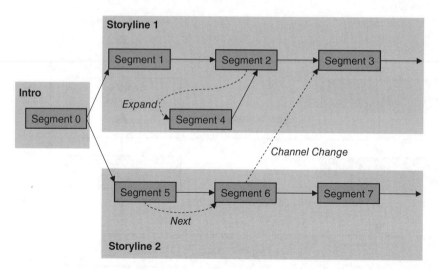

Figure 7.2   Illustration of the playmap model (Huysegems, De Vleeschauwer and De Schepper, 2011).

7.2.3   Delivery Agnostic to Video Reframing – the Spatial Interactivity Case

After having described temporal personalisation, we close the section with its spatial counter-part, which is the core topic of this chapter.

As described in Chapter 3, with recent capturing systems for panoramic and omni-directional video, new types of media experiences are possible where an end user has the possibility to choose his viewing direction and zooming level freely. Even for conventional camera capture, the past few years has seen the resolution of video camera systems increase at a faster pace than the resolution of displays, as depicted in Figure 7.3. All in all, we see that the resolution and field-of-view of many end devices will never catch up with those of the latest acquisition

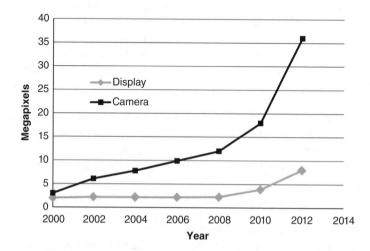

Figure 7.3   Camera and TV set resolution evolution over time. Adapted from (Ip and Varshney, 2011).

systems, hence leaving some room for the end user to interactively control what subset of the video data he wants to see at any point in time.

Many different examples of interactive video delivery have been demonstrated or deployed. In the entertainment sector, companies like Immersive Media offer a web streaming solution to cover events with a 360-degree video camera. However, the solution relies on delivering the complete spherical panorama to the client device, where the final rendering of the interactive viewport takes place. To support the wide-scale delivery over the internet, the source content is severely downscaled, which prevents the end user taking advantage of the high resolution available at the capturing side.

At the other end of the spectrum, companies like KDDI have shown a solution where all rendering takes place on the server side. Client devices, for example, low-powered and low-resolution mobile phones, send a spatial request to the server, requiring the server to reframe and rescale the content accordingly before compression and streaming to the client (KDDI R&D Labs, 2010). In Macq et al. (2011) a similar approach was proposed for spherical content, with support for continuous navigation. Although they allow the end user to access the highest resolution possible, these solutions have the drawback that the servers need to maintain a processing session and a delivery channel for each individual client.

In the following sections, we will propose methods that try to combine the benefits of the two previous approaches: scale to many users and spatial interaction at the highest picture quality available.

7.3 Spatial Random Access in Video Coding

Due to the very high data rate involved with panoramic and ultra-high resolution video, compression is a key aspect for the efficient transmission of interactive video. A brute-force approach would consist of ingesting the whole set of content available at the source as a single monolithic coded stream and delivering it as-is throughout the whole system until final processing at the end user device. As argued in Section 7.2.1, this approach is not able to support the diversity of device capabilities and network capacities. Therefore, in this section, we review the numerous techniques proposed in the literature that allow a more granular representation of the content. Their common objective is to offer an acceptable trade-off between compression performances and spatial random access, that is, the ability of a delivery system to access and forward only a subset of the content.

More precisely, we focus on a new functionality that enables a client device to interactively and efficiently access a portion of ultra-high definition panoramic encoded content. In the literature, this functionality is most often referred to as 'zoomable videos' (Khiem et al., 2010), 'virtual pan-tilt-zoom videos' (Mavlankar and Girod, 2009), 'spatial random access' (Rondão Alface, Macq and Verzijp, 2011), (Niamut et al., 2011) or 'interactive regions of interest' (Lambert and Van de Walle, 2009). Note that the notion of Region-of-Interest (ROI) used in this book refers to a predefined rectangular region in a video frame. In contrast, in the context of this chapter, the region is not known at the time of encoding and is defined by the end user (or the client device). To highlight this difference, we are thus using the term 'Interactive ROI' (IROI) throughout this chapter.

A recurring approach to support spatial random access in panoramic video is to partition the frames into regions, which are independently coded. When these regions are created following

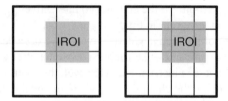

Figure 7.4 Example of 2 × 2 (left) and 4 × 4 (right) tiling grids. Depending on the tile size, a single IROI overlaps multiple tiles, and thus requires multiple tiles for reconstruction.

a regular rectangular pattern, they are called video *tiles*. Figure 7.4 shows two examples of tiling as well as a grey region representing an IROI.

This section starts with a short overview of video compression principles and the adverse notion of random access. Then we review different techniques that offer spatial random access in video, including some based on the tiling approach.

7.3.1 Video Compression and Random Access – A Fundamental Trade-off

The classic objective of video compression is to achieve the best compression performance possible. This means that for a given level of representation fidelity, the compressed stream size should be minimised. Today's video compression standards have achieved great success with regard to this compression objective. The *de facto* video compression standard H.264/AVC and its recent successor H.265/HEVC (Sullivan et al., 2012), reach compression ratios of 150:1 and 300:1 respectively, for typical 1080p TV quality.

However, spatial random access is not directly supported by most video compression standards, even for (non-interactive) ROIs. The reason is that enabling an independent access to a spatial portion of an encoded video usually comes with a loss in compression performance, which is the primary objective of these standards. This loss in compression performance can be explained by the fact that most successful video compression schemes rely deeply on existing redundancy in both the spatial and temporal domain of the video content. This redundancy is removed by progressively predicting the video content from the already encoded content in a way that both the encoder and decoder are able to compute in the same manner, and that requires minimal data transmission. Accessing a spatial region in an encoded stream poses a challenge to the decoder as the necessary predictions for decoding the region will be based on previous frames or other areas of the video content that the user does not want to access. This implies that the encoder must restrict predictions in such a way that all possible ROIs (that should be known in advance) can be easily decoded. Doing so, the redundancy is decreased and the compression performance drops. Adding the requirement of end user interactivity obviously makes the problem worse, as it becomes impossible to isolate every possible IROI.

More precisely, successful video compression schemes such as H.264/AVC and H.265/HEVC make use of hybrid coding, which consists of combining block-based motion compensation and de-correlative transform. Motion compensation and intra-prediction (see Figure 7.5) are tools that enable an encoder to detect and exploit redundancy in time and space when processing a video block (i.e., a macroblock in H.264 and a coding unit in H.265). This block is then predicted from already encoded frames (motion compensation) or from already

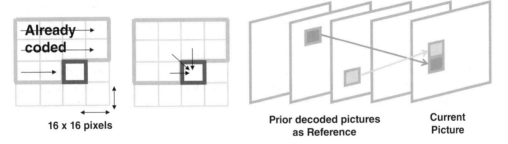

Figure 7.5 From left to right, H.264 macroblock structure, H.264 intra prediction dependencies, H.264 motion compensation dependencies.

encoded blocks of the current frame (intra prediction). The difference signal, also known as residual, is then de-correlated by a transform (the integer cosine transform for both H.264 and H.265) and the resulting signal is made up of transform coefficients. These coefficients are quantised and then sent to an entropy encoder together with the necessary signalling information for reproducing the prediction at the decoder side. A post-processing filter can be also applied on each encoded frame (deblocking filter for H.264 and more generally a set of loop filters for H.265) in order to improve the video reconstruction quality.

All these steps provide compression as a result of signal dependencies because:

- the spatio-temporal signal redundancy is exploited but induces many dependencies (the reference data at the decoder side is required in order to reproduce the same prediction);
- the residual signal energy is compacted by the transform and limits quantisation distortion; and
- statistical redundancy of the quantised coefficients is efficiently reduced by an entropy encoder, but it achieves the best compression by updating a table of symbol probabilities. Since the decoder performs the same operations, avoiding the decoding of previous symbols, which would not be requested by a user, would de-synchronise the tables of probabilities and lead to decoding errors.

Unfortunately, dependencies prevent using spatial random access as much more information than the data related to the spatial region the user is interested in has to be decoded. Similarly, two other video compression functionalities are also impacted by these dependencies: parallel encoding and transmission error resiliency. Error resiliency requires that when some losses occurred in the transmission of the video, the decoder should still be able to decode the subsequent parts of the video stream correctly. Compression dependencies tend to complicate this as, if some necessary information is lost for prediction or for entropy decoding, then the decoder can only perform well if it is able to conceal these losses. In the case of parallel compression, the objective is to distribute the encoding processing (e.g., frame blocks) on different cores while minimising inter-core communication. This communication is directly proportional to the number of compression dependencies between blocks assigned to different cores.

H.264 and H.265 provide some tools to limit these dependencies. The first one is the Group of Picture (GOP) structure length that prevents any dependency between the frames it contains to all other GOPs. The second is the distribution of blocks into slices for H.264 and into slices combined with tiles for H.265. Slices in H.264 can be defined by using Flexible Macroblock Ordering (FMO), which enables defining (non-interactive) ROIs and other compositions of blocks that were initially designed for error resiliency purposes. The intra-prediction and entropy encoding dependencies between blocks of different slices or tiles are limited and enable parallel processing. However, in order to ensure sufficient compression performance, H.264 (and H.265) does not restrict motion compensation of the blocks inside a slice (or a tile) to refer to the entirety of the previous frames inside its GOP. We can conclude here that standard tools do not enable efficient random access per se.

Random Access and Image Compression

In the literature, the first studies on the trade-offs between limiting dependencies and achieving high compression performance have addressed image compression in remote browsing applications with interactive Pan-Tilt-Zoom commands. To enable interactive browsing of TeraPixel images (Guo et al., 2010; Ip and Varshney, 2011), compression techniques with spatial random access capabilities are necessary as all captured pixels cannot be displayed at the same time on a screen. This avoids having to download the entire content by only selecting the requested portions of the encoded stream corresponding to the desired region of the image.

Existing coding strategies for very high resolution images consist of multiscale image representations such as Gaussian or Laplacian pyramids. Usually the scales are independently encoded and redundancy between layers is not exploited for the benefit of easy and simple random access. The use of JPEG2000 has also been explored for interactive browsing (Taubman and Rosenbaum, 2000), as JPEG2000 natively supports tiled representation of the image, where each tile is transformed and encoded separately. Furthermore, the transform itself, which is a rounded version of the bi-orthogonal Cohen-Daubechies-Feauveau (CDF) 5/3 wavelet transform (Adams and Kossentini, 1998), generates coefficients that are encoded independently in blocks. Each block is then encoded in an independent embedded sub-bitstream. These mechanisms are at the core of JPEG2000 capabilities in terms of bitstream scalability and random access. The transmission protocol, called JPEG2000 over Internet Protocol (JPIP) has been designed to support remote interactive browsing of JPEG2000 coded images (Taubman and Prandolini, 2003).

Random Access and Video Compression

Beyond image coding, some video compression schemes also offer random access tools. MPEG-4 Part 2, propose video object coding with background sprites, enabling independent access to video objects (Krutz et al., 2008).

Some works exploit H.264 FMO in order to independently encode and access ROIs (Fernandez et al., 2010). The limitation of these approaches is that the ROIs (and not IROIs) are defined at the encoder side and not at the user side. Moreover, noticeable visual distortions at the borders of the predefined ROIs reduce the applicability to interactive Pan-Tilt-Zoom (PTZ).

H.264/SVC, the scalable extensions of H.264, provides a multiscale video representation in an analogous manner to JPEG2000, although it is not using wavelet transforms. SVC includes some support for slicing, however, the standard does not natively support the extraction of subregions in an interactive manner. In Section 7.3.4, we describe some possible use of multiple SVC encoders to support multi-resolution tiled representation.

7.3.2   Spatial Random Access by Tracking Coding Dependencies

A class of approaches that allow a client decoder to access only a subset of the compressed video proposes to access only the macroblocks containing the relevant pixels and to track dependencies on other macroblocks/video regions created by the encoder.

In Khiem et al. (2010) the concept of Zoomable Videos using a monolithic coding approach (as opposed to a tiled approach) is defined. They propose tracking dependencies at the encoder side and limit the download to the very necessary parts of the encoded stream. Unfortunately, this approach is not standard compliant as it requires the decoder to have knowledge of a tree of dependencies. In addition to the direct dependencies due to intra- and inter-frame prediction, this technique must also track all the coded elements that are required at the client side to avoid any drift in the entropy decoder. Such a drift would create an unrecoverable de-synchronisation between the encoder and the decoder states.

Besides the complexity of the tracking itself, one of the inherent issues with dependency tracking at the macroblock level is the size of the database representing these dependencies. Depending on the nature of the video and the efficiency of the data structure used, the dependency files can be huge (Liu and Ooi, 2012), possibly larger than the original video file!

Lambert et al. (2009) exploited the SVC extension of H.264 for enabling IROIs. Natively, SVC offers support for encoding non-interactive ROIs and tiled representation in a layered-video representation. However, as SVC introduces scalability, it also increases the number of dependencies as shown by Tang and Rondão Alface (2013). Moreover, the standard does not explicitly support the decoding of an isolated ROI. In their approach, Lambert et al. (2009) try to overcome these limitations by building an XML file that tracks all encoding dependencies, so that the decoder is able to decode only the necessary data for the requested IROI. As in Liu and Ooi (2012), the main issue of this approach relates to the size of the XML file and the required processing time for parsing and processing it.

7.3.3   Spatial Random Access by Frame Tiling

More radical approaches have been proposed in order to break dependencies at the cost of the overall compression performance. This is the case for Mavlankar and Girod (2009) who proposed interactive PTZ for videos by using a multilayer tiling approach, as well as by Rondão Alface, Macq and Verzijp (2011) and Niamut et al. (2011). Here tiles are defined as a spatial segmentation of the video content into a regular grid of independent videos that can be encoded and decoded separately. In particular, the partitioning into tiles is temporally consistent. We denote the tiling scheme by 'M x N' where M is the number of columns and N is the number of rows of a regular grid of tiles. Classic video compression corresponds to a 1 × 1 tiling. While this approach nicely removes all dependencies between tiles, the drawbacks of this approach are that synchronisation between tiles (temporally as well as in terms of global encoding rate) must be ensured and that an IROI might require more than one tile to be accessed for reconstructing the view (Rondão Alface, Macq and Verzijp, 2011), as was shown in Figure 7.4. Moreover, the compression performance is also reduced as the existing redundancy between tiles can no longer be exploited.

Therefore, tile-based compression for random access into ultra-high resolution or panoramic video has different objectives than classic video compression. While a good compression performance for the total panoramic video is still desirable at the server side, the objective is

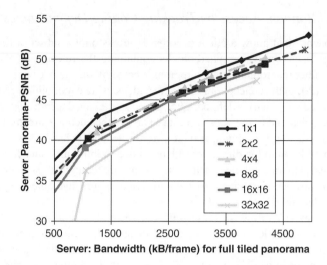

Figure 7.6 Global Rate-Distortion curves for tile-based compression. At the server side, finer tiling decreases the compression performances (Rondão Alface, Macq and Verzijp, 2011).

also to maximise the quality of an IROI reconstructed by an interactive client and minimise the total bitrate of the tiles it needs to decode.

The study by Rondão Alface, Macq and Verzijp (2011) analyses the impact of the tiling scheme on compression performances, in the context of interactive navigation in 4 K spherical panorama. The work is based on an intra-frame coding scheme using the tiling capabilities of JPEG2000 (see Section 7.3.1). Tiling of different granularities are compared, ranging from a 1 × 1 grid (no tiling, the panorama is encoded as a single video) to a 32 × 32 grid (finest tiling). Figure 7.6 and Figure 7.7 present the rate-distortion performance of tile-based encoding, by

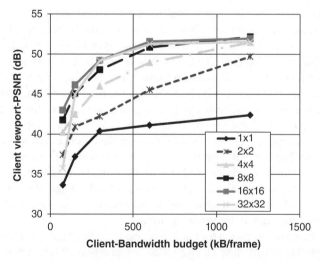

Figure 7.7 Rate-Distortion curves for the rendered viewport of an interactive client. At the client side, finer tiling improves the rate-distortion performances (Rondão Alface, Macq and Verzijp, 2011).

measuring the Peak Signal-to-Noise Ratio (PSNR) as a function of the bandwidth budget at the server and client side respectively. More precisely, Figure 7.6 shows the rate-distortion performance of the full panorama as a function of the bandwidth budget at the server side. Fine tiling schemes decrease the overall compression performances, since the redundancy between tiles cannot be removed by the encoders. At the client side, the PSNR is measured on the final viewport. In the context of spherical video, this means that the measures of PSNR take into account the rendering of the IROI on a spherical surface (see Chapter 8, Section 8.2 for more details on video rendering). For a given IROI size, Figure 7.7 shows the PSNR of the rendered viewport as a function of the bandwidth available at the client side. In this case, finer tilings are beneficial since they allow the client to access a subset of tiles that better approximates the desired IROI. From these two figures, it can be seen that for a given transmission bandwidth budget and a given IROI size and position,

- finer tilings better suit the client-side. Since the resolution of the tiles decreases, the same rate enables to increase quality on the received IROI.
- larger tilings better suit the server-side. When tiles are large, the redundancy is still efficiently removed from the content by the encoder.

It follows that intermediate tilings (e.g., 8×8 in this study) enable to reach a good trade-off for the whole transmission system.

7.3.4 Multi-Resolution Tiling

The aforementioned works mainly proposed tiling the content in a single resolution and recompose the desired IROI by restitching received tiles. However, Mavlankar and Girod (2009) proposed a multi-resolution scheme where lower resolutions could help reduce the compression loss of tiled content at higher resolutions. If the lower resolution tiling is small enough (such as thumbnails) the bitrate overhead of using another resolution layer is affordable. Moreover, multi-resolution tiling enables the quality of user-defined zooming factors on tiles to increase. Once the user zooms into a region of the content, the system will provide the highest resolution tiles that are included in the requested region. As a result of transmission latency or possible data loss, the device reconstructing the IROI may end up in a situation where the currently accessed high resolution tiles do not fully cover the current IROI. In that case, multi-resolution is useful for concealing the missing tiles. For example, tiles may be missing at the borders of the IROI in case of a fast panning from the end user. In that case, lower resolution tiles can be used to approximate those border pixels at a much lower transmission cost. However, the tasks of reconstructing the IROI from multi-resolution tiles is also more complex and might lead to visual quality drops at the border between high resolution tiles and concealed pixel regions.

Encoding such multi-resolution tilings as proposed in Mavlankar and Girod (2009) is not directly compliant with H.264/SVC. However, it is possible to use Extended Spatial Scalability (ESS) as illustrated on Figure 7.8 and nest different enhancement layers altogether in a non-compliant SVC stream. Figure 7.9 shows the rate-distortion performance for such a multi-resolution tiling for 7K content. If inter-layer dependencies and non-compliant enhancement layer nesting are both allowed, then the compression performance increases drastically but this comes at the cost of random access.

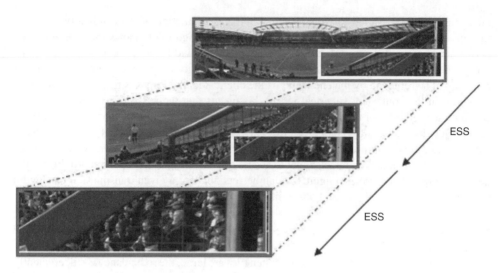

Figure 7.8   Multi-resolution tiling with Extended Spatial Scalability (ESS) in H.264/SVC.

7.3.5   Overlapping Tiling for Low-Powered Devices

When navigating through a panoramic video, each view request from a device translates into a certain IROI with respect to the panoramic video. As was shown in Figure 7.4, when splitting up the panoramic video into multiple tiles, typically more than one tile is required to cover the requested IROI. In the tiled streaming approaches described thus far, each tile should be

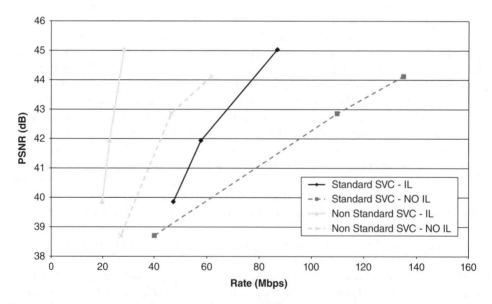

Figure 7.9   Rate-distortion curves for multi-resolution tiling encoding, comparing (i) standard SVC vs. non-standard SVC (nesting ESS layers) and (ii) use of inter-layer (IL) encoding or no IL.

independently decodable and is therefore independently encoded. A direct consequence of this approach is that a receiver needs as many decoders as it needs tiles to cover the requested IROI. On today's desktop and laptop devices this does not pose significant problems since these devices have enough processing power to perform video decoding in software or even on graphic card processing units, and can easily scale the number of simultaneous decoding processes with the number of tiles.

This is not the case for today's mobile devices, such as tablets and smartphones. Apart from the fact that they generally have less processing power available, battery life is a very important consideration in these devices. For this reason, mobile devices are almost always equipped with a hardware-based video decoder, for example, a dedicated H.264 decoder. The downside of hardware-based decoding is that a device is limited by the number of physical decoding chips in a device. Hence, if multiple video streams need to be decoded at once, the device must be equipped with multiple hardware decoding chips. While some advanced hardware decoding chips feature the possibility of decoding multiple videos at once using a single chip, using either time-division multiplexing or multiple dedicated decoding lanes in the chip, most mobile devices can only handle between one and four simultaneous video feeds.

This limitation poses a problem for the forms of multi-resolution tiling as discussed previously. For example, if at least four tiles are required to reconstruct any given IROI, then depending on the tile size, the number of simultaneous tiles needed for a given IROI can be as high as 9, 16 or even 25. Considering the fact that a larger tile size (resulting in fewer tiles) leads to less efficiency (more unnecessary pixels transferred), it is safe to say that tiled streaming is not yet feasible in an efficient way on mobile devices using hardware decoding.

To solve this problem, it is possible to change the tiling scheme in such a way that the client device never needs more than one tile in order to prepare the view to be displayed. This approach uses *overlapping* tiles, that is, tiles having a certain overlap with each other. By using overlapping tiles, the total number of tiles that together cover the requested IROI can be reduced, as shown in Figure 7.10. In this particular case, the total number of tiles required to reconstruct the requested IROI is reduced from four (see Figure 7.4) to one, while the IROI size and total number of tiles remain the same.

The effectiveness of overlapping tiles in reducing the number of tiles required to reconstruct a given IROI is determined by the overlapping factor Z, which gives the relative overlap (per

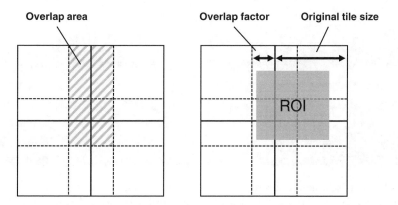

Figure 7.10   Tiling scheme with overlapping tiles (2 × 2 grid) and an overlapping factor of one-third.

Figure 7.11 Regular non-overlapping tiling scheme. An image is split up into four tiles, each having the same size.

planar direction) of a particular tile in relation to its size. Choosing a larger value for Z results in larger overlapping areas, and thus, less tiles are required to reconstruct a given IROI. The downside of this overlap is that it results in more storage on the server, which has to store redundant pixels. Figure 7.11 and Figure 7.12 give an impression of the increase in storage required for an overlap factor Z of 0.5.

A side effect of having overlapping tiles is that tile sizes are no longer equal across a given tiling scheme. That is, for a given overlapping factor Z, a corner or edge tile has an overlap in respectively one or two planar directions. For example, a tile containing the top-left region of a given video stream overlaps with the tile to its immediate right and with the tile directly below it. A tile in the middle of a tiling scheme, however, overlaps in four planar directions and therefore has a larger tile size. This effect is illustrated in Figure 7.13.

7.4 Models for Adaptive Tile-based Representation and Delivery

In the previous section, we described different methods to represent and code ultra-high resolution video in a manner that supports spatial random access. This section shows, with two examples, how the tiling approach gives some lever to optimise the overall coding performance of the panorama, as well as the selection of tiles to be transmitted to an interactive client.

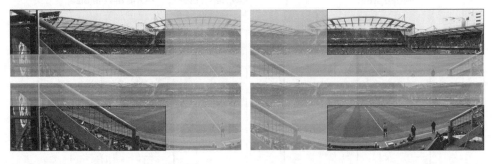

Figure 7.12 Tiling scheme with overlapping tiles. An image is broken up into four tiles, each with overlapping factor Z of 0.5. The clear areas in the figure represent the tiles as they would be without any overlap. The grey areas represent the additional area covered by the tile due to the overlapping factor.

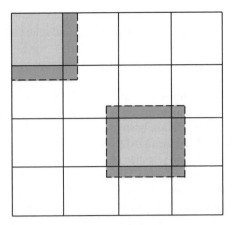

Figure 7.13 Overlapped tiling scheme with Z = 0.25. Overlapping tiles result in heterogeneous tile sizes. In the example the total size of the upper-left corner tile is smaller than the total size of the middle tile.

7.4.1 Saliency-based Adaptive Coding of Tiled Content

Introducing tiling as presented earlier usually comes with a cost in compression performance. This is largely compensated by the fact that it enables the reduction of access bandwidth usage and optimisation of video quality at the user side since only the requested portion of the panorama is received and decoded. However, at the server-side, adaptive encoding of the tiles might be desirable. Indeed, for typical panoramic content with a static multicamera setting, tiles covering static background regions (static tiles) typically only require low rates for achieving high fidelity, while tiles presenting moving objects or people (active or salient tiles) need much higher rates. Since each tile can be encoded independently by a dedicated encoder, some inter-encoder communication might be required for adapting quality and rate based on the tile content activity compared to the total panorama activity. In order to minimise such inter-encoder communication, which is hampering parallel implementations, such as presented in Figure 7.14, a key aspect of coding tiled content relates to an efficient tile rate prediction technique or rate distribution algorithm. Given a global compression objective for the source content, a tile rate predictor can independently inform encoders on the levels of compression to apply for each tile.

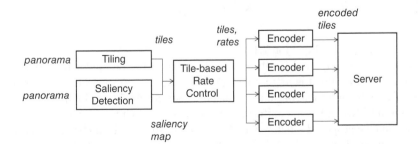

Figure 7.14 Saliency-based adaptive coding of tiled content.

Two main strategies can be used in this context: (a) avoid this issue by encoding each tile multiple times at different rates such that each receiving end selects the best rates with respect to their decoding buffer (as in HTTP Adaptive Streaming – see Section 7.5.1) or (b) encode each tile at a single optimal predicted rate given a fixed bandwidth budget constraint. We focus on the second case in this section.

As explained before, sharing bandwidth equally between all tiles of the panorama content is suboptimal as some parts of the content are static and active tiles require different rates to be encoded with high fidelity. Predicting whether a tile is active or not can be done via priors such as Maximum Absolute Difference (MAD) and visual saliency (see Chapter 5).

Visual saliency is a measure for whether objects or regions stand out relative to their neighbours and is considered a key property for human attention. In the literature, saliency has already been successfully used for rate control in single video H.264 rate control, for example, for Just Noticeable Difference (JND) based H.264 encoding on salient regions (Hadizadeh and Bajic, 2011; Wang and Chiu, 2011). These works actually focus on adapting the bitrate for each macroblock of a video frame. Here, the focus is set on tile total rate prediction with respect to the total panorama content.

The MAD prior, which detects pixel-to-pixel differences between the current frame and the previous frame, has been shown to be efficient at estimating rates for motion predicted frames with a static background (Kim and Hong, 2008). Visual saliency as estimated in Chapter 5 also has the advantage of enabling the efficient prediction of rates for intra-frames.

Given a global bandwidth budget BW, the target rates r_i of each of the $M \times N$ tiles composing the panorama are set according to the following equation:

$$r_i = \frac{\lambda}{M \times N}BW + (1 - \lambda)\alpha s_i \qquad (7.1)$$

where s_i is the aggregated saliency scores for tile i, α is a multiplicative factor that converts saliency values into rate values, which is set in order to meet the bandwidth constraint in Equation (7.2), and λ is a factor that enables tuning the impact of saliency on the target rate. If λ is set to 1, the rates are equal for all tiles. If λ is set to zero, the resulting rates are directly proportional to spatiotemporal saliency. Setting a value of 0.2 for λ leads to better visual results because it ensures a minimal rate for tiles that are not very salient and smoothes inter-tiles quality variations that might result in noticeable artefacts at tile borders at the user side. As stated before, the sum of the rates is subject to the following constraint

$$\sum_{i=1}^{MxN} r_i \leq BW \qquad (7.2)$$

and is used to update α in Equation 7.1.

The bottom part of Figure 7.15 illustrates a heatmap of tile rate prediction for the reconstructed tiled panorama that is represented above it. Figure 7.16 shows differences in PSNR (delta-PSNR) for a non-salient tile and a salient tile respectively. These differences correspond to the offset in PSNR between a uniform rate control tile encoding (all tiles receive the same target rate) and the proposed saliency-based rate control strategy. It can be seen that for larger values of relative saliency of the tile, the PSNR increases for the same total bandwidth budget. For low values of saliency, PSNR values are slightly decreased. The 'percentage saliency'

Figure 7.15   Tile rate allocation heatmap.

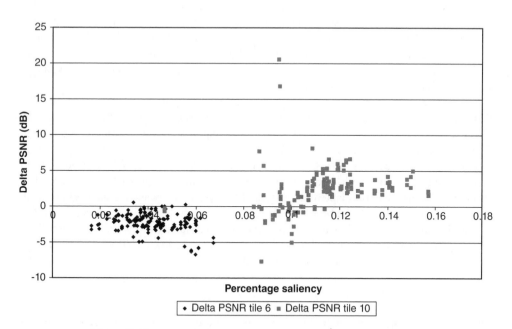

Figure 7.16   Delta-PSNR with adaptive coding for two different tiles.

corresponds to the ratio between the saliency score of the tile and the total saliency of the whole panorama.

On average, for the 7K panoramic video football sequence illustrated in Figure 7.15, when all tiles are used to recompose the panorama, the Delta PSNR gains are equal to 2.3, for a total bandwidth constraint of 50 Mbps and to 3.1 for a bandwidth constraint of 40 Mbps.

7.4.2 Optimisation of Tile Selection Under Delay and Bandwidth Constraints

In a video delivery system supporting continuous spatial navigation, the delay between the submission of IROI requests by the end user commands and the arrival of the correct subset of video data plays naturally a very important role on the QoE. Indeed, at the server side, there is some uncertainty on the requested IROI position due to delay and there is a risk that tiles being sent are no longer the correct ones by the time they are received and processed at the client side.

In tile-based delivery system, there are multiple approaches to deal with this issue. A first approach is to allow displacement only when the required tiles become available (Niamut et al., 2011). This guarantees that the displayed IROI is always reconstructed from tiles at the highest quality available, but at a cost of a larger interactivity delay.

Another approach to reduce the effect of the system latency is to rely on prediction of the end-user behaviour. In Mavlankar and Girod (2009), the IROI trajectory is predicted so that the most probable tiles (with respect to future IROI positions) can already be delivered to the client device. This prediction can be made based on learning the user interaction style that can be modelled as an ARMA process and then fed to a Kalman filter. It is also possible to predict the IROI from the content saliency itself, for example, by analysing lower resolution thumbnails (Mavlankar and Girod, 2009).

In (Rondão Alface, Macq and Verzijp, 2011), the proposed solution relies on a representation of the tiled video at multiple qualities. For a given IROI position, the selection of tiles to be delivered covers an area larger than the one strictly required for reconstructing the current IROI. Each of these tiles is delivered with a quality that decreases with the distance from the current IROI, according to a Gaussian law. With this approach, it is still possible to instantly render the IROI with a best-effort quality that is then progressively enhanced. The Gaussian parameters can be adapted in function of the system delay and the navigation speed. To optimally select the tiles and their quality levels, the approach relies on a combinatorial optimisation model, which aims to maximise the expected quality of the reconstructed IROI under a given bandwidth budget. Figure 7.17 plots the average PSNR of the IROI as a function of the system delay. The performance of the optimisation technique is compared to the following:

- **Perfect forecasting** refers to the ideal situation where only the tiles that cover the current IROI are received on time to the client devices, regardless of the system delay.
- **Constant rate** refers to a case where the full set of tiles (covering the full panoramic video) is sent to the client device at a constant quality, regardless of the IROI requests.
- **Direct IROI mapping** refers to a method where the client requests only the tiles covered by the current IROI position, without trying to compensate for the system delay. The PSNR

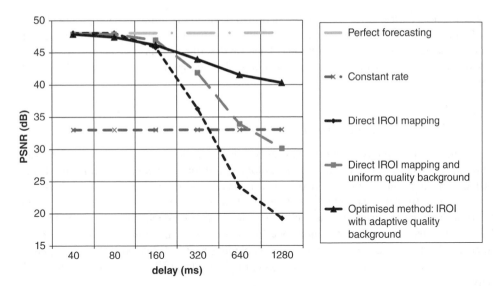

Figure 7.17 Performance of tile selection models against system delay.

for this solution rapidly decreases with delay, since the IROI has usually moved away from the area covered by the received tiles.

- **Direct IROI mapping and uniform quality background** behaves like the previous method, where a low resolution version of the full panorama is always delivered to the client device in order to conceal the mismatch between the current IROI and the delayed tiles.

The performances of the optimised approach described above, *IROI with adaptive quality background*, is the closest to the *perfect forecasting* benchmark. For larger delays, the prediction area of the IROI position progressively covers an area close to the full panorama. Therefore the performances asymptotically approach the PSNR of the *constant rate* solution.

Finally, note that optimisation models have been designed for coding schemes that natively support spatial random access such a JPEG2000. In Naman and Taubman (2011a), the JPEG2000-based Scalable Interactive Video (JSIV) paradigm is introduced as a technique to optimise the interactive consumption of video, where each frame stored on the server is independently coded with JPEG2000. The approach relies on loosely-coupled server and client policies: the server maintains a model of the state of the client decoder and has the freedom to respond to client requests by choosing what subset of the video bitstream to deliver in response. This model is extended in Naman and Taubaum (2011b) where motion compensation is also employed to improve the video data reconstruction. This loose-coupling notion is also a key requirement for designing an adaptive transport system that can scale with many interactive clients, as the next section will show.

7.5 Segment-based Adaptive Transport

This section describes examples of how the previous concepts and techniques can be used in a delivery network in order to support interactive consumption of immersive audio-visual content. We address two approaches to develop a transport mechanism, each targeting different deployment scenarios. After a short introduction to IP video streaming technologies in general, we first describe an evolutionary solution, which aims to extend the adaptive streaming technologies deployed today over the internet. Then we describe a clean-slate approach, based on a publish/subscribe model, allowing a system to realise fine-grained adaptation of the content delivery.

7.5.1   Video Streaming Over IP

Streaming media over IP networks has seen drastic changes and growth in recent years. Through the use of streaming media technologies, millions of people can consume audio-visual content on all of their devices, such as smart and connected TV sets, tablets and smartphones. To realise and understand the challenges of delivering format-agnostic media to all of these devices over an inherently best-effort networking technology, we need to look back at the history of video streaming. We first cover video streaming over the unmanaged internet, also known as 'over-the-top' video streaming or internet video streaming. Then we describe the streaming technologies usually deployed for linear TV services over managed IP networks.

Video Streaming Over the Internet

Media delivery over the internet started out with the HyperText Transfer Protocol (HTTP). Along with IP, HTTP is one of the very basic elements of today's web. It was originally designed for the transfer of linked textual elements and web pages. However, it rapidly became clear that HTTP could be used to transfer and view web-based video, through a number of methods (Kozamernik, 2002). With the *download* method, video files are transferred in their entirety to the end user device or web browser cache. After successful reception, the video file can be viewed in a local player. Full interactivity, such as pause, rewind and seek, is possible. An adaptation of this approach is referred to as *progressive download*; playout can start as soon as part of the video file has been received. Playback can continue as long as the transfer rate is higher than, or equal to, the playout rate. Interactivity is only possible for those parts that have already been received. Last, with *pseudo streaming*, a particular form of progressive download, enhanced interactivity and search is possible through the provision of additional file header information, such that a user can seek portions of a video file that are not yet downloaded.

Initial HTTP-based video transmission came with some significant drawbacks; video files were delivered without regard to the user's actual bandwidth. This led to congestion and video frame drops. Also, low-latency and live video streaming were not possible. This led to the investigation into alternatives to the techniques described above, such as the Real-time Transport Protocol (Schulzrinne, Casner and Jacobson, 2003). This protocol has many interesting properties and provides much more control over the quality of the users' viewing experience, for example, by enabling the switch between video streams of different bandwidths. However, a major drawback for internet delivery is its inability to traverse firewalls and Network Address Translation (NAT) procedures that are present in routers and home gateways.

Recently, a new form of progressive download has emerged, referred to as HTTP Adaptive Streaming (HAS) (Stockhammer, 2011). HAS enables the delivery of (live) video by means

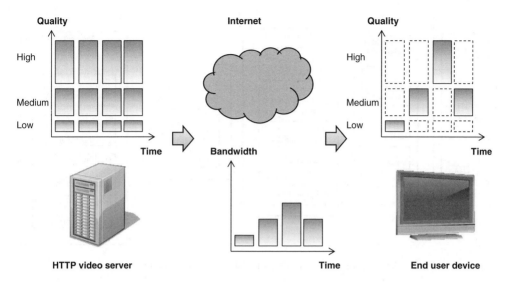

Figure 7.18 HAS allows for switching between different representations when the bandwidth in the network varies. Adapted from (Timmerer and Griwodz, 2012).

of the HTTP protocol, by providing the video in segments that are independently requested by the client from a web server. A video is split into several video segments, which are, in themselves, standalone video files. These segment files can be delivered separately, but when recombined they provide a seamless video stream. A video can be provided in several representations: alternative versions of the same content that differ in resolution, the number of audio channels and/or different bitrate. All representations are temporally aligned such that segments of different representations can be interchanged. As depicted in Figure 7.18, having representations associated with different bitrates allows for seamless adaptation to varying bandwidth conditions.

In most of today's HAS solutions, such as Pantos and May (2013), 3GPP-DASH (2012) and MPEG-DASH (2012), a manifest file also known as a Media Presentation Description (MPD), is used to describe the structure of the media presentation. This manifest includes all information that an HTTP client needs to determine the configuration of the media and to retrieve the media segments corresponding to a media session. This concerns the audio and video codecs, the container format, the media presentation, alternative representations of the media at, for example, low, medium and high quality (as well as associated bitrates), information on the grouping of media, and segment information (e.g., the segment length and resolution). As an example, Figure 7.19 gives the structure of an MPEG-DASH media presentation description.

Video Delivery Over Managed IP Networks – the IPTV Case

As introduced in Section 2.5, many telecom operators and service providers have augmented their service offerings with IP-based TV services. Although audio-visual content is distributed over an IP infrastructure, the so-called IPTV services are to be distinguished from video

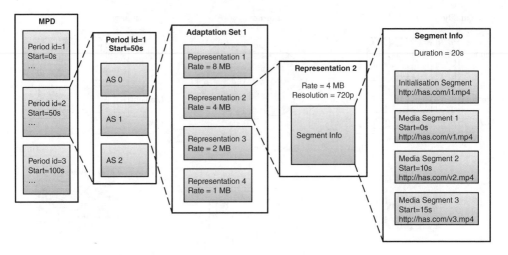

Figure 7.19   Structure of an MPEG-DASH media presentation description.

streaming over the best-effort internet. A key distinction is that IPTV is a controlled service delivered over a managed network. In this aspect it is closer to its cable and satellite equivalents than to the best effort internet.

The possibility for the service provider to fully manage the delivery path from its video streaming servers (the IPTV 'head-end') to the end user allows him to use specific mechanisms that are harder, if not impossible, to deploy over the best effort internet. One of the strongest characteristics of IPTV deployments is the ability to deploy native IP Multicast within the realm of the operator's network. In particular, Source-Specific Multicast (SSM) (Bhattacharyya, 2003; Holbrook and Cain, 2006) is used to deliver a single stream to multiple receivers. The list of subscribers/listeners for a given stream is managed by the Internet Group Management Protocol (IGMP) (Cain et al., 2002) or Multicast Listener Delivery (MLD) (Deering, Fenner and Haberman, 1999) in IPv6. At the transport layer, User Datagram Protocol (UDP) is usually favoured to the reliable Transmission Control Protocol (TCP). This ensures timely delivery of video packets (no acknowledgement is required), at the cost of unreliability (no monitoring and retransmission in case of packet losses). This capability to implement a *push* mechanism from a single source to many receivers has permitted large-scale deployment of linear TV services over IP infrastructure, competing with the traditional broadcasting infrastructure.

Despite its capability to scale to many receivers, push-based delivery is not suited for personalisation or interactivity. Therefore, TCP-based unicast delivery remains mandatory for interactivity features, such a time-shifted TV, Video on Demand (VOD), dynamic user interfaces, Electronic Program Guide (EPG), and so forth. Interestingly, enabling Fast Channel Change (FCC), one of the most basic but essential interactivity features, was one of the original hurdles for IPTV deployment, since the response time provided by analogue tuners is hard to replicate in the digital domain. There the main contributors to channel change delay include the latency of the IP multicast IGMP 'join' process and, particularly, the GOP intervals between I-frames in MPEG-2 and H.264/MPEG-4 streams, which allow the TV display to synchronise with a complete video image.

As a result, many solutions for FCC exist and have been deployed and submitted for standardisation. A solution deployed in practice (Degrande et al., 2008) consists of using dedicated servers to accelerate the channel change speed experienced by the end user: upon a channel change request, a server having buffered the last video frames of the requested channel sends to the client device a unicast stream starting with the last available I-frame, so as to avoid the two main sources of delay listed above (IGMP join and GOP latency). In parallel to that process, the multicast join is also operated, so that the client device can tear down the unicast connection as soon as the multicast stream can be synchronised with the current playout. Many implementation variants exist on how the handover from the unicast to the multicast reception is operated.

This combination of unicast and multicast transport actually allows the system to support massive delivery of video streams to steady-state receivers, as well as a personalised treatment of transient-state receivers, interacting with the system. This observation motivates some of the delivery mechanisms introduced in Section 7.5.3.

7.5.2   Tiled HTTP Adaptive Streaming Over the Internet

In Section 7.3, the concept of spatial video segmentation, or tiling was introduced. It seems natural to extend the temporal segmentation of HAS to spatial segmentation, and indeed, several recent studies, such as Mavlankar et al. (2010), Khiem et al. (2010), Niamut et al. (2011) and Quax et al. (2012), have considered similar approaches. All of these approaches are aimed at encoding, storing and streaming live and/or pre-recorded video in such a way that users can freely zoom and pan into any ROI in a video stream, and in a manner that is scalable to a large number of users by being efficient in terms of bandwidth, computation cost and storage. The generic concept behind these approaches, referred to as tiled HAS, is first described.

With tiled HAS, each video tile is individually encoded and then temporally segmented according to any of the common HAS solutions. All HAS tiles are temporally aligned such that segments from different tiles can be recombined to create the reassembled picture. This process is depicted in Figure 7.20. An advantage of using HAS for the delivery of spatial tiles is that the inherent time-segmentation makes it relatively easy to resynchronise different spatial tiles when recombining tiles into a single picture or frame. As long as the time segmentation process makes sure that time segments between different spatial tiles have exactly the same length, the relative position of a frame within a time segment can be used as a measure for the position of that frame within the overall timeline. For example, frame number n within time segment s of tile A is synchronised with frame number n within time segment s of tile B. On the client side, exact seeking within each segment stream is required, to ensure perfect synchronisation between the segments that make up the final viewport to be rendered on the screen of the end user.

For tiled HAS streaming, we may reuse the MPD from, for example, MPEG-DASH, to describe the structure of a tiled HAS media presentation. As reported by Khiem, Ravindra and Ooi (2012) tiled streaming would require support for new dimensions under the DASH framework, namely those associated with zoom levels and the spatial coordinates of the tiles. Khiem, Ravindra and Ooi (2012) propose that available zoom levels are described via extensions to the stream information tag and grouping the segments belonging to a set of tiles

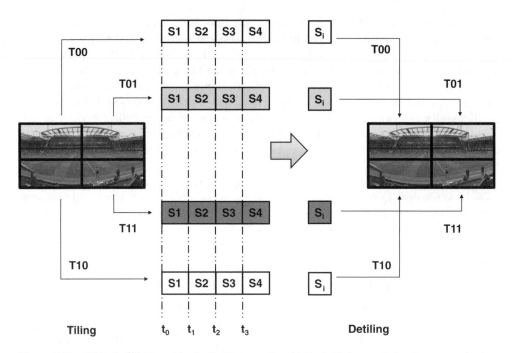

Figure 7.20 With tiled HAS, a video is tiled in a certain grid. Each tile is encoded and segmented using HAS segments. In this example, a 2×2 grid is employed.

as a single playlist element. A potential alternative consists of using the *role* element from the MPD syntax by extending it with a new scheme in order to describe tiled videos. Each tile is then described as a *representation* in its own *adaptation set*.

We now discuss and compare the different approaches for tiled HAS. In many ways, the notion of IROI streaming for today's networks and devices was first explored in-depth by Mavlankar et al. (2010) and Mavlankar and Girod (2011). Mavlankar developed various methods in the context of an IROI streaming system for online lecture viewing, selecting tiled streaming as the best compromise between bandwidth, storage, processing and device requirements. This system, referred to as ClassX (ClassX, 2011; Pang et al. 2011a, 2011b), allows for capture and interactive streaming of online lectures. In building this system, the authors set out to solve three challenges. First, the system had to deal with heterogeneous, potentially wireless, delivery networks and a high diversity of end user devices. Second, it had to deal with the limited processing options in end user devices, in particular mobile devices. Finally, it had to cope with the delays when switching between tiles. Tiled HAS was selected as a way of dealing with the first challenge, as shown in Figure 7.21. The complexity of retiling on low-powered mobile devices was reduced by using an overlapping tiling scheme as described in Section 7.3.5, such that only one tile is needed to satisfy any IROI position. To handle different display sizes and network conditions, each tile was encoded at different spatial and quality resolutions. Tile switching delay was lowered by enabling a form of crowd-driven data prefetching, where viewing patterns from different users collected at the server were exploited in the prediction of IROI requests. Additional

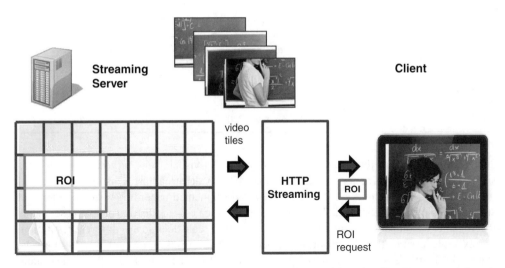

Figure 7.21 The tiled HTTP adaptive streaming approach underlying the ClassX system. Adapted from Mavlankar and Girod (2011).

features of the ClassX system include an automatic ROI selector and the option for showing additional slides in a synchronised manner.

Some of the first comparisons between regular encoding and coding for tiled streaming were investigated by Khiem et al. (2010). In particular, they compared monolithic streaming, as introduced in Section 7.3.2, and tiled streaming. Their results indicated that a monolithic stream with proper choice of parameters achieves better bandwidth efficiency than tiled streams. The research, part of the Jiku project (Jiku, 2011) on interactive and immersive media experiences, was later extended with studies of user access patterns in Khiem, Ravindra and Ooi (2012).

The efficient and interactive distribution of panoramic video sequences was explored by Quax et al. (2012) in the context of the Flemish eXplorative Television project (xTV, 2011). Their focus was on enabling low-delay interaction with high-resolution and high-quality video, with constraints on the available bandwidth and processing capabilities, as encountered in current network technologies and devices. They studied bandwidth requirements for tiled streaming as well the performance of media containers. With a tiled streaming approach, they were able to achieve a bandwidth reduction from 45 Mbps to 4 Mbps for 4k video. They noted that the required bandwidth varied significantly over the scene. For example, in a panoramic video of a live outdoor concert, the least amount of required bandwidth would occur if a user looked at the sky, as the minimal motion would lead to the best compression results. In contrast, the highest amount of required bandwidth would occur when a user looked at the audience. The authors also noted that since rapid seeking is required for the tiled HAS approach, the choice for a particular media container has an impact on the seeking performance. Several codec and container implementations only support seeking to the nearest I-frame only. This results in increased switching delays, as the client waits until all decoded tile frames can be synchronised.

Distribution of panoramic video sequences was further investigated in (Brandenburg et al., 2011; Niamut, 2011), in the context of the European-funded project FascinatE (FascinatE,

Table 7.1 Relation between tiling grid, scale resolutions and number of tiled streams

Tile layout	Scale resolution	Number of streams
1×1	872×240	1
2×2	1744×480	4
4×4	3488×960	16
8×8	6976×1920	64

2010). Here, built into their system, a multilayer audio-visual scene representation is considered, including both panoramic and regular HD broadcast capture sensors, as described in Chapter 3. The captured scene is then tiled and segmented according to the hierarchical structure depicted in Figure 7.22. In this structure, each scale is a collection of spatial segments that together encompass the entire video. However, each scale does so in a different resolution. For example, assume that the panoramic video layer has a resolution of 6976×1920 pixels. With a dyadic tiling approach with tiling grids ranging from 1×1 up to 8×8, the panoramic video is provided in four different scales, as shown in Table 7.1.

The system described in Brandenburg et al. (2011) and Niamut (2011) consists of a number of components, as shown in Figure 7.23. The *Segment Server* is a regular HTTP server or CDN cache node, from which the tiled HAS segments are served. On the client side, the *Video Renderer* includes both a *Segment Client* and a *Frame Combiner*. The Segment Client uses the tiled HAS manifest file to determine which tiles are needed for the current view, retrieves the associated segments and decodes them. The Frame Combiner then recombines the segments and provides the requested view to the user, possibly by cropping out pixels that are unnecessary in the requested IROI.

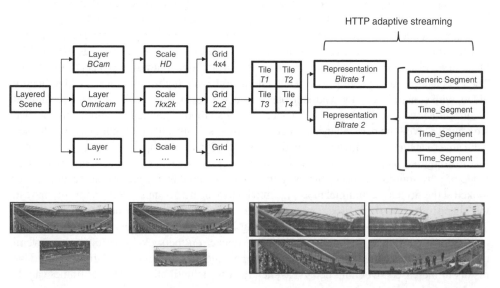

Figure 7.22 Hierarchical structure of spatially tiled content as described in Brandenburg et al. (2011) and Niamut (2011). For every layer, multiple resolution scales are created. These scales are tiled according to a grid layout. Each tile is associated with an adaptive stream, and thus temporally segmented.

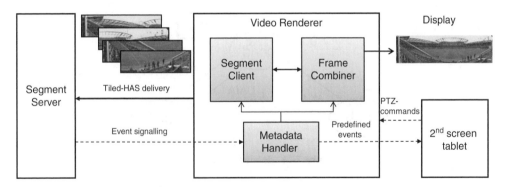

Figure 7.23 Tiled HAS framework as described in Niamut (2011a, 2011b).

An additional component, the *Metadata Handler*, is responsible for event signalling. This allows for asynchronous real-time communication triggered by the Segment Server, for instance notifications of new content layers becoming available. Within today's HAS-based mechanisms like MPEG-DASH, data and control path are tightly coupled, making HAS a mostly purely client-driven technique whereby session updates can only be signalled by updates in the manifest file. Therefore, the authors extend the tiled HAS framework with an event-signalling framework based on the *WebSocket* protocol (Fette and Melnikov, 2011), a web technology that allows for low-latency two-way communications over a single TCP connection. Scene navigation is performed on a second screen device, such as a tablet or mobile phone. In an extension to this system, the overlapped tiling approach, as described in Section 7.3.5, is used for direct streaming to mobile devices. The main advantage here is that only a single adaptive stream is required to meet a requested IROI at any time. However, in order to reduce switching delays, a fallback layer covering the entire scene can be used.

7.5.3 Publish/Subscribe System for Interactive Video Streaming

In this section, we investigate how novel transmission paradigms recently proposed in the literature can be applied to the delivery of interactive video. As pointed out in Section 7.5.1, an ideal system would need to combine the power of push-based transmission and the flexibility of pull-based transmission. Several systems deployed today have somehow implemented this mix at different levels and for different use cases. HBB TV for instance complements broadcast delivery of TV content with personalised unicast streams over the internet. IPTV systems support hand-over mechanisms between multicast and unicast, essentially to support FCC (see Section 7.5.1). However, these examples are adding new complexity on top of existing transport mechanisms. In contrast, several research initiatives are now looking at ways to create delivery networks that put the content publishing and subscribing mechanisms at the basic layers of the system.

Publish/subscribe (Pub/Sub) is an information paradigm that shifts the power away from the data sender. Data consumers express their interest in specific pieces of information explicitly, which are forwarded to them by the network when they become available. The sender and receiver become a loosely coupled pair, in comparison to the traditional client-server model.

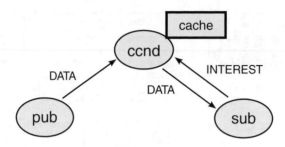

Figure 7.24   Basic CCNx set-up.

Pub/Sub systems are particularly suitable for large-scale information dissemination. It provides the flexibility for users to subscribe to information of interest, without being intimately tied to who makes the information available or even to when that information is made available.

Today, the Pub/Sub paradigm is used for systems like stock quotes or RSS (rich site summary) feeds (Pubsubhubbub, 2013), where selected parts of a large pool of quotes/topics need to be delivered in real-time to different end users. The Pub/Sub paradigm is also actively researched in a broader networking context: for example, DONA (Koponen et al., 2007), PSIRP (Fitiou, Trossen and Polyzos, 2010), COPPS (Chen et al., 2011). Several names were given to such networking systems: Named Data Networking (NDN), Information-Centric Networking (ICN) or Content-Centric Networking (CCN). These systems differ in their data naming scheme and how the data routing is actually handled, but they all adhere to the Pub/Sub paradigm. In the following, we will use the term CCN consistently.

Pub/Sub Video Streaming

We are interested in applying the Pub/Sub paradigm for interactive video streaming. Starting from the tiled representation introduced in Section 7.3, we have multiresolution tiled video content that we want to consume interactively and in real-time. As a way to introduce the concepts of Pub/Sub systems, we present a very basic example that will be helpful in highlighting some of the problems arising when applying these concepts to interactive video streaming. To that end we use PARC's CCNx prototype (CCNx) in combination with a Gstreamer-based publisher and subscriber (Letourneau, 2010). The setup is depicted in Figure 7.24.

Centrally, we see the CCNx router (ccnd), which is equipped with a data cache. The CCNx router clearly decouples the publisher and the subscriber from each other, as is required by the Pub/Sub paradigm. CCNx only knows two types of packets: DATA and INTEREST. The publisher captures video from for example, a webcam, encodes it and wraps the encoded data inside a media container. The webcam produces a continuous feed and, as we are interested in live streaming, we do not want to record a huge file first before making it available to the network. Therefore, we need to chunk it. CCNx expects chunked data and the size of each chunk is determined at the discretion of the publisher. In our example we use a chunk size of 4 KB, which is the default value in CCNx. As such, each data packet between the publisher and the ccnd will contain a chunk of 4 KB data. This data chunk is complemented with a CCNx header that contains the unique name of this data chunk (in addition to other fields used, for example, for data authentication, which are irrelevant for the discussion here). This unique naming is critical and CCNx uses a hierarchical naming scheme for it. In our example, the naming scheme can take the following form: *ccnx://publisher.com/camera-id/chunk-id/checksum*.

The ccnd caches all incoming data, independent of any interest (or any subscriber) for it. As a result, incoming and outgoing data at a CCNx router can differ, in contrast to traditional IP routers. In principle, each data chunk is stored in each cache along the path for as long as possible. Once a cache is full, the oldest chunks will be overwritten.

A subscriber that wishes to receive the webcam stream needs to express its interest for it. In fact, it sends an interest packet for every data packet that it needs. If the ccnd has the packet cached, it will send it from the cache. If not, it will forward the interest, based on some configured forwarding rules. Data will follow the reverse path where interests have been indicated.

Problems with Pub/Sub Video Streaming

Each CCN chunk has its own unique name. Since there are no source and destination addresses in the chunk, the forwarding is based on a lookup of this name. Fast name lookup is the most critical problem for scalable CCN forwarding (Yuan, Song and Crowley, 2012). For the discussion here, we will not dive further into the forwarding problem. Instead we focus on aspects that are more tightly related to the streaming of interactive video content.

The CCN system as presented above does not allow for random access to the video content. The subscriber may start by requesting the first chunk of the video stream. As long as this chunk is present in the ccnd cache, everything works fine. If it is not present anymore, the subscriber can express its interest for the latest chunk and continue onwards from there. Unfortunately this does not work due to the specific nature of video (see Section 7.4.1). This problem may be partly solved by using a specific format of the video and a better chunking, for example, using fragmented MP4, as is done for HAS, can allow random access to each fragment. We can map each such fragment to one CCN chunk. Thus, each chunk will be independently accessible. However, we have now created very large chunks (compared to the CCNx implementation default of 4 KB) spanning an integral number of video GOPs. Large chunks minimise the header overhead (which is in the order of 650 bytes per chunk in the CCNx implementation), but are not effective in implementing transport level flow control (Salsano et al., 2012). Salsano and colleagues propose a two level content segmentation and argue that reasonable chunk sizes can be up to a few MBs.

More problematic for video streaming is the pure pull-based nature of CCNx. Indeed, each chunk needs to be asked for separately via an interest request. For streaming data, this is wasteful. Some solutions to the problem have been proposed in literature. VoCCN (Jacobson et al., 2009) suggests employing pipelining by sending interests for multiple future media chunks. COPPS (Chen et al., 2011) presents the use of a subscription table inside the ccnd, complemented with IP multicast. As such, subscription state can be kept in the ccnd and a push-based multicast transmission can be used.

A Pub/Sub System for Interactive Video Streaming

In this section we give some details on the Pub/Sub mechanism implemented for the delivery system built in the FascinatE project (FascinatE, 2013) where a system for live delivery of

Figure 7.25   Basic Pub/Sub model.

panoramic video to interactive users has been built. In this system, the forwarding part is not included, leading to a system as presented in Figure 7.25. The publisher makes a set of live streams available. This set is composed of spatially tiled content, in different resolution layers (see Section 7.3.4). In the context of live delivery, the content caching aspect does not need to be taken into account, although one can imagine that the publisher keeps its own proper cache.

This setup may seem trivial at first, but, here, we want to focus on the following elements: scalability, low delay and random access.

Scalability

A set of streams is published that are intimately tied to one another, from which each subscriber will request a subset at any moment in time. For example, each subscriber can request its own region of interest, or can request the most appropriate resolution that makes the best fit with its device capabilities. As the number of subscribers becomes larger, more and more duplicate media streams will occur. Clearly this situation would benefit from IP multicast, as proposed by Chen et al. (2011). Alternatively, using IP multicast for all streams all the time would require a huge pool of multicast addresses, which may not be available. As an example, the FascinatE prototype requires 85 multicast addresses. This is still manageable for a single content production, but becomes problematic if all content would be offered in that way. We can benefit from a system where the publisher can dynamically change the used transport for the chunks transmitted. The FascinatE prototype uses a basic algorithm that employs unicast for as long as there is only a single subscriber and multicast if there is more than one subscriber. Ravindra and Ooi (2012) analysed several other algorithms that can be used at the sender side to determine the transport switch.

Of course, this change in the underlying transport protocol must happen seamlessly from the subscriber. During a live session, if one of the required chunks changes in the way it is transported, the subscriber should be able to handle that without any glitches in the video. The FascinatE system handles this by publishing control information as a separate stream. Each subscriber requests this control stream. It contains the list of available tile streams (different spatial tiles, different resolution levels) and the transport location on which they can be requested. Each time the transport selection algorithm that runs at the publisher, decides to change the underlying transport for a specific tile, it will update the control information and push it towards all current subscribers. Once a subscriber is notified, it will start to listen on the new transport, but in parallel keeps listening for data chunks on the current transport. Once it receives the first data chunk on the new transport, it can close its connection towards the old transport. The publisher does not get explicit feedback from all subscribers (as this would break the Pub/Sub paradigm). Therefore it will wait a certain fixed amount of time (e.g., 10 frames) before sending data chunks on the new transport.

Low Delay and Random Access

When navigating in the video content, the subscriber requires real-time interaction and this is expressed by sending IROI requests. The resulting trajectory of the IROI will result in the subscription/unsubscription towards different tiled streams. Keeping the delay as low as possible between the IROI request and the visual feedback on the subscriber display is crucial. Although this minimisation is partially addressed by the architecture and delivery models (see Section 7.4), for the Pub/Sub system, we do not want to wait for a very large piece of data to come in, before we can start to process it (i.e., decode, crop, etc.). The minimal amount of data to decode is a single frame, so the Pub/Sub data chunk contains a single video frame.

With this chunk definition, it is not possible to have random access to each data chunk. Indeed, some data chunks (e.g., video frames) are independently decodeable, while others are not. In order to stick with the data centric approach, a subscriber needs to be able to detect easily if it can start decoding a chunk, or if it needs to wait for further chunks. This clearly has a large impact on the delay as perceived by the end user. It is therefore important to keep this delay, and hence the video GOP size, small to ensure a good level of temporal random access (see Section 7.3.1).

The format of the data chunks is presented in Protobuf (Protobuf, 2013) serialisation format. It contains the data of a single encoded video frame, commonly known as an access unit. Appended to that we have header elements that allow identification of each data chunk uniquely. In order to keep the description simple, we assume only a single production, which means that a spatial identifier (sid) and a temporal identifier (pts) are sufficient. We could write both in CCNx name format as *ccnx://sid/pts*. If multiple productions from multiple organisations need to be supported, we can easily add these to our chunk definition, and obtain the equivalent of *ccnx://organization/production/sid/pts*. The chunk header also contains the key flag, indicating if this data chunk is independently decodeable. For codecs that require a different decoding order versus presentation order, we also have a dts field.

```
message AccessUnit {
required uint32 sid = 1;
required bool key = 2;
required uint64 pts = 3;
required uint64 dts = 4;
required bytes data = 5;
}
```

With the above chunk definition we can synchronise different spatial tiles. The chunk header pts can be used directly for this. As such, a client can independently subscribe to different chunk (or tile) streams, and carry out spatial reconstruction of the content based on this timestamp. In the FascinatE prototype a small synchronisation buffer of a few frames was sufficient for this spatial reconstruction.

In conclusion, the Pub/Sub mechanism for video delivery allows the system to very flexibly adapt the video flows across the end-to-end chain. On the one hand, the data granularity (at frame-level) is considerably reduced (in comparison to segments of several seconds used, e.g., in HAS). On the other hand, the underlying transport can easily support massive data push to numerous clients in combination to on-demand data pull from interactive clients.

7.6 Conclusion

In this chapter, our objective was to review the impact of the format-agnostic concept on the end-to-end content delivery chain. In Section 7.2 we reviewed the different interpretations of the concept, ranging from the basic concerns of supporting multiple resolutions and aspect ratios, to the more demanding case of spatial navigation in ultra-high resolution panoramic video.

As we have focused on this latter case in the remainder of the chapter, we have naturally addressed the problem of representing the video data in a manner that combines good compression results with the ability to interactively access spatial subsets of the content. This has led to the introduction of the spatial random access notion in Section 7.3.

There we have seen that maintaining the highest level of compression usually leads to unpractical solutions for spatial random access. Therefore, we have investigated a broad range of solutions based on the idea of tiling. By making the video tiles independent from an encoding/decoding point-of-view, one can design a system that optimally trades off overall compression performance at the sender side and optimal interactive experience for bandwidth-constrained clients. Details on a selection of optimisation models were given in Section 7.4.

Finally Section 7.5 addressed the actual transport mechanisms. We described how the techniques largely deployed today for bitrate-adaptive video streaming can be extended to support spatial interactivity. Then we presented an outlook on the future content delivery mechanisms, which inherently support very flexible content representation and a combination of one-to-many and one-to-one delivery patterns.

Ultimately, we envision a situation where the acquisition of full audio-visual scenes at ultra-high fidelity will be completely decoupled from the way end users access and consume this content. Future 3D and 'free viewpoint video' production systems, as well as an always increasing range of consumer devices are key trends that can only make the concept of a format-agnostic delivery chain more relevant in the coming years.

References

3GPP-DASH (2012) '3GPP TS26.247, Transparent End-to-End Packet-switched Streaming Service (PSS); Progressive Download and Dynamic Adaptive Streaming over HTTP', accessed 20 August 2013 at: http://www.etsi.org/deliver/etsi_ts/126200_126299/126247/10.00.00_60/ts_126247v100000p.pdf

Adams, M. and Kossentini, F. (1998) 'Performance Evaluation of Different Reversible Decorrelating Transforms in the JPEG-2000 Baseline System,' *1998 IEEE Symposium on Advances in Digital Filtering and Signal Processing*, Victoria, BC 5–6 June, 20–24.

APIDIS (2008) 'Apidis Overview', accessed 20 August 2013 at: http://www.apidis.org

Bhattacharyya, S. (2003) 'An Overview of Source-Specific Multicast (SSM),' RFC 3569 (Informational), Jul. 2003, accessed 20 August 2013 at: http://www.ietf.org/rfc/rfc3569.txt

Brandenburg, R. van, Niamut, O., Prins, M. and Stokking, H. (2011) 'Spatial Segmentation For Immersive Media Delivery'. *Proceedings of the 15th International Conference on Intelligence in Next Generation Networks (ICIN)*, Berlin, 4–7 October, 151–156.

Cain, B., Deering, S., Kouvelas, I., Fenner, B. and Thyagarajan, A. (2002) 'Internet Group Management Protocol, Version 3,' RFC 3376 (Proposed Standard), Oct. 2002, updated by RFC 4604, accessed 20 August 2013 at: http://www.ietf.org/rfc/rfc3376.txt

CCNx, (2013) Homepage, accessed 20 August 2013 at: http://www.ccnx.org

Changuel, N., Mastronarde, N., Schaar, M. van der, Sayadi, B. and Kieffer, M. (2010) 'End-to-end Stochastic Scheduling of Scalable Video Overtime-Varying Channels'. *Proceedings of the 18th International Conference on Multimedia 2010*, Firenze, Italy, 25–29 October, 731–734.

Chen, F. and De Vleeeschouwer, C. (2010) 'Automatic Production of Personalized Basketball Video Summaries from Multi-sensored Data'. *Proceedings of the IEEE International Conference on Image Processing (ICIP10)*, Hong Kong, 26–29 September.

Chen, F. and De Vleeschouwer, C. (2011) 'Formulating Team-Sport Video Summarization as a Resource Allocation Problem'. *IEEE Transactions on Circuits and Systems for Video Technology* 21(2), 193–205.

Chen, J., Arumaithurai, M., Jiao, L., Fu, X. and Ramakrishnan, K. K. (2011) 'COPSS: An Efficient Content Oriented Publish/Subscribe System', *Proceedings of ANCS 2011*, Brooklyn, NY, 3-4 October, 99–110

ClassX (2011) ClassX – Interactive Lecture Streaming, accessed 20 August 2013 at: http://classx.stanford.edu/ClassX/

De Schepper, K., De Vleeschauwer, B., Hawinkel, C., Van Leekwijck, W., Famaey, J., Van de Meerssche, W. and De Turck, F. (2012) 'Shared Content Addressing Protocol (SCAP): Optimizing Multimedia Content Distribution at the Transport Layer'. *Proceedings of the IEEE Network Operations and Management Symposium (NOMS)*, Maui, HI, 16–20 April, 302–310.

Deering, S., Fenner, W. and Haberman, B. (1999) 'Multicast Listener Discovery (MLD) for IPv6,' RFC 2710 (Proposed Standard), Oct. 1999, updated by RFCs 3590, 3810, accessed 20 August 2013 at: http://www.ietf.org/rfc/rfc2710.txt

Degrande, N., Laevens, K., De Vleeschauwer, D. and Sharpe, R. (2008) 'Increasing the User Perceived Quality for IPTV Services', *IEEE Communications Magazine*, 46(2), 94–100.

FascinatE (2010) 'Format Agnostic SCript-based INterAcTive Experience', accessed 20 August 2013 at: http://www.fascinate-project.eu/

FascinatE (2013) 'Deliverable D4.2.2c – Delivery Network Architecture', accessed 20 August 2013 at: http://www.fascinate-project.eu/

Fernandez, I.A., Rondão Alface, P., Tong, G., Lauwereins, R. and De Vleeschouwer, C. (2010) 'Integrated H.264 Region-Of-Interest Detection, Tracking and Compression for Surveillance Scenes'. Packet Video Workshop (PV), 2010 18th International, Hong Kong, 13-14 December, 17–24.

Fette, I. and Melnikov, A. (2011) 'The WebSocket Protocol', IETF RFC 6455. 2710 (Proposed Standard), December 2011, accessed 20 August 2013 at: http://tools.ietf.org/html/rfc6455

Fotiou, N., Trossen, D. and Polyzos, G.C. (2010) 'Illustrating a Publish-Subscribe Internet Architecture', TR10-002 PSIRP project, January 2010

Google Maps (2013) Accessed 20 August 2013 at: http://maps.google.com

Guo, D. and Poulain, C. (2010) 'Terapixel, A spherical image of the sky'. Microsoft Environmental Research Workshop

Hadizadeh, H. and Bajic, I.V. (2011) 'Saliency preserving video compression', *IEEE International Conference on Multimedia and Expo (ICME)*, Barcelona, 11–15 July, 1–6.

He, Y., Ostermann, J., Tanimoto, M. and Smolic, A. (2007) 'Introduction to the Special Section on Multiview Video Coding', *IEEE Transactions on Circuits and Systems for Video Technology* 17(11), 1433–1435.

Holbrook, H. and Cain, B. (2006) 'Source-Specific Multicast for IP,' RFC 4607 (Proposed Standard), August, accessed 20 August 2013 at: http://www.ietf.org/rfc/rfc4607.txt

Huysegems, R., De Vleeschauwer, B. and De Schepper, K. (2011) 'Enablers for Non-Linear Video Distribution', *Bell Labs Technical Journal* 16(1), 77–90.

Ip, C.Y. and Varshney, A. (2011) 'Saliency-Assisted Navigation of Very Large Landscape Images'. *IEEE Transactions on Visualisations and Computer Graphics* 17(12), 1737–1746.

Jacobson, V., Smetters, D., Briggs, N.H., Plass, M.F., Stewart, P., Thornton, J.D. and Braynard, R.L. (2009) 'VoCCN: Voice-over Content-Centric Networks', *Proceedings of the 2009 workshop on Re-architecting the Internet (ReArch) Co-NEXT*, Rome, 1–4 December, 1–6.

Jiku, (2011) 'Sharing Event Experience, Instantly', accessed 20 August at: http://liubei.ddns.comp.nus.edu.sg/jiku/

KDDI R&D Labs (2010) 'Three Screen Service Platform', accessed 20 August at: http://www.youtube.com/watch?v=urjQjR5VK_Q

Khiem, N., Ravindra, G., Carlier, A. and Ooi, W.T. (2010) 'Supporting Zoomable Video Streams via Dynamic Region-of-Interest Cropping'. *Proceedings of the 1st ACM Multimedia Systems Conference*, Scottsdale, AZ, 22–23 February, 259–270.

Khiem, N.Q.M., Ravindra, G. and Ooi, W.T. (2012) 'Adaptive Encoding Of Zoomable Video Streams Based On User Access Pattern'. *Signal Processing: Image Communication* 27(2012), 360–377.

Kim, M.-J. and Hong, M.-C. (2008) 'Adaptive Rate Control in Frame-Layer for Real-Time H.264/AVC'. 14th Asia-Pacific Conference on Communications, Tokyo, 14–16 October.

Koponen, T., Chawla, M., Chun, B.-G., Ermolinskiy, A., Kim, K.H., Shenker, S. and Stoica, I. (2007) 'A Data-Oriented (and beyond) Network Architecture'. *Proceedings of SIGCOMM'07*, Kyoto, Japan, 27-31, August.

Kozamernik, F. (2002) 'Streaming Media over the Internet — An Overview of Delivery Technologies'. *EBU Technical Review*, October.

Krutz, A., Glantz, A., Sikora, T., Nunes, P. and Pereira, F. (2008) 'Automatic Object Segmentation Algorithms for Sprite Coding Using MPEG-4'. ELMAR, 2008, 50th International Symposium, Zadar, Croatia, 10–12 September, vol.2, 459, 462.

Lambert, P., Debevere, P., Moens, S., Van de Walle, R. and Macq, J.-F. (2009) 'Optimizing IPTV Video Delivery Using SVC Spatial Scalability', *Proceedings of the 10th Workshop on Image Analysis for Multimedia Interactive Services*, 2009 (WIAMIS '09), London, 6–8 May, 89–92.

Lambert, P. and Van de Walle, R. (2009) 'Real-time interactive regions of interest in H.264/AVC'. *Journal of Real-Time Image Processing* 4(1): 67–77.

Letourneau, J. (2010) 'GStreamer Plug-In Used to Transport Media Traffic Over a CCNx Network', accessed 20 August at: https://github.com/johnlet/gstreamer-ccnx

Liu, F. and Ooi, W.T. (2012) 'Zoomable Video Playback on Mobile Devices by Selective Decoding'. *Proceedings of the 2012 Pacific-Rim Conference on Multimedia (PCM'12)*, Singapore, 4–6 December, 251–262.

Macq, J.-F., Verzijp, N., Aerts, M., Vandeputte, F. and Six, E. (2011) 'Demo: Omnidirectional Video Navigation on a Tablet PC using a Camera-Based Orientation Tracker'. *Fifth ACM/IEEE International Conference on Distributed Smart Cameras (ICDSC)*, Ghent, Belgium, 22–25 August 2011.

Mavlankar, A., Agrawal, P., Pang, D., Halawa, S., Cheung, N.-M. and Girod, B. (2010) An Interactive Region-Of-Interest Video Streaming System For Online Lecture Viewing. Special Session on Advanced Interactive Multimedia Streaming, *Proceedings of 18th International Packet Video Workshop (PV)*, Hong Kong, 13–14 December.

Mavlankar, A. and Girod, B. (2009) 'Background Extraction and Long-Term Memory Motion-Compensated Prediction for Spatial-Random-Access-Enabled Video Coding'. Picture Coding Symposium PCS 2009, Chicago, 6–8 May, 1–4.

Mavlankar, A. and Girod, B. (2011) 'Spatial-Random-Access-Enabled Video Coding for Interactive Virtual Pan/Tilt/Zoom Functionality'. *IEEE Transactions on Circuits and Systems for Video Technology (CSVT)* 21(5), 577–588.

Merkel, K. (2011) 'Hybrid Broadcast Broadband TV, The New Way to a Comprehensive TV Experience'. 14th ITG Conference on Electronic Media Technology (CEMT), Dortmund, Germany, 23-24 March, 1–4.

MPEG-DASH, (2012) ISO/IEC 23009-1:2012 Information technology – Dynamic adaptive streaming over HTTP (DASH).

Naman, A.T. and Taubman, D.S. (2011a) 'JPEG2000-Based Scalable Interactive Video (JSIV)'. *IEEE Transactions on Image Processing* 20(5), 1435–1449.

Naman, A.T. and Taubman, D.S. (2011b) 'JPEG2000-Based Scalable Interactive Video (JSIV) with Motion Compensation'. *IEEE Transactions on Image Processing* 20(9), 2650–2663.

Niamut, O.A., Prins, M.J., van Brandenburg, R. and Havekes, A. (2011) 'Spatial Tiling and Streaming in an Immersive Media Delivery Network', EuroITV 2011, Lisbon, Portugal, 29 June.

Pang, D., Halawa, S., Cheung, N.-M. and Girod, B. (2011a) 'ClassX Mobile: Region-of-Interest Video Streaming to Mobile Devices with Multi-Touch Interaction'. *Proceedings of ACM Multimedia (MM'11)*, Scottsdale, AZ, 28 November–1 December.

Pang, D., Halawa, S., Cheung, N.-M. and Girod, B. (2011b) 'Mobile Interactive Region-of-Interest Video Streaming with Crowd-Driven Prefetching'. International ACM Workshop on Interactive Multimedia on Mobile and Portable Devices, ACM Multimedia (MM'11), Scottsdale, AZ, 28 November–1 December.

Pantos, R. and May, W. (2013) HTTP Live Streaming, IETF Internet-Draft, draft-pantos-http-live-streaming-11, April 16, accessed 20 August 2013 at: http://datatracker.ietf.org/doc/draft-pantos-http-live-streaming/

Protobuf, (2013) 'Protocol Buffers', accessed 20 August 2013 at: https://code.google.com/p/protobuf/

Pubsubhubbub (2013) 'pubsubhubbub A simple, open, web-hook-based pubsub protocol & open source reference implementation', accessed 20 August 2013 at: http://code.google.com/p/pubsubhubbub/

Quax, P., Issaris, P., Vanmontfort, W. and Lamotte, W. (2012) 'Evaluation of Distribution of Panoramic Video Sequences in the eXplorative Television Project'. NOSSDAV'12, Toronto, 7–8 June.

Ravindra, G. and Ooi, W.T. (2012) 'On Tile Assignment for Region-Of-Interest Video Streaming in a Wireless LAN'. *Proceedings of the 22nd International Workshop on Network and Operating System Support for Digital Audio and Video (NOSSDAV)*, Toronto, 7–8 June.

Rondão Alface, P., Macq, J.-F. and Verzijp, N. (2011) 'Evaluation of Bandwidth Performance for Interactive Spherical Video'. *Proceedings of IEEE ICME Workshop on Multimedia-Aware Networking*, Barcelona, 11–15 July.

RUBENS (2009) 'Re-thinking the Use of Broadband Access for Experience Optimized Networks and Services', accessed 20 August 2013 at: http://wiki-rubens.celtic-initiative.org/index.php/Main_Page

Salsano, S., Detti, A., Cancellieri, M., Pomposini, M. and Blefari-Mellazzi, N. (2012) 'Transport-Layer issues in Information Centric Networks', ACM SIGCOMM Workshop on Information Centric Networking (ICN-2012), Helsinki, Finland, 17 August.

Schierl, T., Stockhammer, T. and Wiegand, T. (2007) 'Mobile Video Transmission Using Scalable Video Coding', *IEEE Transactions on Circuits and Systems for Video Technology* 17(9), 1204–1217.

Schulzrinne, H., Casner, S., Frederick, R. and Jacobson, V. (2003) 'RTP: A Transport Protocol for Real-Time Applications'. IETF RFC 3550, accessed 20 August 2013 at: http://datatracker.ietf.org/doc/rfc3550/

Segall, C.A. and Sullivan, G.J. (2007) 'Spatial Scalability Within the H.264/AVC Scalable Video Coding Extension'. *IEEE Transactions on Circuits and Systems for Video Technology* 17(9), 1121–1135.

Stockhammer, T. (2011) 'Dynamic Adaptive Streaming over HTTP – Standards and Design Principles', *MMSys '11 Proceedings of the 2nd Annual ACM Conference on Multimedia Systems*, San Jose, CA, 23–25 February, 133–144.

Sullivan, G.J., Ohm, J., Han, W.-J. and Wiegand, T. (2012) 'Overview of the High Efficiency Video Coding (HEVC) Standard'. *IEEE Transactions on Circuits and Systems for Video Technology* 22(12), 1649–1668.

Tang, S. and Rondão Alface, P. (2013) 'Impact of Packet Loss on H.264 Scalable Video Coding'. *Proceedings of 5th International Conference on Advances in Multimedia*, Venice, Italy, 21–26 April.

Taubman, D. and Prandolini, R. (2003) 'Architecture, Philosophy and Performance of JPIP: Internet Protocol Standard for JPEG2000'. *Proceedings of the SPIE International Symposium on Visual Communications and Image Processing (VCIP'03)*, vol. 5150, Lugano, Switzerland, July, 649–663.

Taubman, D. and Rosenbaum, R. (2000) 'Rate-Distortion Optimized Interactive Browsing of JPEG2000 Images'. *Proceedings of the IEEE International Conference on Image Processing (ICIP'00)*, Barcelona, 14–17 September, 765–768.

Timmerer, Ch. and Griwodz, C. (2012) 'Dynamic Adaptive Streaming over HTTP: From Content Creation to Consumption'. Tutorial at ACM Multimedia 2012, Nara, Japan, 29 October 29.

Vanhastel, S. and Hernandez, R. (2008) 'Enabling IPTV: What's Needed in the Access Network'. *Communications Magazine* 46(8), 90–95.

Wang, R.J. and Chiu, C.T. (2011) 'Saliency Prediction using Scene Motion for JND-based Video Compression'. *Proceedings of the IEEE Signal Processing Systems*, Beirut, 4–7 October, 73–77.

xTV (2011) 'eXplorative TV', accessed 20 August 2013 at: http://www.iminds.be/en/research/overview-projects/p/detail/xtv-2

Yuan, H., Song, T. and Crowley, P. (2012) 'Scalable NDN Forwarding: Concepts, Issues and Principles'. *Proceedings of the 21st International Conference on Computer Communications and Networks (ICCCN)*, Munich, Germany, 30 July–2 August.

8

Interactive Rendering

Javier Ruiz-Hidalgo[1], Malte Borsum[2], Axel Kochale[2] and Goranka Zorić[3]

[1] *Universitat Politècnica de Catalunya (UPC), Barcelona, Spain*
[2] *Deutsche Thomson, Hannover, Germany*
[3] *Interactive Institute, Stockholm, Sweden*

8.1 Introduction

The appearance of richer forms of video content has opened the door to the need for new, also richer, visual rendering to present this content and interfaces to interact with it. Traditionally, video production is designed to support a particular *script* that the director is following, which is targeted to a specific video format and display system (e.g., HD video on widescreen TVs). Currently, viewers expect to enjoy video content on a variety of displays, such as high-resolution cinema screens, TVs, tablet, laptops and smart-phones. Furthermore, they are increasingly expecting to be able to control their audio-visual experience by themselves, moving away from a fixed *script* and selecting one of several scripts suggested or even by freely exploring the audio-visual scene.

Conventional production has limited support for such functionality. One fundamental approach to overcome the limitations is to adapt the video production step for a format-agnostic approach. This leads to the concept of a layered format-agnostic representation, as developed in Chapter 2, Section 2.8.4, where several cameras with different spatial resolutions and fields-of-view can be used to represent the view of the scene from a given viewpoint. The views from these cameras provide a base-layer panoramic image, with enhancement layers from one or more cameras more tightly focussed on key areas of the scene.

In this chapter, the layered format-agnostic representation is exploited in order to create an advanced interactive rendering tool that: (a) allows the scene to be rendered to multiple display systems, from ultra-high resolution screens to mobile devices; and (b) can be controlled by more natural user interactions such as a gesture recognition system for the pan, zoom, framing and viewpoint selection functionalities available to the viewer. Combining the rendering and

Media Production, Delivery and Interaction for Platform Independent Systems: Format-Agnostic Media, First Edition. Edited by Oliver Schreer, Jean-François Macq, Omar Aziz Niamut, Javier Ruiz-Hidalgo, Ben Shirley, Georg Thallinger and Graham Thomas.
© 2014 John Wiley & Sons, Ltd. Published 2014 by John Wiley & Sons, Ltd.

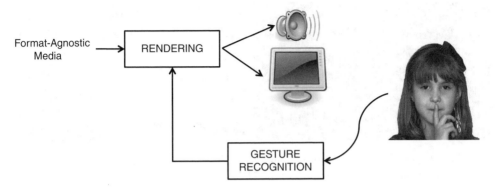

Figure 8.1 Interactive rendering of the format-agnostic media using a device-less gesture recognition system.

gesture recognition tools provides the viewer a large degree of freedom to select the way the contents are displayed and to freely exploit the scene in a natural and non-intrusive way.

Figure 8.1 shows a general scheme of the chapter. The rendering module is responsible of adapting the layered format-agnostic representation to be rendered into the specific user screen and loudspeakers. At the same time, the gesture recognition module is monitoring users, their gestures recognised and their interactions transmitted to the rendering module. This allows, for instance, the user to navigate through the panoramic video.

The organisation of the chapter is as follows. Section 8.2 describes the challenges of rendering of a format-agnostic representation. The section is focused in the video aspects of the rendering. Section 8.3 is devoted to an overview of the techniques and algorithms needed for device-less gesture recognition. Both sections also include a description of a technological solution. Finally, a user study of the gesture recognition is given in Section 8.3.6.

8.2 Format-Agnostic Rendering

End-user devices presenting the visual content from an interactive service need significant dedicated processing. Traditionally this processing aims to render a perfect image on 2D surfaces that are displays of various sizes and capabilities. As the display size on the consumer side may vary from small to large, the content producers assume an average dimension for framing the captured video. With the bottleneck of limited transmission bandwidth video services typically provide pre-rendered material dedicated to this average display, hence fracturing and limiting the customer base. New applications raise the challenge to address a customer group large enough for a profitable business. In the following such applications are discussed and a solution for tackling the challenge of customer diversification by rendering format-agnostic video is presented.

8.2.1 Available Rendering Solutions in End Terminals

This section revisits available solutions of rendering for interactive video presentation. Here, the user interaction in the traditional consumer electronic devices is offering multiple viewing

Figure 8.2   Rendering for end terminals (Multi-Angle Player (Youswitch Inc, 2013)), left: concert, right: sports. Reproduced by permission of Youswitch Inc.

positions (or angles, see Figure 8.2) by switching to different video streams. These are either provided 'live' with the main programme or 'online' via a feedback channel requesting data from additional internet channels or prerecorded on storage media (Blu-Ray discs for sports or concert features etc.).

Video rendering is considered in this chapter as translation of a model of the real world for projection on 2D displays (Borsum et al., 2010). This is basically the process to generate device dependent pixel data from device independent image data, which is the rasterisation of geometric information (Poynton, 2003). However, the term rendering, originally used in computer graphic modelling (Ulichney, 1993), is nowadays combined with techniques from computer vision and also includes operations such as scaling/filtering, colour adjustment, quantisation and colour space conversion. This also applies to operations in the data pipeline preparing the raw video data for the rendering process. As a consequence, elements of the complete rendering process are proposed to be distributed across the distribution and presentation chain balancing the computational load described in Wang et al. (2011), for example.

While computer generated imagery (CGI) is becoming common for film and television production to create visual effects (VFX) or to create unique scenes, the increased level of detail required for flawless presentation on large screens still needs significant effort to produce content that looks natural. Here, the model that represents the video data and the transmission of this data to the renderer (data pipeline) are the weak points for broadcast video production. Furthermore, how the user can access this data for smooth content personalisation such as navigating a virtual camera, depends largely on swift updates of the model data required for the individual perspective selection.

Rendering Techniques Applied in Interactive Video Applications

Techniques offering such individual perspectives are researched in image-based rendering (IBR) (Szeliski, 2012). In IBR new photorealistic perspectives will be generated that also employ prior knowledge about the scene's geometry. By adding this geometric information the demands on realistic rendering, compactness of the representation and rendering speed can be eased and its performance improved. Shum and Kang (2000) suggest an IBR continuum classifying the various techniques according to their level of geometry modelling

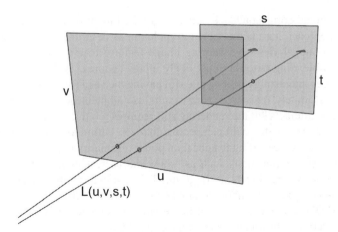

Figure 8.3   LF parameterisation (Levoy and Hanrahan, 1996).

(none/implicit/explicit), which was later extended to include the type of rendering such as a pixel or layer (Shum, Chan and Kang, 2007).

Recently pushed by advances in camera technology (Lytro, 2013; Raytrix, 2013), Light Field (LF) rendering is an example where the scene geometry is unknown (Levoy and Hanrahan, 1996). Here, the LF as radiance of a set of light rays is considered as a 4D function getting sliced by a number of 2D planes for parameterisation (Figure 8.3) given that there is no light dispersion. These planes (or light slabs) are measured ray intensities, for example, at the camera plane and focal plane. An LF enables several features beyond multiperspective rendering almost independent from the scene complexity such as access to depth of field or dynamic focus and allows SNR (signal to noise ratio) or dynamic range improvements. Its acquisition is challenging but rendering an LF is as easy as using the plane parameters in a modified ray tracer or simply by texture mapping supported by current graphics hardware. However, creating LFs require a host of input images and changes in the scene lighting are limited. The available camera technology with micro lens arrays still is restricted in available perspectives and resolution. Tackling issues like ghosting artefacts requires expensive computations – too costly for all of today's end-user devices. The LF discussed here is analogous to the Wave Field (WF) used in spatial audio rendering presented in Chapter 4.

Panoramic image mosaics such as described in Szeliski and Shum (1997) belong to the same IBR category. Here, a plenoptic function as intensity and chromaticity of light observed from every position and direction in 3D space can be constructed with a limited number of registered images. Thus images can be provided, for example, by omni-directional cameras or cameras with fish eye lenses following a motion track. This reduces the perspective change to a simple translation within the panoramic image making the rendering in the terminal easy to implement. The general idea was also extended to register multiple images to a panoramic mosaic without the need to project them on a common surface such as a cylinder. However, tackling registration errors to prevent accumulation and distortions by motion parallax introduced by camera motion are challenging issues for the terminal.

Adding a positional correspondence between the used input images means using implicit geometry. One approach here is the view interpolation as described in (Chen and Williams, 1993) that renders novel views by warping image features to estimate the positions of the features and then blending onto the most likely image textures. Such a process, also called morphing, produces interesting results while still having the need for a large number of sample images that are accurately registered. New results use ambient point clouds improving the handling of visible artefacts but the rendering still requires too much processing performance to be put in end-user devices for swift perspective changes. Offline versions allowing morphing between various images are starting to appear on powerful smartphones (mostly used to morph between faces), hence applications such as 'show the perfect image setting from the best (virtual) vantage point' are already at the brink. This also will apply to technology evolved in the research of computational photography (Zhou and Nayar, 2011) as object masking and occlusion filling requires extensive use of processing, which later allows the most attractive elements in the picture to be seen (Cohen and Szelisk, 2006).

Having access to the explicit geometry like a 3D model with depth information such as supplied by laser scanned environments allows texture mapped modelling as described in Heckbert (1986). Here, 2D textures are projected onto the model represented by triangular meshes. Depending on the perspective of the model, the visibility of the polygon from the mesh has to be computed and a view map needs to be maintained (Magnor, 2005). The rendered view is easily determined through view projection. Current GPU hardware strongly supports these operations and texture mapping is common in today's computer games. The remaining challenge is the support for dynamic scenes that requires fast computation on large textures. The bottleneck here is the upload speed of large data chunks onto the texture memory to allow those fast computations (Owens et al., 2008; Cozzi, 2012).

Finally, video-based rendering (VBR) combines IBR with the notion of motion capture with the objective to render photorealistic, arbitrary views of dynamic, real world events (Magnor, 2005). For VBR, a dynamic scene should ideally be captured with a dense multiview camera arrangement to create sufficient perspective data for rendered scenes with limited or no occlusions and other artefacts. For solving this, several approaches are already proposed demanding prior scene knowledge, dynamic light field sampling, scene approximations by visual hulls or regularisation across several captured or interpolated images. These operations are far from being integrated into an end terminal and reasonable simplifications are researched such as example-based video texturing on 3D models (Wei et al., 2009). The goal to allow free view video in the future is already backed by an MPEG standard on multiview coding (MPEG MVC, 2008). Nevertheless, challenges of video material encoded for VBR such as how to allow editing and how to compress for reasonable transmission remain open.

However, the benchmarks for assessing the rendering results are rated by their creation of visual realism with regard to motion, image resolution and colour perception. Therefore, most of the researched rendering technologies are still incomplete in one sense or another. As the rendering speed strongly depends on the complexity of the scene, the result might only be useful for offline processing (example use case: cinema production). An application featuring such advanced rendering technology for broadcast video needs to maintain at least the level of image and service quality the end user knows today; otherwise it will not support a business that creates return on investment. Changes in production and distribution infrastructure demand requirements ensuring this (see Section 8.2.2).

Applications for Interactive Perspective Rendering

The previous paragraphs concentrated on the potential techniques included in an interactive rendering application, independent of the service model or use case. The remaining sections will focus on the available services and their relation to end-user devices. Following that is a summary of important differentiators of the technology and a classification of the end-user devices using the technologies.

Nowadays, before deploying such new technology in end-user devices certain features can be offered faster on an internet service. Here the end user accepts, more easily, content that is 'strange' in look and feel (such as YouTube videos) and does not fit perfectly with the target display. Interactive TV web sites such as Hulu, Netflix, Amazon Prime and other provide video-on-demand (VOD) or Web TV services that can easily be extended to offer 3D or other immersive content stretching the notion of media portals. Such services make use of existing infrastructures and are referred to as over-the-top (OTT) services. Firmware components such as widgets, gadgets and apps extend existing media platforms to provide new functionality (e.g., Watch ESPN (Entertainment and Sports Programming Network), BBC's iPlayer or Mark Levoy's SmartCam (2013) for multipoint and tilt-shift refocusing).

Broadcast video requires live rendering for complex scenes. Current rendering farms, such as, Rebusfarm (2013), Render Rocket, (2013) and Fox Renderfarm (2013), still concentrate on offline rendering functions for cinema or games production using software such as Vray, Mental Ray, Maxwell or VUE. Such tools also allow virtual tours to real architecture or geological places and are combined with virtual reality tools such as Quicktime VR to allow movement through complex scenes and the presentation of panoramic images.

In contrast, the online games industry introduced cloud gaming in 2010. It performs the rendering of complex scenes that have already been encoded as audio-visual stream for connected clients (Onlive, Gaikai (now Sony), Otoy, StreamMyGame, Playcast, G-Cluster, Ciinow, CoreOnline and a host of others) in a way that is similar to the way movies are delivered by Netflix and others. This on-demand gaming complements the idea of videos on demand to pass interactions and to connect a community of users in a multiplayer environment. The online community is already huge, providing business value, but there is competition with gaming services, such as, Zynga, that provide socially-connected entertainment with simpler multiplayer strategy games. The benefits of cloud games are that they can support cross-platform rendering while reducing the hardware requirements of the connected end device. This means that new titles, which start competing with movie productions in film festivals ('L.A. Noire'), can be accessed by popular platforms such as tablets or smartphones. These services – categorised into 3D graphics streaming and video streaming with and without post rendering – still suffer from insufficient high-speed broadband access to distribute the content. Thus, several issues such as compression artefacts, interaction latency and content/privacy protection still need to match the quality the end users accept (Choy et al., 2012; Jarschel et al., 2013). The new cloud services offered by Amazon (2013) and Google (Google Compute Engine, 2013) promise to improve content distribution on a variety of end-user platforms.

Media libraries for Pay TV or VOD still do not offer complex modelling of produced content but stick to common HD and SD video formats. Rarely do they offer anything beyond IMAX (Vudu, 2013) or 3D, for example. The challenges for TV broadcast remains the realism of the content, the response of the service to user interactions and the ease of accessing the content with standardised formats and an intuitive user interface.

A rendering application on today's end-user devices needs a dedicated use-case for meaningful deployment. The available technology can therefore be differentiated in the following way:

- The platform supported (mobile, home, cinema) and how the content scales to this (e.g., reframing, repurposing).
- How this platform connects to the service and to other users (multi player/user).
- The kind of immersion achieved (extension of field of view, augmented or virtual views) and whether the user is part of this immersion (interaction).
- Allowing the content production a directed view for specific storytelling or whether the user can decide on or control an own story or perspective.

For broadcast video the end-user devices can be classified basically into *two* groups – television and edutainment:

Television
Interactive TV: The rendering here is mostly understood as mapping/overlaying content with various resolutions onto a display (YouTube Videos, Internet Browser, User-Generated Content). The interaction offered can be categorised into local, feedback and active. Local interaction relates to selecting from program mosaics (AT&T's U-Verse Multiview, 2013; DirecTV, 2013) used by ESPN or plain EPGs and controlling time-shifted presentation. Feedback interaction is considered to be voting reaction or advertisement/tag triggered shopping, for example, Triviala's quizshows (Triviala, 2013). Direct involvement for active interaction is still limited to game shows (e.g., MoPa-TV allows controlling of self-created avatars). However, all these do not demand sophisticated rendering techniques in the terminal.

First use cases to enhance the interactive choices for the users are multifeed channels providing different views (EVS's C-Cast used by Canal+, Sky showing parallel helmet camera feeds at Formula 1 broadcasts). Other services concentrate on complementing the traditional video programs with additional data in a synchronised second device such as Emmy® award winning Fourth Wall Studio's 'Dirty Work' (2013) and XBOX 'SmartGlass' (2013) or ESPN's 'Sportscenter Feeds'. This fits the requirements of repurposing content for mobile use cases ('content to go', Sky's 'skygo'). Recent developments on Smart TVs add more functionality mainly to control access to content from local or streamed sources or via cloud services (e.g., LG integrating the OnLive Cloud Gaming into their GoogleTV based Smart TVs). Applications to manipulate own content such as photos, organising content as a media centre, recognising the user for profile management, functions for surveillance and access to games all demand more processing power in the local device but challenge existing pay-TV business models (Future Source, 2013). However, personalisation is basically understood as collecting user data and preparing recommendations/playlists. Rendering features still do not combine decently the 3D graphics with the photorealistic video world but are already adopted in the production of previsualisation content for program creation (Ichikari et al., 2007).

3DTV: The rendering for 3D reproduction requires a significant amount of effort to produce images in real time. Hence, recent developments in technology for 3D displays created devices with panels featuring refresh rates of 400 fps or higher (Didyk et al., 2009; Holliman et al., 2011). Common LCD panels have 5ms response time or less and use contrast push signalling. While these high refresh rates were introduced to reduce motion artefacts it also allows widespread deployment of comfortable 3D or 3D like rendering in TVs and set top boxes despite the concerns raised in Woods and Sehic (2009). The recent HDMI (2010) specification (HDMI 1.4a) includes formats (frame packed, over/under, side-by-side) that ensure device connectivity and require various rendering techniques to be included in the end device. This and the processing for automatic 2D/3D conversions need to be common in nowadays 3DTVs and home media servers (e.g., Technicolor's DXI807). However, 3D is still no major seller, more the appeal to buy a future-proof TV (PwC 3D, 2012).

Free or multiview video as described in Kubota et al. (2007) and specified in MPEG MVC has been demonstrated in case studies but not released to the wider broadcast market. What can be seen are virtual enhancements of offline rendered sports broadcasts as a tool to indicate certain situations (Vizrt, 2013).

Entertainment/Edutainment

Video games: Traditionally, the performance demand for rendering of video games is high and strongly depends on the capability of the latest platforms (XBOX, PlayStation and Wii). Here, the trend towards photorealistic rendering of videos has almost reached movie production quality and is supported by cloud services and media streaming services such as Netflix that push the content to a wide range of gaming platforms (NetFlix Devices, 2013). Multidisplay rendering as promoted by various gaming simulations, already requires high power processing platforms hence combining monitor mosaics to render a wider view or visualising more information. Again, the available DisplayPort specification (Vesa, 2010) ensures only local connectivity.

Reality augmentation/virtualisation: The useful combination of graphical and real visual information is researched in areas beyond the gaming industry. Some examples are for driving simulation or driver assistance (e.g., Roth and Calapoglu, 2012; NVidia Driver Assistance, 2013), or a transparent backseat (Keio University, 2013), medical imaging (Balsa Rodriguez et al., 2013) or virtual factories/logistics (Takahashi et al., 2007). Here, the rendering platforms are still expensive and highly customised but a potential platform in the automotive industry can easily push solutions to a wider audience.

The need for dedicated rendering is pushed by the demand to create realistic models for entertaining or for supporting our daily life. Other aspects not connected to main stream devices having influence are:

Niche applications/market: Corporate video production or hotel TV solutions along with video conferencing systems are use cases where higher prices are accepted and complex rendering workstations (centralised in hosting server or in dedicated client systems) allow sophisticated processing.

Innovative content production: To promote interest in individual rendering per connected end device, content needs to be created for a wider audience. First models are virtual tours such as the Smithsonian (2013) or AirPano (2013) that are 360° panoramas of world attractions or second screen applications complementing the broadcast channel. However, pushing CGI animations or even partial VFX rendering to individual machines requires a common API (Application Programming Interface) or framework in the entertainment industry, which is not imminently available.

Finally, the rendering performance and quality is strongly influenced by the platform components supporting the streaming access or performing the rendering operations. The most recent platforms in home entertainment adopt common components from mainstream personal computer domains in order to make use of graphical processors (GPUs) and open source software and APIs (Linux/Android, OpenGL ES).

8.2.2 Requirements for Format-Agnostic Video Rendering

The goal of format-agnostic video rendering is to provide optimal visual quality on all kind of end-user devices, even when they have different capabilities. Final rendering results shall provide visual experience regarding motion of objects or virtual cameras, available resolution and colour perception comparable to video material produced dedicated for that type of end device.

While rendering a new visual perspective for an individual end device (called user view), it is desirable to use pre-rendered material whenever possible. Other material has to be re-targeted to match the device capabilities. Obviously all necessary rendering steps have to be real-time capable, for which some processing power has to be available directly in the end device or provided by the transmission network dedicated to each device.

Transmission and storage of the available material has to ensure that the whole scene is accessible with only minimal latency. This allows maximum freedom, for the user view, in interaction as all necessary parts can be requested. Very fine granular restrictions on availability or quality of the material can be based upon geographical content rights or user groups which have subscribed to, for example, premium services.

Interaction on the end device provides selection from predefined professional views and free navigation inside the scene. Navigation being defined as switching available camera sources and panning, tilting and zooming the virtual camera.

Layered Scene and Render Categories

To enable this flexibility in rendering a concept that offers access to the content of the whole scene is required as currently only necessary parts are transmitted. A layered scene representation as introduced in Chapter 2, Section 2.9 is a promising candidate. By subscribing to only the currently necessary parts of the LSR (layered scene representation) all material for rendering is available to the device without the need for transmission of the whole scene content.

All elements eventually to be rendered on the end device can be categorised into three major groups, which are defined as rendering categories (Figure 8.4):

> **Video scene:** All traditional video sources that build the LSR. These are ultra-high resolution panoramic images that build the base layer or background of the scene and images from all satellite cameras.
>
> **Scene related:** All artificial elements that enrich the scene that need to be positioned relative to the scene geometry. These can be highlighting of objects, artificial commercials or parts of the user interface. This would typically be implemented as a list of vector graphics objects, but bitmaps for photos or logos are also possible.
>
> **Screen related:** All artificial elements that build the GUI (graphical use interface). These are logos, menus and so on that have to be positioned relative to the screen and not to the scene.

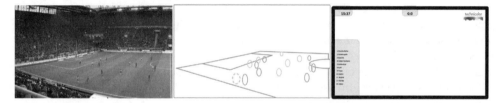

Figure 8.4 Categories in scene composition, from left to right: video scene, scene related, screen related.

Rendering Framework

The general rendering framework can be divided into two main categories of processing steps:

Composition
This encompasses all processing steps that are independent from the physical device capabilities. These can easily be moved into the transmission network and can mostly be shared by a larger number of end-user devices. The output of a composition processing step can replace layers of the LSR or even add new ones. Composition steps are all post-production tasks for image enhancement and correction such as removing lens distortion or adapting the colour and luminance. They include all post-production tasks that allow the scripting engine to generate metadata and additional information layers, for example, registration of all video signals to the world coordinate system, object tracking and recognition, ROI (Region-of-Interest) definition and framing proposals for different screen aspect ratios. Next, image scaling and tiling for better network utilisation and bandwidth reduction and also recoding, for example, from raw or lossless formats to typical broadcasting formats. It also includes caching of all or parts of the scene for later use as on-demand service for time-shift or replays and selection/reduction of image material based on the business model, for example, based on content rights or user-level service agreements. Finally, adding commercials and other additional data, even from other content providing sources. This is especially useful for adding or replacing commercials based on geographical location or user preferences.

Presentation
Presentation encompasses all processing steps that change the material according to physical limitations of specific devices. In some cases these steps can also be moved into the transmission network, for example, processing for generic device classes like standard HD displays. In most cases these steps have to be done inside the end device or at least by rendering units dedicated to a single end device. Examples of presentation steps are: The determination of visible elements/streams out of the scene description and thereby reduction of current frame scene graph by unused parts. This decision is mainly based on what can be seen on the currently selected user view, but also on some binary switches if the user enables/disables additional information layers. Further more, all perspective correction, stitching and blending of video elements from the LSR belonging to the 'video scene' category need to be performed. The current perspective to be used is defined by the device parameters of the connected displays and the viewport the user has selected. It also includes the generation of artificial scene parts.

This means perspective correction and alpha blending of all elements of the 'scene related graphics' category from the scene description. This processing is also based on the individual user's view. It is also used to render special GUI elements that have to be placed relative to the scene and not the screen. Finally other presentation tasks include: 2D-rendering, bitmap scaling and font rendering of all elements of the 'screen related graphics' category from the scene description. This is also used to render the typical GUI elements to interact with, and feedback information towards the user.

Platform Requirements

The platform requirements for video rendering depend highly on the selected scenario. In order to target all cases from large cinemas with nearly any hardware restrictions, set top boxes in the living room to mobile devices with the lowest capabilities, the achievable features also have to be adopted. Some general requirement areas can be identified.

Overall **processing performance** is mainly driven by perspective correction and video format conversion. Even if this is an extreme processing workload, it is a typical use case for dedicated GPU processing and scaling with display resolution. Nowadays GPUs with sufficient performance are available even in low-end smartphones.

The **network access bandwidth** can be very high if the complete LSR has to be available. It is easily achievable with optical fibre connection to a cinema. At home optical fibre will become more common. With special network management, tiling and multicast transmission it is possible to provide only the necessary parts of the video material to home and mobile devices.

Internal **data throughput** can be a problem for existing devices if their internal buses are designed only for HD video data rates. As the LSR needs larger video sources to render individual views some excess is necessary.

The **scalability** of the architecture is mainly a software and control problem. As multiple video renderers can split the work across the network, much more control messaging is necessary to negotiate the current render pipeline.

Low **power consumption** is a major consideration for mobile devices, as their run time from battery is limited. Set top boxes at home should also have low power consumption to appeal to the purchaser and to avoid overheating.

The **level of user interaction** offered is a matter of display size and available sensors measuring the user interaction. The possible methods of interaction change with the distance and the capabilities of the device. But user interactions are also determined by the network latency and therefore responsiveness to commands and the possibly limited choices in the case of reduced LSR transmission due to network constraints.

The effort to support **immersive live-event experience** includes the display size and loud-speaker capabilities. While home scenarios can allow an immersive media consumption within a large screen and complex loudspeaker setting, a mobile scenario can offer the wider 'community factor' of experiencing content in a group.

8.2.3   Description of a Technical Solution

This section describes the technical solution for format-agnostic video rendering. It is mainly targeted at a living room scenario but generalisation to cinema and mobile scenarios are given whenever applicable.

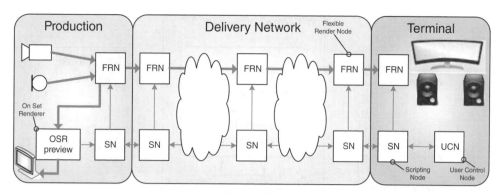

Figure 8.5   Flexible placement of Flexible Rendering Nodes (FRN).

Distributed Rendering

As a main concept of distributed rendering the processing steps of audio-visual rendering can be distributed across the complete chain from production via network up to the end terminal. Therefore multiple Flexible Rendering Nodes (FRNs) could be used as indicated in Figure 8.5. Obviously these FRNs are targeting different processing steps and are also implemented on different platforms. FRNs located in the production or network domain are more likely to implement the composing steps as defined in the previous section, whereas the terminal FRN will implement the most presentation steps.

As one example implementation the FRN placed at the output of the production is responsible for all post-processing tasks (e.g., colour correction) while the first FRN of the delivery network acts as ingest node and encodes all raw video to compressed streams. These blocks will typically be integrated on-set or close to the event location.

Other FRNs in the network are responsible for transmission and shaping the LSR, first for national needs, then for regional and, as a last step, for individual user needs. They may use tiling and quality/resolution changes due to network needs and may even add new content for specific user groups.

Both FRNs located at the network terminal boundary work in close cooperation to transform the multiuser streams to an individual rendered experience on the display. Depending on the selected scenario varying from cinema via home to mobile, implemented processing steps are moving from the terminal side towards the network side. In extreme cases everything is integrated in a big rendering PC next to the cinema and the mobile terminal receives only a pre-rendered video stream and is only responsible for rendering the GUI for user interaction.

Other elements, depicted in Figure 8.5, relate to scripting nodes (SN) and user control nodes (UCN) or production site (aka on-site) rendering (OSR). These were suggested in this sample implementation to support production automatism and user interaction and to monitor the rendering results at a local or remote production site.

Architecture and Interfaces

As the implementation described in this section is targeting a technology demonstrator, that has to integrate modules from different sources and targeting several demonstrations with

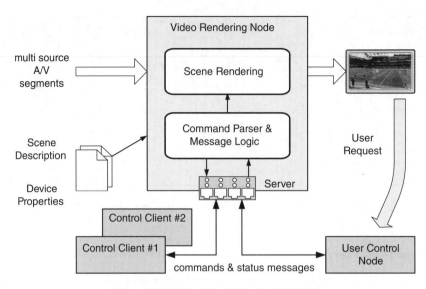

Figure 8.6   Video Rendering Node (VRN).

contrary requirements, a network based communication and control architecture was chosen. A real implementation would integrate these modules much further into one application or firmware. Therefore it would be possible to use application internal communication instead of a network based one.

The FRN can contain functions for rendering video as well as audio. While this chapter concentrates on video rendering aspects performed in a video rendering node (VRN) the audio rendering aspects are discussed in Chapter 4, Section 4.7.2.

Figure 8.6 shows the major interfaces of a Video Rendering Node (VRN).

Initial configuration is done via loading profile files. The *device properties* like display resolution, speaker positions and other capabilities are stored in these files and are loaded once at start-up. This could also include user profiles that can be modified by the user. For demo purposes an initial *scene description* can be loaded to define the initial LSR and location of sources and streams.

Ingest of video data comes either directly from a panorama stitching module or the distribution network. This interface is implemented as InfiniBand (InfiniBand, 2013) network to allow raw data transmission, but 10 GBit-Ethernet (IEEE 802.3ae 2002) or other high bandwidth networks could also be used. For compressed streams a standard Ethernet network would be sufficient. For offline demos, the ingest can also be configured to come from local disc, either as raw images or compressed videos.

To allow traffic management in the network, *region requests* are sent from the VRN towards the network. Based on that information the LSR can be reshaped to transmit only the necessary data for the current view.

The normal output goes to the screen attached to the GPU in the terminal platform. In case of a network-located FRN the output has to be encoded again to build a new layer in the LSR.

Multiple control clients can connect to the control parser to concurrently control the rendering process. Built-in mouse and keyboard control would allow basic user control, but typically a user control (as introduced in the next section) with more natural interactions will connect to implement the user interaction. A scripting node can also provide automatically generated user views and other predefined views from the production.

All connected clients receive regular status messages with all relevant rendering parameters to update their controlling. This interface is also used to synchronise the video with audio rendering nodes (ARNs).

Video Rendering

To allow easy rendering of arbitrary captured panoramic content, as a first step a virtual model of the screen is generated inside the VRN. This virtual model is shaped according to the camera model used for capturing. It is possible to use flat or spherical models, but a half cylinder with 180° viewing angle is used in most cases. With this model the inverse warping process is simplified, as any shape can easily be adopted with a different shaped virtual screen. This is discussed in Chapter 3, Section 3.2.5.

To render the user defined view from that generated model, a virtual camera is defined and modified according to the view changes the user transmits. Virtual cameras and textured mesh models of virtual worlds are the core features of state of the art computer and gaming graphics solutions like OpenGL or DirectX. Therefore, it imposes an implementation inside one of these frameworks. OpenGL is a good choice in order to stay platform independent. Even typical smartphone processors are supporting with OpenGL-ES the necessary feature set to render appropriate panoramic views.

Typical video codecs are using the colour space coding for luminance (Y) and colour difference (U/V or Cr/Cb) with 4:2:0 pixel sampling, whereas the colour space RGB and its sampling 4:4:4 is used for computer graphics. So some conversion has to be done inside the rendering terminal. OpenGL provides fragment shaders to solve that problem via GPU processing without any additional performance needs. Therefore the CPU processing is only used for controlling purposes and feeding the video data to the GPU.

An example of a rendered 2D image from the VRN is illustrated in Figure 8.7. The background showing the rendered 2D image segment from a cylindrical panorama is overlaid by several additional views into such a panorama that the user can select as a main view. To highlight the mapping of the optional image regions, several marker textures are rendered onto the existing image (rectangular shapes).

The next section provides an overview of the algorithms used to provide device-less gesture recognition and a description of a technical solution. These gesture recognition systems will provide a natural way to interact with the rendered content explained in previous sections.

8.3   Device-less Interaction for Rendering Control

Until recent years, the only interaction between humans and rendered TV content has been driven through specific devices such remote controls. The previous section has shown how the rendering of a layered format-agnostic representation increases the possible interactions presented to the end user, such as freely navigating through a panorama image. More recently, the wide acceptance of tablets in the consumer market and the adoption of Smart TVs (Kovach, 2010) has improved the user experience when interacting with multimedia content by

Figure 8.7 Rendered 2D image.

combining simple movements with finger configurations (i.e., pinch to zoom). Furthermore, device-less interaction has recently experienced an exponential growth. The growth in the acceptance of device-less visual systems may be mainly attributed to two factors: First, the emergence of new sensors that facilitate the recognition of human pose, hand and finger gestures. Second, the last advances in image processing algorithms that have opened the possibility of implementing real-time systems that recognise human gestures with high fidelity. Both factors have had a key role in the acceptance and emergence of gesture recognition interfaces to control and interact with TV content.

Even though one can consider multimodal systems (visual and audio) applied to TV control interaction, audio-based systems have not been really used as extensively as video. This is mainly due to the fact that audio uses the same communication channel as the TV audio. Therefore, saying commands might prevent you from hearing the TV. On the other hand, visual-based device-less interaction does not have these problems, allowing the user to interact with the system without clashing with the audio or blocking the screen.

However, when considering gesture recognition interfaces to interact with richer format-agnostic content, several requirements and issues must be considered to avoid some of the classic pitfalls when translating standard remote control paradigms to a gesture-based interface.

Gesture recognition interfaces should ensure that the interface is intuitive by providing gestures that are easy to learn, remember and repeat over time. It should guarantee a good responsiveness of the interface by ensuring a reasonable latency of the system (Nielsen, 1993). It is important to note that there are two different levels of this responsiveness of the interface. For instance, visual feedback of some gestures (e.g., zooming or raising volume) must be fast enough to allow fluid interaction (i.e., less than 0.2 seconds). On the other hand, other interactions such as changing audio or video channels are less restrictive in latency (i.e., less than 1 second) and greater latencies can be allowed.

The system must provide an unobtrusive interface where users can interact without the need for additional devices or complex marker systems. This unobtrusiveness is important to ensure that a natural and immersive interface is provided to the user. The system should also be designed to avoid the fatigue typically caused by gesture recognition interfaces. One of the main drawbacks of these new interfaces is the fatigue that the user may experience after interacting with the system for some time. The interface must be comfortable and convenient by using a set of natural gestures that do not force the user into difficult or tiresome positions, resulting in the so called 'gorilla arm' syndrome (Pogue, 2013). This means that the GUIs associated with the interfaces must be simple and not very precise in order to allow simple gesture to reach GUI elements easily as, traditionally, gesture recognition interfaces do not have the precision of other devices such as mice or keyboards.

Gesture recognition interfaces must contemplate the social context of TV viewing, especially at home. The social aspect is complex and varies over time: family and friends watching together; public shared space; negotiation with regards to who is in control of the interface. Therefore, the interface should be immediately available and instantly shareable among the group (Vatavu, 2010).

Finally, the gesture recognition interfaces must provide a proper feedback to the user. When using these interfaces, it is very important that users know what the system is recognising or whether they are in control of the interface. For instance, the user must always know whether or not he is being tracked, where he is located in the interface so he knows how and where to move his hand in order to get to his desired location.

Gesture recognition systems can be decomposed as a pipeline of three basic components (Figure 8.8). The first component corresponds to the capture device, that is, one sensor or multiple sensors. They capture a representation of the gesture (which could be a colour image, a video, a depth map or a complete 3D representation). The second corresponds to the gesture analysis where the needed segmentation and tracking (in the case of motion capture) are done. This component is also responsible of extracting the features, reducing the dimensionality that will define the gesture. The last component labels or classifies the performed gesture. Both, the feature extraction and the recognition stages may influence the previous blocks by predicting the values at the next time instance.

When dealing with systems that recognise human gestures in human-computer interaction, it is important to take into account that human gestures have a very high variability between individuals and even for the same individual between different instances. Furthermore, when designing gesture recognition systems, there exist many-to-one mappings from concepts or semantic meanings to gestures. Hence, gestures are ambiguously and incompletely specified. For example, to indicate the concept of *silence*, a finger can be placed across the lips, a hand can cover the mouth completely or both hands can be used to cover the ears. Moreover, gestures are often culture specific and the same gesture could symbolise completely different things.

Human gestures can be static or dynamic. In static gestures the user assumes a certain pose and configuration, and this pose and location is crucial for recognition. Dynamic gestures are composed of pre-stroke, stroke and post-stroke phases and in this case, also the motion and evolution of the gesture (between the pre-stroke pose and the post-stroke pose) must be evaluated for recognition (Mitra and Acharya, 2007).

Depending on the body part, which performs the gesture, they can be classified into *hand and arm gestures* such as sign language, hand numbers. *Head and face gestures* such as,

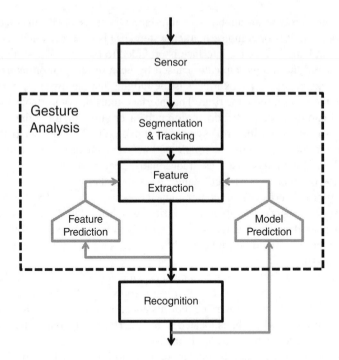

Figure 8.8 Components of a gesture recognition system.

nodding or shaking the head and *body gestures* that involve the full body motion such as distinction between seating, standing, walking and running.

Gesture recognition systems must take all these concepts into account when designing and defining a human-computer interface using gestures. The following sections further discuss each one of the components, shown in Figure 8.8, needed in a gesture recognition system highlighting their importance inside a gesture recognition interface. Several algorithms and approaches followed by the literature are given as examples of each of the components of a gesture recognition system.

8.3.1 Sensors for Gesture Recognition

In order to start recognising gestures, the human body position (hands, arms and legs), configuration (angles, rotations) and movement (velocity, acceleration) must be sensed. This can be done either using devices attached to the user, such as gloves, accelerometers, magnetic trackers, or by device-less vision-based sensors. Attached sensors are usually cumbersome to wear and, for the most part, hinder the ease and naturalness of the user's interaction with the computer. Vision-based device-less sensors overcome this problem although they tend to be more complex and, until now, less robust when recognising gestures.

Classically, vision-based gesture recognition systems have used normal colour cameras as input data for the gesture analysis. However, recently, there has been a quick switch to 3D based sensors. Even though 3D data has been extracted previously from stereo or multiview setups, the recent appearance of new, low budget 3D sensors, has facilitated the use of these sensors in gesture analysis. In general, the different sensors used in gesture recognition systems could be classified in colour cameras, infrared cameras and 3D cameras (Berman and Stern, 2012).

Colour cameras were the first sensors to be used in gesture analysis systems. Although a single-camera sensor provides an inherent lack of available information, very interesting works have studied how to extract gestures from single colour cameras. For example, Yan and Pollefeys (2008) recovered the articulated structure of a body from single images with no prior information. Guan et al. (2009) obtained a synthetised shaded body. Body pose is estimated by searching into the learned poses, reflectance and scene lighting, which most probably produced the observed pose. Brubaker, Fleet and Hertzmann (2009) used a simple model of the lower-body, based on physical walking movement called *Antropomorphic Walker*, proposed by Kuo (2002). Even though some reasonably good results in unconstrained scenarios have been achieved using skin colour based segmentation (Lenkiewicz, 2011), single colour cameras inherently provide poor information, due to information loss originating from perspective projection to a single viewpoint. Single-camera based methods are usually very specific and do not effectively generalise to different kinds of movement, scenes and viewpoints. Furthermore, the segmentation of arms and hands in colour cameras is difficult and without a controlled environment (where the background is known) becomes an ill-posed problem.

Infrared cameras have also been employed as data for gesture recognition systems. Thermal imaging provides useful signatures of flesh objects that are sensitive to ambient lightning by measuring the heat energy radiating from the object. The use of thermal imaging can improve recognition in low illumination conditions and can help locate hands or arms. Larson et al. (2011) presented a system called *HeatWave* that uses digital thermal imaging cameras to detect, track and support user interaction on arbitrary surfaces. However, thermal imaging systems present a very high cost and are not currently suitable for home scenarios.

While classically conventional 2D colour cameras have been used for gesture recognition systems, there is a tendency to replace them with *3D cameras* as they facilitate some gesture analysis task such as segmentation and they are very tolerant of illumination changes. Existing 3D cameras can be categorised into two classes: passive or active (Zhu et al., 2011). Among the passive methods, stereovision and multiview can be considered the most widely used. Among the active methods, scanning based, time-of-flight (TOF) and structured light sensors have been used as inputs for gesture recognition systems.

Stereovision systems are composed of two cameras or a simple camera with dual lenses and they can be used to obtain stereo images. The depth information provided by the stereo images can be exploited to recognise gesture. For instance, Li et al. (2011) proposed a system that recognises five hand gestures from a stereo camera by means of a thinning algorithm to obtain the angle and distance of feature points in the hand. Despite these advances, the fundamental problems of stereo systems, such as occlusions, remain unsolved and are difficult to overcome in gesture analysis (Zhu et al., 2011).

Multiview-based systems offer more precise tracking, but non-portable and costly setups are required. Results in this area are excellent since complete 3D movement information is available. Relevant studies using multiview data include Gall et al. (2009) that estimate pose and cover possible non-rigid deformations of the body, such as moving clothes. Sundaresan and Chellappa (2009) predict pose estimation from silhouettes and combined 2D/3D motion queues and Corazza et al. (2009) propose a system that generates a person-wise model updated through Iterative Closest Point (ICP) measures on 3D data. Nevertheless, 3D capture environments are very expensive and cumbersome to setup, since they require precise calibration and, usually, controlled illumination conditions. In addition, the computational cost of multiview methods is prohibitive and real-time is hardly achieved.

Scanning-based systems such as Radio Detection and Ranging (RADAR) or Light Detection and Ranging (LIDAR) could also be used as input for gesture recognition. The use of these systems has been limited mainly because of the scanning speed limit, the maintenance of their mechanical parts and the high cost. An initial work on gestures can be found in Perrin, Cassinelli and Ishikawa (2004) that performed gesture recognition tracking fingers based on a wide-angle photo-detector and a collimated laser beam. Newer systems, such as the radarTOUCH (2013) have recently appeared, these also rely on rotating laser scanners to perform gesture recognition.

Time-of-flight cameras based on the TOF principle (Kolb et al., 2010) are able to deliver a depth estimation of the recorded scene. TOF cameras currently provide low-resolution images with resolutions of 176×144 (SR4000) (MESA, 2013) and 204×204 (PMD CamCube) (PMD, 2013). Figure 8.9 shows an example of a data captured by the SR4000 sensor. Among the works using TOF cameras to be noted are: Knoop, Vacek and Dillmann (2006) who propose a fitting of the depth data with a 3D model by means of ICP; Grest, Krüger and Koch (2007) who use a non-linear least squares estimation based on silhouette edges; and Lehment et al. (2010) propose a model-based annealing particle filter approach on data coming from a TOF camera.

Structured light technology is another method for obtaining depth. Recently, the Microsoft Kinect (Kinect, 2013) a low-cost structured light image sensor combined with a colour camera is commercially available. The Kinect sensor has a resolution of 640×480, a frame rate of 30fps and an average depth estimation error of 1cm (similar to TOF cameras). However, Kinect provides more stable depth estimation, with fewer outliers and artefacts at depth edges than TOF cameras (Figure 8.10). These advantages, together with the low price of Kinect, have revolutionised gesture recognition using these sensors. For instance, Ganapathi et al. (2012) extract full body pose by filtering the depth data using the body parts' locations and

Figure 8.9 Images captured by a SR4000 sensor: left, the confidence of the captured pixels; right, the depth information.

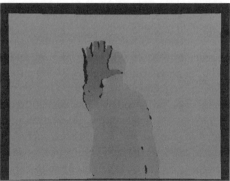

Figure 8.10   Images captured by a Kinect sensor: left, the colour image; right, the depth information.

Shotton et al. (2011) estimate the pose of humans exploiting an enormous dataset to perform a pixel-wise classification into different body parts.

It should be stressed that the appearance of TOF and structured light sensors such as Kinect has really pushed the limits on the gesture recognition technologies providing a low-cost sensor where gesture analysis algorithms can work. However, they are still bulky and too big for mobile applications, for instance. More recently, new, smaller sensors are appearing that may facilitate further the advancements in gesture recognition systems. Among them are LeapMotion (LeapMotion, 2013) that combines visible and infrared imaging, the PMD CardBoard nano (PMD, 2013) or a new prototype by PrimeSense (the company behind the creation of Microsoft Kinect) that claim reducing the size of the Kinect sensor by a factor of ten (PrimeSense, 2013).

8.3.2   Gesture Analysis Techniques

One of the most important components of all gesture recognition systems corresponds to the gesture analysis (middle component of Figure 8.8). This component analyses the data given by any of the sensors reviewed in the past section and extracts a sets of features that will help directly identify or classify, from a previously learned database, the gesture performed by the user and recorded by the sensors.

A set of blocks common to most gesture analysis systems is depicted in Figure 8.8. They usually start by a segmentation and tracking step (only if recognising dynamic gestures) that will help the subsequent feature detection step. Initially, the segmentation step, where body parts where extracted from the background, has been one of the most difficult steps in the gesture recognition process. However, due to the use of depth-based sensors, using the depth data directly or the associated 3D points in space basically solves the segmentation problem. An example can be seen in Figure 8.10, where segmenting the user hand in the left image (colour information) could pose a challenging problem but it is easy to segment based on depth information from the image on the right.

After the segmentation is performed, a feature extraction step is responsible for reducing the dimensionality of the problem by extracting certain features that distinguish the different gestures to be recognised. The features used depend greatly on the spatial model chosen to represent the gestures.

The literature differentiates two main approaches in gesture analysis: a 3D-model-based or an appearance-based (Pavlovic, Sharma and Huang, 1997; Zhou and Nayar, 2003).

The *3D-model-base* makes use of 3D information to model directly the 3D appearance of the body part performing the gesture. They create kinematic 3D models of the user body that can later infer the entire body posture used to create the gesture. 3D model-based algorithms can use volumetric or skeletal models (or even a combination of both). Volumetric models are generally created of complicated 3D surfaces, such as polygon meshes (Heap and Hogg, 1996; Stenger et al., 2006). Even though volumetric models realistically represent the human body, they are often too complex and their computational complexity renders them unfeasible to be used in real-time scenarios. Other volumetric approaches try to reduce this computational complexity by using simple 3D geometric structures to model the human body such as, cylinders, spheres, ellipsoids. Initial studies such as Azarbayejani, Wren and Pentland (1996) or Clergue et al. (1995) have successfully reduced the complexity of full volumetric models by using ellipsis to model human arms. Instead of dealing with all the parameters of a volumetric body part, models with a reduced set of equivalent joint angle parameters together with segment lengths are often used. Such models are known as skeletal models and have been and are still used extensively in gesture analysis. For example, Figure 8.11 shows a skeletal representation of an upper body. The skeletal model approximates the anatomical structure of the body in a fixed numbe. of links (nodes) and edges.

The most common visual features extracted from the image sensors in 3D-model-based approaches vary greatly. For instance in Lu et al. (2003) and de la Gorce (2011), a combination of silhouettes, edges, colour and shading are used in monocular and multiview images to create a 3D model of hands. More recently, depth sensors have been used to find image cues. Oikonomidis et al. (2011) built an efficient 3D model for arms based on the depth map provided by a Kinect camera. Another, more sophisticated approach includes techniques using 3D salience (Suau, Ruiz-Hidaglo and Casas, 2012) to detect extremities of the human body.

Figure 8.11   Skeleton representation of an upper body part.

Nowadays, the best performing techniques combine both approaches where a skeleton representation of a body is further refined by a surface or volumetric representation. For instance, Wu, Varanasi and Theobalt (2012) proposed a technique to capture full body pose using shading cues. Similarly, Ballan et al. (2012) created a combined skeletal and volumetric 3D hand model using discriminative salient points from multiview images.

3D-model-based approaches do offer a very rich description of the human body. However, the main drawbacks include the difficulty to cope with occlusions and singularities that arise from ambiguous views and the large image database needed to train the multiview system covering all the characteristic shapes of the human body.

Appearance-based models use image features to model the visual appearance of the human body. These models do not use a spatial representation of the body as they derive the parameters directly by relating the appearance of any gesture with a set of predefined template gestures. Some initial works used skin information (Sånchez-Nielsen et al., 2004) to create blob regions modelling the hands. Other approaches use an Eigen space to represent body parts using a small set of basis vectors. An example of this approach is given by Black and Jepson (1998) where hand gestures are recognised using 25 basic images. Similar techniques use a large set of training data to learn a direct mapping from image features to specific body poses (Rosales, Athitsos and Sclaroff, 2001).

Many approaches work with local invariant features. In Wang and Wang (2008), a learning algorithm is applied to Scale-invariant Feature Transform (SIFT) features (Lowe, 2004). Similarly, Suau, Ruiz-Hidalgo and Casas (2102) propose the use of an oriented radial distribution to locate salient areas in the depth map and to predict user extremities. Several approaches use Haar like features for gesture recognition. Haar features focus more on the information within a certain area of the image rather than each single pixel. For instance, Bergh, Koller-Meier and Gool (2009) retrieve body pose by using Average Neighbourhood Margin Maximisation to train 2D and 3D Haar wavelet-like features. The 2D classification is based on silhouettes, and the 3D classification based on visual hulls. More recently, Local Binary Patterns (LBP) (Ding et al., 2011) and extensions of Histogram of Gaussians (Meng, Lin and Ding, 2012) have been applied to the detection of hand gestures.

Appearance-based models can recover from errors faster than their 3D-model-based techniques. However, their accuracy and type of gestures they are able to recognise depend greatly on the training data. Furthermore, appearance-based approaches fail when accurate pose estimation is needed to recognise the gesture.

8.3.3   Recognition and Classification Techniques

This component uses the features extracted in the previous step to, first, update the model of the human body and then classify the gesture based on the parameters of the model. In the case, for instance, of the 3D-model-based approach, the parameters used will be the joint angles and position of the different elements of the model. In the case of appearance-based models, many different parameters could be used to perform the classification and vary deeply of the appearance model chosen.

In order to update the model, several techniques are used that, basically, perform an optimisation to infer model parameters from the features extracted in the previous step. One of the most popular optimisation methods is local optimisation (de la Gorce, 2011). Local optimisation is a very efficient method but needs a very carefully designed objective function to avoid local minima. Other approaches rely on stochastic optimisation techniques such as particle filters (MacCormick and Isard, 2000) or combinations with local optimisations (Bray, Koller-Meier and Gool, 2007). Other kinds of optimisation techniques have also been explored, such as, belief propagation (Sudderth et al., 2004; Hamer et al., 2009) and particle swarm optimisation (Oikonomidis, Kyriazis and Argyros, 2010).

Once the model parameters have been estimated, they can be used to correctly identify the gesture performed by the user. If a predefined grammar of gestures is used, this step usually requires a classification technique. Again, multiple classification techniques have been employed for the specific task of gesture recognition. Among them, Nearest Neighbour (NN) or Support Vector Machine (SVM) classification techniques have been applied to detect static gestures of hands (Kollorz et al., 2008; Rekha and Majumder, 2011). When dealing with dynamic gestures, techniques such as Hidden Markov Models (HMM) or Finite State Machines (FSM) have been successfully employed for gesture recognition. HMM (Rabiner, 1989) is a double stochastic process governed by: (a) an underlying Markov chain with a finite number of states, and (b) a set of random functions each associated with one state. In discrete time instants, the process is in one of the states and generates an observation symbol. Each transition between the states is defined by: (a) a transition probability, and (b) an output probability that defines the probability of emitting an output symbol from a finite alphabet database when given a current state. HMMs have been applied to gesture recognition, a set of model parameters are employed to train the classifier and similar test data is used for prediction and verification. Yoon et al. (2001) propose a HMM network to recognise gestures based on hand localisation, angle and velocity. Similarly, Ramamoorthy et al. (2003), use a combination of static shape recognition, Kalman-filter based hand tracking and HMM-based temporal characterisation scheme for reliable recognition of single-handed dynamic hand gestures.

In the FSM approach, a gesture is modelled as an ordered sequence of states in a spatio-temporal configuration. Initial works in the gesture recognition field include Bobick and Wilson (1997) and Yeasin and Chaudhuri (2000) where state machines are used to model distinct phases of generic gestures. For instance, in Yeasin and Chaudhuri (2000) the FSM is used to define four directions (up, down, left and right) to control the direction of a robot similar to a panning in a panorama image. Additional approaches include where an adaptive boosting technique is used for gesture classification (Wang and Wang, 2008).

Recently, Random Forests (RF) has been successfully used in gesture recognition systems. Initially proposed by Breiman (2001), RF is an ensemble classifier that consists of a combination of tree predictors such that each tree depends on the values of a random vector sampled independently and with the same distribution for all trees in the forest. RFs combine the idea of bootstrap aggregating and the random selection of features in order to construct a collection of decision trees with controlled variation. As a result it is a very accurate, fast and efficient classification algorithm. Initial approaches using RF in gesture recognition include Demirdjian and Varri (2009) who proposed a temporal extension of random forests for the task of gesture recognition and audio-visual speech recognition. However, automatic temporal segmentation of the gestures was still unsolved. Shotton et al. (2011) proposed to use RF to estimate the pose of humans exploiting an enormous dataset. Gall et al. (2011) propose a Hough forest framework for dealing with action recognition, including object detection and localisation. López-Méndez and Casas (2012) formulate gestures as objects allowing the use of RF for localising the user performing the gesture. Recently, Keskin et al. (2012) extended RFs to a multilayered framework for hand pose estimation.

As previously noted, the various classification techniques are used to perform the last step in the gesture recognition process, where gestures are labelled (and also possibly located) in the image.

8.3.4   Remaining Challenges in Gesture Recognition Systems

Vision-based gesture recognition is not, by a long way, a solved issue. Although, as highlighted in previous sections, new sensors and image processing algorithms have recently helped in the development of new, more robust systems that perform gesture recognition. For instance, recently, consumer examples of device-less interaction between users and TVs have appeared. Samsung offers the Samsung Smart Interaction (Samsung, 2012) that allows users to control the graphical user interface of the TV with their hands. Unfortunately, the hand control is limited to move a mouse pointer around the screen. Panasonic recently presented their Gesture Control (Panasonic, 2013) that allows control of the TV through basic gestures. Users can change channels and access specific content by waving their hand. However, the system only works if the hand is very close (10cm) to the screen. In the past few years, many academic researchers have focused on gesture recognition.

In general, vision-based gesture recognition systems need to overcome some remaining challenges in order to completely solve the human–computer interaction satisfactorily. Usually, common gesture recognition systems understand a very limited gesture database. The robustness and accuracy of the recognition, even though very high, is still a challenge for current systems. For instance, false detections can seriously damage the user experience.

Furthermore, current systems are not capable of dealing with occlusions. Real gesture recognition interfaces experience multiple occlusions: other users, tables or chairs in the field-of-view, and so on. Gesture recognition systems must be able to recover from occlusions and to provide robust detections in these circumstances.

The temporal segmentation of user gestures, for example, locating where the gesture starts and ends, is still an open problem. Some systems solve this by forcing users to go to a *neutral* state between gestures. This sometimes results in a frustrated user as gestures cannot be performed as fast as the user would like or it results in an increase of the fatigue to the user.

Usually, complete gesture recognition systems are too specific to the application and only work in specific environments or with very controlled scenarios. Gesture recognition system should be general enough to work in different environments and to support different users concurrently in many situations.

The next section presents a specific gesture recognition system implemented to control a TV set in a home scenario.

8.3.5   Description of a Technical Solution

This section describes an implementation of a gesture recognition interface to control a format-agnostic representation on a TV set in a home environment. The proposed system works in a device-less and marker-less manner allowing users to control the content on their TV sets

Figure 8.12   Users interacting with the gesture recognition system.

using only hand gestures. It is responsible for locating the people who wish to interact with the system and of understanding and recognising their gestures.

The current setup allows the user to be standing or seated in a chair or on a sofa. Although a single user is able to interact with the system at any given time, the interface is multiuser capable in the sense that several users can ask for control of the system and interact with it while the others might still be present in the scene (Figure 8.12).

In order to allow the user to interact with the TV content, the system supports the following functionalities:

- **Menu selection:** A menu is overlaid on the current screen and the user can select any button of the menus by pointing at it (Figure 8.13).
- **Navigation:** The user is able to navigate through the panorama scene by panning, tilting and zooming on the content.
- **Region of interest selection:** The system informs the user of the available ROIs in the current view and the user is able to change between them.

Figure 8.13   Feedback to the user as overlay on the panoramic screen.

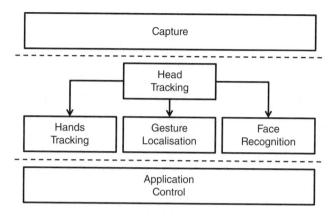

Figure 8.14 Architecture of the proposed gesture recognition system.

- **Pause/resume:** The user is able to pause/play the content at any time.
- **Audio volume:** The user is able to increase or decrease the volume and to mute it.
- **Take control:** The control of the system can be passed between users.

To fulfil all the functionalities, the current system implements the architecture depicted in Figure 8.14. It is divided in three main layers. In the first layer, the *capture* component is responsible for communicating with a single Kinect camera and feeding the images into the system. The second layer is the core of the interface where the processing, detection and recognition algorithms take place. It is composed of a number of components.

The *head tracking*, where the depth image obtained in the capture module is analysed to detect heads (oval areas of the same depth) (Suau, Ruiz-Hidalgo and Casas, 2012). The position of the heads is used in all subsequent components to locate possible users of the system. Other persons in the field of view of the Kinect sensor are not tracked and their gestures do not interfere with the system.

The *face identification* component recognises users. Faces are detected using a modified Viola-Jones detector (Viola and Jones, 2004) and recognised using a temporal fusion of single image identifications. The recognised users will determine the number of people able to control the system.

Once users are detected, hands are tracked by the *hand tracking* component using a 3D virtual box in front of the head of the user with control of the system. 3D blobs in the virtual box are segmented and treated as hands (Suau, Ruiz-Hidalgo and Casas, 2012).

The *gesture localisation* component is responsible for detecting and classifying static gestures (gestures performed with still hands). In this case, the classification of the gesture is purely done using the shape and position of the hand, extracted using only the depth data provided by the capture component. The classification is based on random forests (Breiman, 2001), aiming to accurately localise gesture and object classes in highly unbalanced problems.

Finally the third layer of the architecture contains the *application control* component. This module is responsible for acquiring all the detections and tracking information obtained by the components in the middle block (head, hands, user recognised and classified gestures) and mapping them into the functionality listed in the previous section. It communicates with

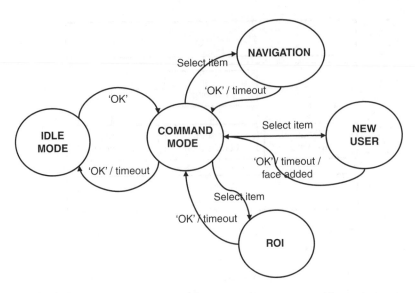

Figure 8.15   A state diagram of the proposed gesture recognition system.

the TV to perform all the needed interactions (select regions of interests, change volume, pan-tilt-zoom the panorama, etc.). It controls the user interface and provides all the needed feedback through the GUI overlay on the TV (Figure 8.13).

The gesture recognition interface is controlled as in the state diagram depicted in Figure 8.15. Each state in the diagram allows users to perform specific gestures to control the system. The *application control* module is responsible for controlling the general status of the system and performing the transition between states. The end user does not need to be aware of the general state diagram or the current state as the application control component informs the user about the available commands at each stage through feedback in the GUI overlay. Figure 8.15 shows the different states possible. The arrows and boxes indicate the gesture or timeout that will transition the interface from one state to the other one.

The system always starts in *idle mode*. In this state no menu is shown in the screen and the only gestures available are the *OK* or *take control* gestures (Figure 8.16). In this state, the gesture localisation component only recognises the *OK* gesture and all other components are suspended. If the *OK* gesture is detected, the user performing the gesture takes control and the interface transitions to the next state. In all following states only the user in control is recognised and other users might be present in the scene without affecting the system.

The *command mode* is the principal state where users interact with the system. In this state all the components (head and hand tracking, gesture localisation and face identification) are started and report to the application control. The head tracking ensures that the user is followed and that all subsequent components only focus on the user controlling the system. The face identification recognises users and allows them to access specific options or interactions with the system. The hand tracking tracks the position of a single hand that is used to access the menus of the GUI overlay. The selection of menu items in this mode is performed by pointing a single hand to a menu item and opening your hand to select it.

(a) (b) (c)

(d) (e)

Figure 8.16 Proposed static gestures recognised by the system in command mode: (a) *OK* gesture to take control of the system, (b) Shush gesture to lower the volume, (c) Hand on ear to increase the volume, (d) Cross to mute and (e) Parallel-hands to pause the video.

Additionally, the gesture localisation component recognises five static gestures (Figure 8.16) and the application control maps them to the following functionalities:

- The *OK* gesture (the same one the user used to take control of the system) is used to go back to *idle mode* (a).
- The volume can be increased by locating the finger on top of the mouth (b) and lowered by putting the hand on the ear (c).
- The volume can be completely muted by doing a cross sign in front of the mouth (d).
- The video can be paused or resumed by keeping the hands parallel as if about to clap (e).

By selecting any of the menus shown in the GUI overlay, the interface can transition to the states of navigation, region of interest selection or new user. From all these states the user can go back to command mode by doing the *OK* gesture or by waiting several seconds without doing anything to trigger a timeout.

Figure 8.17   Recognised gestures for menu selection and pan/tilt navigation.

The *navigation mode* allows the user to navigate freely through the panorama. In this mode no menus are shown in the screen and only the hands of the user are overlaid on the screen. The user can then pan and tilt the panorama by using one hand (Figure 8.17) and 'grabbing' the scene and zoom in or out (Figure 8.18) by using both hands (such as done in maps applications on tablets). The hand tracking component is responsible of tracking these movements and, in this state, no gesture localisation is performed as experimental results showed an increase of false/positive detections.

The *ROI mode* allows the user to navigate through the available regions of interest by moving the hand left to the right or the right hand to the left. The hand tracking algorithm is also responsible for tracking the hands of the user to detect grabbing and movements in this state with no possibility of using static gestures.

Figure 8.18   Recognised gestures for performing zoom navigation.

Finally, the *new user mode* is used to add new users to the system. This mode creates new user models for use in the face identification component.

8.3.6   User Study of the Gesture Interface

This section describes a user study with the gesture interaction, evaluation results and gives implications for the design. The study has been performed using the gesture control system with included interaction commands and gestures as described in Section 8.3.5.

New layered format-agnostic representations (such as high-resolution panoramic video) bring viewers new opportunities, rich content and possibilities to choose their own view of an event. This is in contrast to traditional TV viewing, where only preselected views are offered to the viewer. The question is what it actually means for TV viewers to get the possibility to select views and the level of detail of an event. In the past decade, there has been a lot of research on interactive TV (iTV) resulting in novel systems offering advanced viewing features. Consequently, a range of interactive functions has been suggested for viewers enabling much more than traditional channel and volume changing, while at the same time less effort has been invested in studying viewers' needs. Live video material additionally brings new challenges – the viewing of live events requires real-time interaction and mistakes in handling the content could lead to a possible loss of important information.

This new type of the content, apart from offering various viewing opportunities, also brings interactional challenges and a need for appropriate interactional methods. The aim here is to explore ways of approaching this type of high resolution content – while the traditional remote control might be readily available, it might not be the best way to interact with this new presentation of content. The focus of this section is on the novel gesture-based interaction described in the previous section, and how this interactional method might influence television experiences. More specifically exploration includes, (1) user study of the acceptance of the gesture based interface, and (2) their applicability for the use with this new type of the rich content. These steps are necessary in order to be able to design appealing user interfaces and to design for interaction with mass content.

Laboratory Study of the Gesture Control System

A full laboratory-based study of the gesture interaction with panoramic TV was conducted during 3 days in early November 2012 in Stockholm. The first day was reserved for setting up and checking the system and environment, and the other two days for the evaluation. The study set up was mimicking the home, that is, as a living room with a sofa bed, coffee table, flowers and candles. The study included two types of content – the panoramic video of the football match between Chelsea and Wolverhampton Wanderers played in 2010, and panoramic video of a dance show where the Berlin Philharmonic Orchestra accompanied youth dancers from Sascha Waltz at the Arena Berlin in May 2012 in the show Carmen. The study was conducted with 20 people consisting of seven pairs and six singles. This enabled us to see any differences in dynamics when users watched and interacted with the system when they were alone as opposed to when they were in a more social situation. All the pair users knew each other well. Most of participants have had experiences with panoramic videos before, either on big cinema screens, on tablets or mobile devices. At the beginning of the session participants

Figure 8.19 Detail from the final study setup.

were given instructions how to interact with the system, using the described gestures to zoom, pan, tilt and control the volume. The sessions were video recorded focusing on interactions with the content from two directions, front and back where it was possible to see the screen and the resulting interactions. First, the participants had been interacting with the system for about 10 minutes for each of the two available contents, though in some cases even longer. Afterwards, the researcher asked them a set of questions about their immediate experiences and impressions. These were audio recorded and later transcribed. Figure 8.19 shows a detail from this study setup.

Study Results

This section continues by describing observations and findings from the gesture interaction evaluation. The evaluation is based on the analysis of video recordings of participants' interaction with the gesture control prototype, but also on the interviews with each of the participants carried out at the end of the sessions.

The study confirmed that different people have different learning times – the ability to control the video with hand gestures differed, ranging from learning very fast, to having a long learning process and problems with memorising gestures and available commands. A good example of learning is one participant that was trying to control the system using gestures. If the gesture would not work, the participant showed fast adaptation, by trying a modified version of the gesture. In another example, trying to use the same 'wrong' gesture was causing confusion for some participants. In this case, no adaptation occurred and finally it was the reason for the user to stop interacting with the content. Differences were also visible in learning styles – a set of people preferred to learn by observing, that is, to first spend time just looking at the interface, and then trying, while others took a more 'hands-on' approach.

When it comes to the gesture form and size, the study showed that different people have different styles of the interaction, in terms of the size of the gesture, parts of the body involved (just fingers or the whole arm etc.), as well as body posture (leaning back or forward). In some cases, although the participants performed the gestures slightly differently to that expected by the system, the system still interpreted these gestures correctly. For example, the pause gesture was explained as two parallel hands, but two flat hands in front of the body also worked and was used by one participant seamlessly.

In the gesture control system two modes are available, the navigation mode and the ROI mode. In each of these modes, specific gestures could be used. This however was not clear to a few of our participants despite receiving explanations before the interaction session. They needed additional instructions to be able to go back to the desired mode.

Some gestures were more problematic than others. Zooming out was very difficult for all the participants; the gesture of using pointing fingers to zoom and then open hands (or taking the hands down and out again) seemed unnatural. Instead of zooming out this resulted in constantly zooming in and out. It did not seem clear how the interaction should be made. The reason is that the gestures for zooming in and zooming out are too similar, that is, it is actually the same movement for both, just in the opposite direction. Another reason why these gestures were found to be problematic is because they are repetitive gestures and are too tiring. For the user to get to the desired level of zooming, it takes too many steps. Panning also created similar difficulties. It was hard for many participants to physically achieve the movements required. However, this problem would not appear if participants used two hands instead of one when navigating.

When one looks to the individual gestures, the selection of different channels and region of interests was the easiest to perform and was achieved easily. This can be explained by the familiarity with swiping the screen paradigm taken from the smartphone interaction.

In the gesture interaction the participants preferred one-handed gestures. In most of the cases, they would use just one hand, even if that resulted in almost awkward arm movements where they would put one arm in front of themselves. It was not specified that only one hand or arm should be used for pointing and selecting, in fact it was possible to seamlessly shift from one hand to another according to the pilot study carried out prior to the user testing described.

The guiding illustrations of gestures were interpreted widely. However, the importance of immediate feedback was clear. When people were not getting feedback immediately they often attempted to change the gesture to gain a response.

Although the participants were trying out the system's gesture interaction and navigation in the video image for the first time, they were not active at all times throughout the test sessions. Users shifted between active and passive modes, and between interacting with the system and each other. Video recordings show that the participants were looking mostly at the screen and occasionally to each other, though there was no need to look for a control device. In this social setting two problems were identified. First, 'big' gestures like panning are impractical in social watching. Second, fights for taking control are more easily resolved in the case when a remote control is used.

Interviews with participants showed the need and appreciation for having the overview of the whole event. That is something that panoramic picture uniquely offers, and is especially useful when it comes to having a freedom in exploring the content. Participants loved that freedom and the possibility to explore, but still they needed an easy way to find their 'way back' in the case of getting lost. Although the participants loved the possibility of zooming in and exploring, they were critical of how it should be used – zooming in should not go so deep that it allows for low quality pictures. 'The resolution is the biggest killer of the application, if I zoom I want a proper resolution, otherwise it doesn't make sense. If I could get detailed view I would zoom in more', one of them commented.

Most of the participants in this study said that interacting with the gesture control system was fun, even if interaction didn't go completely smooth. This was visible on video recordings, particularly in the cases when two people were 'watching' together.

Discussion

The development of high-resolution panoramic content comes with challenges such as how to show all viewing possibilities and how to interact with the content shown. This section has focused on interactional challenges and investigated: a new interactional method – gesture device-less interaction, its acceptance among test participants and usability; and gesture interaction in the context of high-resolution panoramic content.

In general, participants were positive and open toward this new interaction method. Learning for each of the participants followed a different path, but all managed to interact successfully with the system. Compared to the traditional remote control, there were few issues with this new method. First, most of the participants tend to use only one hand, which made it difficult to perform some commands, like long panning. Second, gesture interaction, and interactive content in general add active components to the traditionally passive TV viewing. Participants did not have problems with performing other gestures other than with zooming and to some extent panning. In the study it was clear that interacting with the system and getting used to the gesture control was taking a lot of the participants' attention. However, pairs were managing to communicate between themselves. It was apparent that the gestures themselves were a conversation topic and something that added to an enjoyable social viewing experience. Users actively commented on the gestures both while doing them and while observing the other. This communication seemed to serve the dual purpose – social interaction and learning the gesture techniques by putting words to them. This may be partly due to the novelty of the system in the testing situation, but it suggests that social interaction and gesture control of the system do not necessarily need to be treated as two separate and possibly conflicting activities. Rather, the two modes contributed to the novel viewing experience.

In the gesture interaction prototype, a specific gesture is assigned to the command taking the control. Although the gesture was individual, participants enacted it socially, for example, mimicking the movement of giving the remote control, for example, leaning towards the person with the control (see Figure 8.20 right). The space around the user that movements within gestures were taking, like in panning the panorama, varied between participants. Bigger gestures were sometimes experienced as tiring, but participants also stated that they were taking up the space of the person sitting next to them, as shown on Figure 8.20 left. These observations suggest that the design of multiuser gesture systems should allow for unintrusive socially adapted gestures for controlling and navigating in rich video content, that is, social needs should be combined with the gesture control in a natural way.

These studies showed that the gesture interaction fits well with this new type of interactive content. For most of the participants it felt natural to navigate through the high-resolution panoramic picture, zoom in and out, look for details, but also to choose between predefined views. However, the only confusion was in not knowing which interaction mode they were in at a given moment and therefore which commands were possible. This is not a specific gesture interaction problem, but an interface design problem – a task for the future.

Figure 8.20   Social interaction with the gesture recognition prototype.

Although interaction with such TV content was perceived as fun for most of the participants, we must not forget that interaction with the traditional TV mainly means changing channels or volume, and that viewers are used to TV as a lean back medium. Viewers at the same time want to have the opportunity to lean back and passively watch the action, and actively work with broadcast content. Thus the design of future TV services must support both types of behaviour.

8.4   Conclusions

This chapter has discussed the challenges that a layered format-agnostic representation brings to the end-user device. In particular, both the challenges in the rendering and the user interaction have been investigated and discussed. Combining the rendering of format-agnostic representations together with innovative gesture recognition tools allows the viewer larger degrees of freedom to select the way the contents are displayed. Furthermore, the combination allows the user to freely exploit the scene in a natural and non-intrusive way.

The rendering module is responsible for adapting the layered scene representation into a 2D virtual rendered view tailored to a specific display system at a particular resolution, and rendered at a specific viewport selected by the user. The selection allows the user to interact and navigate within a cylindrical panorama by selecting pan, tilt or zoom factors for a specific perspective. Available solutions present in the literature for rendering interactive video have been visited and several applications for interactive video rendering have been reviewed. Furthermore, the architecture of a technical solution used to render a layered format-agnostic representation into a high-definition TV screen in a living room scenario has been detailed.

The gesture control system allows the viewer to interact as naturally as possible, allowing the user to perform simple interactions to control and navigate through the layered format-agnostic representation. The chapter has examined the sensors used in gesture recognition systems together with a review of several state-of-the-art techniques and algorithms found in the literature. Moreover, a particular integrated gesture recognition system has been presented. The gesture recognition system proposed works in a device-less and marker-less manner allowing users to control the content of their TVs using a small set of natural gestures.

Additionally, a usability study has been performed on the proposed interactive rendering system. An evaluation of the gesture recognition system has been given and several results of the evaluation have been discussed. It has shown that, even though some problematic gestures are considered difficult and non-natural for some users, most participants in the study were positive and open towards these kinds of new interaction methods.

The interactive video rendering approach presented, combining a format-agnostic video rendering together with a device-less gesture recognition system, allows the scene to be rendered in multiple device systems and be controlled by natural user interactions. The combination overcomes the limited support of conventional video production providing viewers a large degree of freedom to select the way visual content is displayed.

Nevertheless, the challenges in interactive video rendering still remain open and interesting. It is expected that they will remain of central importance to future rendering systems in the forthcoming years. Many improvements can be foreseen for the interactive video rendering discussed. For the video rendering node, the applicability of multicore architectures and distribution of processing power should be assessed to bring more complex rendering algorithms to cost effective end-user devices. For the gesture based interaction, more natural gestures should be investigated improving the overall robustness of the system against additional, more challenging, scenarios.

References

AirPano (2013) Home page, accessed 21 August 2013 at: www.airpano.com

Amazon (2013) 'Amazon Elastic Transcoder', accessed 21 August 2013 at: http://aws.amazon.com/elastictranscoder/

AT&T U-Verse Multiview (2013) Home page, accessed 21 August 2013 at:
    http://uverseonline.att.net/uverse/applications

Azarbayejani, A., Wren, C. and Pentland, A. (1996) 'Real-Time 3D Tracking of the Human Body'. *Proceedings of IMAGE'CON*, Bordeaux, France, May, pp. 780–785.

Ballan, L., Taneja, A., Gall, J., Van Gool, L. and Pollefeys, M. (2012) 'Motion Capture of Hands in Action using Discriminative Salient Points'. *Proceedings of the 12th European conference on Computer Vision*, Florence, Italy, 7–13 October, accessed 28 August 2013 at: http://cvg.ethz.ch/research/ih-mocap/ballanECCV12.pdf

Balsa Rodriguez, M., Gobbetti, E., Iglesias Guitian, J.A., Makhinya, M., Marton, F., Pajarola, R. and Suter, S.K. (2013) 'A Survey of Compressed GPU-Based Direct Volume Rendering' *Eurographics 2013*, accessed 21 August 2013 at: http://www.crs4.it/vic/cgi-bin/bib-page.cgi?id=%27Balsa:2013:SCG%27

Bergh, M.V., Koller-Meie, E. and Gool, L.V. (2009) eal-Time Body Pose Recognition Using 2D or 3D Haarlets'. *International Journal on Computer Vision* 83(1), 72–84.

Berman, S. and Stern, H. (2012) 'Sensors for Gesture Recognition Systems'. *IEEE Transactions on Systems, Man and Cybernetics* 42(3), 277–289.

Black, M. and Jepson, A. (1996) 'EigenTracking: Robust Matching and Tracking of Articulated Objects Using a View-Based Representation'. *Journal of Computer Vision* 26(1), 63–84.

Bobick, A.F. and Wilson, A.D. (1997) 'A State-Based Approach to the Representation and Recognition of Gesture'. *IEEE Transactions on Pattern Analysis and Machine Intelligence* 19(12), 1235–1337.

Borsum, M., Spille, J., Kochale, A., Önnevall, E., Zoric, G. and Ruiz, J. (2010) 'AV Renderer Specification and Basic Characterisation of Audience Interaction'. *FascinatE Report D5.1.1.*

Bray, M., Koller-Meier, E. and Gool, L.V. (2007) 'Smart Particle Filtering for High-Dimensional Tracking'. *Computer Vision and Image Understanding* 106(1), 116–129.

Breiman, L. (2001) 'Random Forests'. *Machine Learning* 45(1), 5–32.

Brubaker, M.A., Fleet, D.J. and Hertzmann, A. (2009) 'Physics-Based Person Tracking Using the Anthropomorphic Walker'. *International Journal of Computer Vision* 87(1–2), 140–155.

Chen, S. and Williams, L. (1993) 'View Interpolation for Image Synthesis'. *ACM SIGGRAPH Computer Graphics, Proceedings of the 20th Annual Conference on Computer Graphics and Interactive Techniques*, Anaheim, CA, 2–6 August.

Choy, S., Wong, B., Simon, G. and Rosenberg, C. (2012) 'The Brewing Storm in Cloud Gaming: A Measurement Study on Cloud to End-User Latency'. *Workshop on Network and Systems Support for Games (NetGames)*, Venice, 22–23 November.

Clergue, E., Goldberg, M., Madrane, N. and Merialdo, B. (1995) 'Automatic Face and Gestural Recognition for Video Indexing'. *Proceedings of International Workshop on Automatic Face and Gesture Recognition*, Zurich, Switzerland, 26–28 June, pp. 110-115.

Cohen, M. and Szelisk, R. (2006) 'The Moment Camera'. *IEEE Computer Society* 39(8), 40–45.

Corazza, S., Mundermann, L., Gambaretto, E., Ferrigno, G. and Andriacchi, T.P. (2009) 'Markerless Motion Capture through Visual Hull, Articulated ICP and Subject Specific Model Generation'. *Journal of Computer Vision* 87(1–2), 156–169.

Cozzi, P. (2012) *OpenGL Insights*. Boca Raton, FL: CRC Press.

de la Gorce, M., Fleet, D.J. and Paragios, N. (2011) 'Model-Based 3D Hand Pose Estimation From Monocular Video'. *IEEE Transactions on Pattern Analysis and Machine Intelligence* 33(9), 1793–1805.

Demirdjian, D. and Varri, C. (2009) 'Recognizing Events with Temporal Random Forests'. *Proceedings of International Conference on Multimodal Interfaces*, New York, USA, Cambridge, MA, 2–4 November, 293–296.

Didyk, P., Eisemann, E., Ritschel, T., Myszkowski, K. and Seidel, H.P. (2009) 'A Question of Time: Importance and Possibilities of High Refresh-rates'. *Visual Computing Research Conference*, Saarbrücken, Germany, 8–10 December.

Ding, Y., Pang, H., Wu, X. and Lan, J. (2011) 'Recognition of Hand-Gestures Using Improved Local Binary Pattern'. *International Conference on Multimedia Technology*, Barcelona, 11–15 July, 3171–3174.

DirectTV (2013) 'Tips and Tricks', accessed 21 August 2013 at:
www.directv.com/DTVAPP/global/contentPage.jsp?assetId=1200065

Fourth Wall Studio (2013) 'Dirty Work', accessed 21 August 2013 at: http://rides.tv/dirty-work

Fox Renderfarm (2013) Home page, accessed 21 August 2013 at: www.foxrenderfarm.com

Future Source (2013) 'CES Show Highlights'. Las Vegas.

Gall, J., Stoll, C., De Aguiar, E., Theobalt, C., Rosenhahn, B., and Seidel, H.P. (2009) 'Motion Capture Using Joint Skeleton Tracking and Surface Estimation'. *IEEE Conference on Computer Vision and Pattern Recognition*, Miami, FL, 20–25 June, 1746–1753.

Gall, J. Yao, A., Razavi, N., Gool, L.V. and Lempitsky, V. (2011) 'Hough Forests for Object Detection, Tracking, and Action Recognition'. *IEEE Transactions in Pattern Analysis and Machine Intelligence* 33(11), 2188 –2202.

Ganapathi, V., Plagemann, C., Koller, D. and Thrun, S. (2012) 'Real-Time Human Pose Tracking from Range Data'. *European Conference in Computer Vision*, Firenze, Italy, 7–13 October, pp. 1–14.

Google Compute Engine (2013) Home page, accessed 21 August 2013 at:
https://cloud.google.com/products/compute-engine

Grest, D., Krüger, V., and Koch, R. (2007) 'Single View Motion Tracking by Depth and Silhouette Information'. *Lecture Notes in Computer Science*, 4522, 719–729.

Guan, P., Weiss, A., Balan, A., and Black, M. (2009) 'Estimating Human Shape and Pose from a Single Image'. *IEEE International Conference on Computer Vision*, Kyoto, Japan, 27 September–2 October, 1381–1388.

Hamer, H., Schindler, K., Koller-Meier, E. and Van Gool, L. (2009) 'Tracking a Hand Manipulating an Object'. *Proceedings of International Conference in Computer Vision*, Kyoto, Japan, 27 September–2 October, 1475–1482.

HDMI (2010) High Definition Multimedia Interface (HDMI) version 1.4a supporting 3D and 4k.

Heap, A.J. and Hogg, D.C. (1996) 'Towards 3D Hand Tracking Using a Deformable Model'. *International Face and Gesture Recognition Conference*, Killington, VT, 14–16 October, 140–145.

Heckbert, P.S. (1986) 'Survey of Texture Mapping'. *IEEE Computer Graphics and Applications* 6(11), 56–67.

Holliman, N.S., Dodgson, N.A., Favalora, G.E. and Pockett, L. (2011) 'Three Dimensional Displays: A Review and Applications Analysis'. *IEEE Transactions on Broadcasting* 57(2), 362–371.

Ichikari, R., Tenmoku, R., Shibata, F., Ohshima, T. and Tamura, H. (2007) 'MR-Based PreViz Systems for Filmmaking: On-set Camera-Work Authoring and Action Rehearsal'. International Symposium on Mixed and Augmented Reality, Nara, Japan, 13–16 November.

Infiniband (2013) Infiband Trade Organization Home page, accessed 21 August 2013 at: http://www.infinibandta.org

Jarschel, M., Schlosser, D., Scheuring, S. and Hoßfeld, T. (2011) 'An Evaluation of QoE in Cloud Gaming Based on Subjective Tests'. *5th International Conference on Innovative Mobile and Internet Services in Ubiquitous Computing (IMIS)*, Seoul, 30 June–2 July, 330–335.

Keio University (2013) DigInfoTV homepage, accessed 21 August 2013, www.diginfo.tv/v/12-0204-r-en.php

Keskin, C., Kiraç, F., Kara, Y.E. and Akaru, L. (2012) 'Hand Pose Estimation and Hand Shape Classification Using Multi-layered Randomized Decision Forests'. *Lecture Notes in Computer Science* 7575, 852–863.

Kinect (2013) 'Introducing Kinect for Xbox 360', accessed 21 August 2013 at: http://www.xbox.com/en-US/kinect

Knoop, S. Vacek, S. and Dillmann, R. (2006) 'Sensor Fusion for 3D Human Body Tracking with an Articulated 3D Body Model'. *International Conference on Robotics and Automation*, Orlando, FL, 1686–1691.

Kolb, A., Barth, E., Koch, R. and Larsen, R. (2010) 'Time-of-Flight Cameras in Computer Graphics'. *Computer Graphics Forum* 29(1), 141–159.

Kollorz, E., Penne, J., Hornegger, J. and Barke, A. (2008) 'Gesture Recognition with a Time-Of-Flight Camera'. *International Journal of Intelligent Systems Technologies and Applications* 5(3), 334–343.

Kovach, S. (2010) 'What Is A Smart TV?'. *Business Insider*, accessed 21 August 2013 at: http://www.businessinsider.com/what-is-a-smart-tv-2010-12

Kubota, A., Smolic, A., Magnor, M., Tanimoto, M., Chen, T. and Zhang, C. (2007) Multiview Imaging and 3DTV, *IEEE Signal Processing Magazine* 24(6), 10–21.

Kuo, A.D. (2002) 'Energetics of Actively Powered Locomotion Using the Simplest Walking Model'. *Journal of Biomechanical Engineering* 124(1), 113–121.

Larson, E., Cohn, G., Gupta, S., Ren, X., Harrison, B., Fox, D. and Patel, S.N. (2011) 'HeatWave: Thermal Imaging for Surface User Interaction'. *ACM CHI* Conference on Human Factors in Computing Systems, Vancouver, Canada, 7–12 May, 2565–2574.

Leap Motion (2013) Home page, accessed 21 Auguts 2013 at: https://www.leapmotion.com/

Lehment, N.H., Kaiser, M., Arsic, D. and Rigoll, G. (2010) 'Cue-Independent Extending Inverse Kinematics For Robust Pose Estimation in 3D Point Clouds'. *International Conference on Image Processing*, Hong Kong, 26–29 September, 2465–2468.

Lenkiewicz, P., Wittenburg, P., Masneri, S., Schreer, O., Gebre, B.G. and Lenkiewicz, A. (2011) '*Application of Video Processing Methods for Language Preservation 5th Language & Technology Conference (LTC)*', Poznan, Poland, November, 25–27.

Levoy, M. and Hanrahan, P. (1996) 'Light Field Rendering'. *ACM SIGGRAPH 96*, New Orleans, LA, 4–9 August.

Li, X., An, J.-H., Min, J.-H. and Hong, K.-S. (2011) 'Hand Gesture Recognition by Stereo Camera Using the Thinning Method'. International Conference on Multimedia Technology, Hangzhou, China, 26–28 July, 3077–3080.

López-Méndez, A. and Casas, J.R. (2012) 'Can our TV Robustly Understand Human Gestures?: Real-Time Gesture Localization in Range Data'. *Proceedings of European Conference on Visual Media Production*, London, 5–6 December, 18–25.

Lowe, D.G. (2004) 'Distinctive Image Features from Scale-Invariant Keypoints'. *International Journal of Computer Vision* 60(2), 91–110.

Lu, S., Metaxas, D., Samaras, D. and Oliensis, J. (2003) 'Using Multiple Cues for Hand Tracking and Model Refinement'. *Proceedings of Computer Vision and Pattern Recognition Conference, Madison*, WI, 16–22 June, accessed 28 August 2013 at: https://cvc.cs.stonybrook.edu/Publications/2003/LMS03/192cvpr03-hand.pdf

Lytro (2013) Home page, accessed 21 August 2013 at: www.lytro.com

MacCormick, J. and Isard, M. (2000) 'Partitioned Sampling, Articulated Objects, and Interface-Quality Hand Tracking'. *Proceedings on European Conference in Computer Vision*, Dublin, Ireland, 26 June–1 July, 3–19.

Magnor, M. (2005) *Video Based Rendering*. Wellesley, MA: AK Peters.

Meng, X., Lin, J. and Ding, Y. (2012) 'An Extended HOG Model: SCHOG for Human Hand Detection'. *IEEE International Conference on Systems and Informatics*, Yantai, China, 19–21 May, 2593-2596 (2012).

MESA Imaging SR4000, (2013) 'SwissRanger^{TM} SR4000', accessed 21 August at: http://www.mesa-imaging.ch/prodview4k.php

Mitra, S. and Acharya, T. (2007) 'Gesture Recognition: A Survey'. *IEEE Transactions on systems. Man and Cybernetics, Part C: Applications and Reviews* 37(3), 311–324.

MPEG MVC (2008) MPEG Multiview Video Coding ISO/IEC 14496-10:2008 Amendment 1.

NetFlix Devices (2013) 'Streaming Devices', accessed 21 August at: https://signup.netflix.com/MediaCenter/DeviceImages

Nielsen, J. (1993) *Usability Engineering*, San Fransisco: Morgan Kaufmann.

NVidia Driver Assistance (2013) 'Automotive', accessed 21 August 2013 at: www.nvidia.com/object/advanced-driver-assistance-systems.html

Oikonomidis, I., Kyriazis, N. and Argyros, A. (2010) 'Markerless and Efficient 26-Dof hand Pose Recovery'. *Lecture Notes in Computer Science* 6494, 744–757.

Owens, J.D., Houston, M., Luebke, D., Green, S., Stone, J.E. and Phillips, J.C. (2008) 'GPU Computing'. *Proceedings of IEEE* 96(5), 879–899.

Panasonic (2013) 'Omek Interactive and Panasonic Electric Works Bring Complete 3D Gesture Recognition and Interactive Solutions to Advertisers and Retailers', accessed 21 August at: http://pewa.panasonic.com/news/press-releases/omek-interactive-panasonic-3d-gesture-recognition

Pavlovic, V.I., Sharma, R. and Huang, T.S. (1997) 'Visual Interpretation of Hand Gestures for Human-Computer Interaction: A Review'. *IEEE Transactions of Pattern Analysis and Machine Intelligence* 19(7), 677–695.

Perrin, S., Cassinelli, A. and Ishikawa, M. (2004) 'Gesture Recognition Using Laser-Based Tracking System'. *Proceedings of the IEEE International Conference on Automatic Face and Gesture Recognition*, Seoul, Korea, 17–19 May, 541–546.

PMD (2013) Homepage, accessed 21 August 213 at: http://www.pmdtec.com

Pogue, D. (2013) 'Why Touch Screens Will Not Take Over', *Scientific American*, 3 January, accessed 21 August 2013 at: https://www.scientificamerican.com/article.cfm?id=why-touch-screens-will-not-take-over

Poynton, C. (2003) *Digital Video and HDTV: Algorithms and Interfaces*. Amsterdam: Elsevier.

Prime Sense (2013) Home page, accessed 21 August 2013 at: http://www.primesense.com/

PwC (2012) 'Waiting for the Next Wave – 3D Entertainment'. PricewaterhouseCoopers Surveys, accessed 21 August 2013 at: http://www.pwc.com/gx/en/entertainment-media/publications/waiting-for-the-next-wave-3d-entertainment-2012.jhtml

Rabiner, L.R. (1989) 'A Tutorial on Hidden Markov Models and Selected Applications in Speech Recognition'. *Proceedings of the IEEE* 77(2), 257–285.

radarTOUCH (2013) Homepage, accessed 21 August 2013 at: http://www.radar-touch.com

Ramamoorthy, A., Vaswani, N., Chaudhury, S. and Banerjee, S. (2003) 'Recognition of Dynamic Hand Gestures'. *Pattern Recognition* 36(9), 2069–2081.

Raytrix (2013) '3d Light Field Camera Technology', accessed 21 August 2013 at: www.raytrix.de

Rebusfarm (2013) Home page, accessed 21 August 2013 at: www.rebusfarm.net

Rekha, J. and Majumder, S. (2011) 'Hand Gesture Recognition for Sign Language: A New Hybrid Approach'. *Proceeding of International Conference on Image Processing, Computer Vision and Pattern Recognition*, Las Vegas, pp. 80–86.

Render Rocket (2013) 'Meet Render Rocket', accessed 21 August 2013 at: www.renderrocket.com

Rosales, R., Athitsos, V. and Sclaroff, S. (2001) '3D Hand Pose Reconstruction Using Specialized Mappings'. *Proceedings of IEEE International Conference on Computer Vision*, Vancouver, 7–14 July, 378–387.

Roth, E. and Calapoglu, T. (2012) 'Advanced Driver Assistance System Testing using OptiX'. GPU Technology Conference, San Jose, accessed 21 August 2013 at: http://on-demand.gputechconf.com/gtc/2012/presentations/S0319-Advcanced-Driver-Assistance.pdf

Samsung (2013) Home page, accessed 21 August 2013 at: http://www.samsung.com/us/2012-smart-tv

Sánchez-Nielsen, E. Antón-Canalís, L. and Hernández-Tejera, M. (2004) 'Hand Getsure recognition for Human Machine Interaction'. *Proceedings of International Conference on Computer Graphics, Visualization and Computer Vision*, Plzen-Bory, Czech Republic, 2–6 February, 1–8.

Shotton, J., Fitzgibbon, A, Cook, M., Sharp, T., Finocchio, M., Moore, R. et al. (2011) 'Real-Time Human Pose Recognition in Parts from Single Depth Images'. *Computer Vision and Pattern Recognition*, Colarado Springs, 21–25 June, 1297–1304.

Shum, H.Y., Chan, S.C. and Kang, S.B., (2007) Image-Based-Rendering. Berlin: Springer.

Shum, H.Y. and Kang, S.B. (2000) 'A Review of Image-based Rendering Techniques'. *SPIE Proceedings of Visual Communication and Image Processing*, Perth, Australia, June.

SmartCam (213) Home page, accessed 21 August 2013 at: https://sites.google.com/site/marclevoy/

Smithsonean (2013) 'Panoramic Virtual Tour', accessed 21 August 2013 at: http://www.mnh.si.edu/panoramas/

Stenger, B., Thayananthan, A., Torr, P.H.S. and Cipolla, R. (2006) 'Model-Based Hand Tracking using a Hierarchical Bayesian Filter'. *IEEE Transactions on Pattern Analysis and Machine Integillence* 28(9), 1372–1384.

Suau, X., Ruiz-Hidalgo, J. and Casas, J.R. (2012) 'Oriented Radial Distribution on Depth Data: Application to the Detection of End-Effectors'. *Proceedings of IEEE International Conference on Acoustics Speech and Image Processing*, Kyoto, Japan, 25–30 March, 789–792.

Sudderth, E., Mandel, M., Freeman, W. and Willsky, A. (2004) 'Visual Hand Tracking Using Nonparametric Belief Propagation'. *Proceedings of Workshop on Generative Model Based Vision*, Washington, DC, 2 July, 189.

Sundaresan, A. and Chellappa, R. (2009) 'Multicamera Tracking of Articulated Human Motion Using Shape and Motion Cues'. *IEEE Transactions on Image Processing* 18(9), 2114–2126.

Szeliski, R. (2012) *Computer Vision – Algorithms and Applications*. Berlin: Springer.

Szeliski, R. and Shum, H.Y. (1997) 'Creating Full View Panoramic Image Mosaics and Environment Maps'. *Proceedings of the 24th Annual Conference on Computer Graphics and Interactive Techniques*, Los Angeles, 3–8 August, 251–258.

Takahashi, H., Tamura, N., Furue, T. and Yoshie, O. (2007) 'The Implementation of Virtual Factory with Advanced Low Cost Image-Based Rendering', accessed 21 August 2013 at: http://www.icpr19.cl/mswl/Papers/238.pdf

Triviala (2013) 'Triviala LIVE TV Show', accessed 21 August 2013 at: www.triviala.com/pages/interactive-dual-screen-participation-tv

Ulichney, R. (1993) 'Video Rendering'. *Digital Technical Journal* 5(2), 9–18.

Vatavu, R.D. (2010) 'Creativity in Interactive TV: Personalize, Share, and Invent Interfaces'. In A. Marcus, A. Cereijo Roibas and R. Sala (Eds) *Mobile TV: Customizing Content and Experience*. London: Springer, pp. 121–139.

Vesa (2013) 'DisplayPort specification v1.2 Overview', accessed 21 August 2013 at: http://www.vesa.org/wp-content/uploads/2010/12/DisplayPort-DevCon-Presentation-DP-1.2-Dec-2010-rev-2b.pdf

Viola, P. and Jones, M.J. (2004) 'Robust Real-Time Face Detection'. *International Journal of Computer Vision* 57(2), 137–154.

Vizrt (2013) Home page, accessed 21 August 2013 at: www.vizrt.com

Vudu (2013) Vudu Movies Home page, accessed 21 August at: www.vudu.com/movies/}search/imax

Wang, C.C. and Wang, K.C. (2008) 'Hand Posture Recognition Using Adaboost with SIFT for Human Robot Interaction'. *Springer Lecture Notes in Control and Information Sciences* 370(1), 317–329.

Wang, L., Chen, D., Deng, Z. and Huang, F. (2011) 'Large scale Distributed Visualization on Computational Grids: A Review', *Computers & Electrical Engineering* 37(4), 403–416.

Wei, L., Lefebvre, S., Kwatra, V. and Turk, G. (2009) 'State of Art in Example-based Texture Synthesis'. *Eurographics 2009*, Munich, 30 March–3 April.

Woods, A. and Sehic, A. (2009) 'The Compatibility of LCD TVs with the Time Sequential Stereoscopic 3D Visualisation'. *Proceedings of Electronic Imaging, Proc SPIE Vol. 7237*, San Jose, CA, 19–21 January.

Wu, C., Varanasi, K. and Theobalt, C. (2012) 'Full Body Performance Capture under Uncontrolled and Varying Illumination: A Shading-Based Approach'. *ECCV'12 Proceedings of the 12th European conference on Computer Vision*, Florence, Italy, 7–13 October, vol. IV, 757–770.

XBOX (2013) 'SmartGlass', accessed 21 August 2013 at: http://www.xbox.com/en-US/smartglass

Yan, J. and Pollefeys, M. (2008) 'A Factorization-Based Approach for articulated Non-Rigid Shape, Motion and Kinematic Chain Recovery from Video'. *IEEE Transactions on Pattern Analysis and Machine Intelligence* 30(5), 865–877.

Yeasin, M. and Chaudhuri, S. (2000) 'Visual Understanding of Dynamic Hand Gestures'. *Pattern Recognition* 33(11), 1805–1817.

Yoon, H.S., Soh, J., Bae, Y.J. and Yang, H.S. (2001) 'Hand Gesture Recognition Using Combined Features of Location, Angle and Velocity'. *Pattern Recognition* 34(7), 1491–1501.

Youswitch Inc. (2013) Home page, accessed 21 August 2013 at: www.youswitch.tv/Home

Wu, C., Varanasi, K. and Theobalt, C. (2012) 'Full Body Performance Capture Under Uncontrolled and Varying Illumination: A Shading-Based Approach'. *Lecture Notes in Computer Science* 7575, 757–770.

Zhou, C. and Nayar, S. K. (2011) 'Computational Cameras: Convergence of Optics and Processing'. *IEEE Transactions on Image Processing* 20(12), 3322–3340.

Zhou, H. and Huang, T.S. (2003) 'Tracking Articulated Hand Motion with Eigen Dynamic Analysis'. *Proceedings of International Conference on Computer Vision* 2(1), 1102–1109.

Zhu, J., Wang, L., Yang, R., Davis, J.E. and Pan, Z. (2011) 'Reliability Fusion of Time-of-Flight Depth and Stereo Geometry for High Quality Depth Maps', *IEEE Transactions in Pattern Analysis and Machine Intelligence*, 33(7), 1400–1414.

9

Application Scenarios and Deployment Domains

Omar Aziz Niamut[1], Arvid Engström[2], Axel Kochale[3], Jean-François Macq[4], Graham Thomas[5] and Goranka Zorić[2]

[1]*TNO, Delft, The Netherlands*
[2]*Interactive Institute, Stockholm, Sweden*
[3]*Technicolor, Hannover, Germany*
[4]*Alcatel-Lucent, Bell Labs, Antwerp, Belgium*
[5]*BBC Research & Development, London, UK*

9.1 Introduction

Chapters 3 to 8 gave details about the underlying technologies that together constitute the format-agnostic media approach. In these chapters, significant technological advances have been described compared to the state-of-the art in production, delivery, rendering of, and interactivity with, live audio-visual media, as described in Chapter 2. We argue that this new approach of format-agnostic media allows content to be produced in a more flexible way, better suiting the interests of the viewer and the capabilities of their device. However, the introduction of new technology is driven by the articulation of end user demand and their willingness to accept these new technologies, as well as the logical fit of these technologies within the domains where they will typically be deployed. Hence, it is worth considering how the approach of format-agnostic media can be introduced into existing production, delivery and device domains. Studying the techniques that underlie the approach through a variety of application scenarios and use cases helps to better understand their added value as well as their impact on the respective deployment domains.

This chapter describes the impact of deploying the format-agnostic media approach, along the lines of three deployment domains and the end user perspective. First, in section 9.2 we elaborate on a series of potential application scenarios. We start from state-of-the-art production

Media Production, Delivery and Interaction for Platform Independent Systems: Format-Agnostic Media, First Edition. Edited by Oliver Schreer, Jean-François Macq, Omar Aziz Niamut, Javier Ruiz-Hidalgo, Ben Shirley, Georg Thallinger and Graham Thomas.
© 2014 John Wiley & Sons, Ltd. Published 2014 by John Wiley & Sons, Ltd.

of TV channels and gradually increase the level of end user interaction that is offered. Then, we focus on the impact of deployment in Sections 9.3, 9.4 and 9.5, respectively. In the production domain, we provide an outlook on new production practices and consider the changes in the roles of a production team. In the network domain, we consider inherent network limitations and the impact of the application scenarios on the network. In the device domain, we consider the limitations of device capabilities and how to take these into account when deploying the format-agnostic media approach. These views are then completed in Section 9.6 with an end user perspective on how this introduction impacts user behaviour.

9.2   Application Scenarios

The following application scenarios describe features of a potential format-agnostic system from an end user perspective, in order to develop understanding of how technology enables and fits within the overall audio-visual media experience. An application scenario describes these features typically in the form of a short story. Such a description increases the understanding of how technology enables and fits within the overall audio-visual media experience. While the format-agnostic media approach can be broadly applied, we constrain the application scenarios in a number of ways, in order to arrive at some basic examples. First, scenarios are focussed on live events, such as watching soccer matches or musical performances. Second, we expect that users will consume format-agnostic media in a cinema, at home on a TV screen and/or on mobile devices such as tablets and smartphones.

We also make several assumptions about systems that implement the format-agnostic media approach. Audio and video will be captured using a selection of cameras and microphones. Specifically, in line with the video acquisition technologies described in Chapter 3, there will be one or more fixed extreme wide-angle cameras and multiple conventional broadcast cameras with the ability to pan, tilt and zoom. Similarly, in line with the technologies from Chapter 4, microphones for capturing the sound field at one or more points, as well as individual sound sources are present. A mechanism is provided to combine these audio-visual sources into a hierarchical representation, for example, the *layered scene representation* introduced in Chapter 2. It is possible to produce a range of different views or regions-of-interest (ROIs) of the captured scene by selecting different viewpoints and fields-of-view, to suit different viewer preferences and device capabilities, for example, making fields-of-view appropriate for the screen size of the device. Metadata is generated that describes how to create a particular view based on the hierarchical representation. Such metadata could be generated at the production side, for example, analogous to the shot framing and selection decisions made by a cameraman and vision mixer, or at the end user side, for example, by a user choosing the part of the scene he wants to examine in detail, or some combination of the two. One embodiment of this type of metadata is the production script, described in Chapter 6.

With these constraints and assumptions in mind, we construct the following three example application scenarios.

9.2.1   Digital Cinema: A Totally Immersive Experience

Imagine that you are a big fan of the opera diva Anna Netrebko. Unfortunately, she does not perform in the small town where you live. However, there is a premium event place in your town offering an immersive experience in a cinema equipped with a 180° panoramic screen providing ultra-high resolution video quality and 3D sound. This cinema presents a live transmission of a performance of Anna Netrebko at the arena in Verona. Arriving in the cinema, you are one of 30 premium guests, enjoying the opera 'La Traviata'. You watch the opera in crystal clear image quality with 180° field of view. The spatial audio reproduction sound system is based on Wave Field Synthesis or Higher Order Ambisonics. Thus, you are able to watch the performance as if you had the best and most expensive seat in the arena in Verona – right in the middle in front of the stage. From time to time some close-up views of the diva and other performers are shown in different parts of the panoramic screen. Thus, you are able to see details of the show, for example, facial expressions of the artists – which you would never have seen from any seat at the actual performance.

9.2.2   Home Theatre: In Control

You arrive home late to watch your favourite football team Barcelona in tonight's match against Chelsea. The match has already started and you realise that you just missed a goal, as football players are already celebrating it. Even though several streams and channels are available to you, that show the same football match from different angles, none of them is showing any replay you like. A bit frustrated, you continue watching the match but, this time, you decide to select a personal view of the football field by using your second screen device to interact with a video stream coming from the fixed extreme wide-angle camera. Later on, the system signals you by a small icon in the top-left side of the screen that a complementary channel is streaming a view automatically following your favourite player, who is playing a fantastic match. You quickly change channels so you can follow him, displaying this dedicated player view on your TV set.

9.2.3   Mobile: Navigation and Magnification

You are at a concert of the Rolling Stones, and listen entranced to your favourite music. You point your phone camera to the stage and select the drummer to be in the centre of the picture. Your mobile initially shows the picture from its own camera on its screen. You wait until a connection has been established with the in-stadium video service. After pressing the OK button the picture is replaced by a high-quality close-up live stream of the drummer, as recorded by the camera system and repurposed for mobile usage. You can now see the wrinkles of Charlie Watts! You pan a bit by using the touch screen on your mobile and watch Charlie performing. After a while you get bored and select Mick Jagger in the same manner. You press the button 'Follow' on the screen to make sure Mick will not walk out of the viewing frame on the mobile, as he jumps up and down on the stage. When the concert has finished, you are informed that an edited version of the concert is available. You can watch it on your mobile during your trip home, just to enjoy the concert again. However, you are disappointed with the directors' choices for scene cuts and framing, so you activate the free navigation mode to gain access to the whole set of recorded views and navigate freely (in both time and viewing window) to watch your favourite parts of the concert again.

   The above application scenarios provide a glimpse of what is on offer if the format-agnostic media approach becomes part of our media experiences. However, the deployment of technology, required for a particular application scenario, must be further analysed from production-, network- and delivery-centric perspectives. The three scenarios described above gradually

Table 9.1 Overview of the deployment domains with respect to the degree of end user interactivity. This overview assumes *no interaction*, for example, in the case of state of the art production of a linear TV programme

	Low Impact	Medium Impact	High Impact
Production	Works as today. Extra tools allow for novel shots and immersive audio.	Automation is split between production and network domains. Content is automatically rendered and adapted within the network to device capabilities.	All production is automated. Content is delivered to the end device including rendering instructions. All reproduction of content is fixed from the production or network side.
Network	Normal bandwidth, only delivers linear video streams/TV channels from production onwards.	Requires high bandwidth to production, but reduced bandwidth to the end user.	Requires high bandwidth towards the user.
Device	Low-end device, only receives linear video streams/TV channels from production onwards.	Regular device, only receives linear video streams/TV channels from network proxy onwards.	High-end device, capable of receiving very high bandwidth streams, interpreting rendering instructions and rendering of multiple high-resolution video streams, as well as 3D audio.

increase the impact in each deployment domain as well as the level of end user interactivity, as offered by the underlying format-agnostic media system. Tables 9.1, 9.2 and 9.3 provide an overview of the impact of increasing end user interactivity on the deployment domains.

The following sections each contain a more in-depth domain-centric analysis, including a more detailed take on the end user perspective.

9.3   Deployment in the Production Domain

9.3.1   Outlook on New Production Practices

As explained in Chapter 2, current production practices are usually aimed at creating a single version of a programme, with minimal support for re-purposing or interaction. This applies both to the technical standards used, for example, 16:9 HDTV with 5.1 sound, and in terms of the selection and composition of shots. Some specific examples include:

- Shot framing is optimised for a particular aspect ratio, possibly with 'safe areas' to allow parts to be discarded when producing a version for a different-shaped screen (Section 2.2.4).
- Audio signals are mixed down to a particular format, for example, ignoring the height component that may be available from a Soundfield microphone when producing a 5.1 mix (Section 2.3.3).
- A single edited version of a live programme is produced under the guidance of the director, who will be aiming to produce a programme that will please the majority of the audience (Section 2.4).

Table 9.2 Overview of the deployment domains with respect to the degree of end user interactivity. This overview assumes *medium interaction*, for example, where a user can choose between predefined streams of content

	Low Impact	Medium Impact	High Impact
Production	In order to generate multiple streams, more staff are required. Some of this work could be automated.	Generating multiple streams can be split between production and network domains.	All production is automated or automated/supervised mark-up and script generation.
Network	Must be capable of delivering a large number of different streams.	High ingest volume on the interface between production and network, including capability for selecting parts of the captured media.	Must be capable of delivering all captured media to the end user.
Device	Regular device, only receives linear video streams/TV channels from production onwards.	Regular device, only receives linear video streams/TV channels from network proxy onwards.	High-end device, capable of receiving very high bandwidth streams and rendering of multiple high-resolution video streams, as well as 3D audio. Additionally offers selection of streams by the end user.

Table 9.3 Overview of the deployment domains with respect to the degree of end user interactivity. This overview assumes *full interaction*, for example, where a user can freely navigate around the scene

	Low Impact	Medium Impact	High Impact
Production	Lots of scripting, both automatic and supervised in gallery.	Production and provider galleries cooperate on script creation and metadata generation.	Provider galleries are responsible for script creation and metadata generation.
Network	Interaction is limited to stream selection, a simulation of 'full interaction' can only be realised by delivering more channels/streams.	Must deliver all captured media in tiled format to the end user including capability to render specific ROIs.	Must be capable of delivering all captured media to the end user.
Device	Regular device, only receives linear video streams/TV channels from production onwards.	Mid-to-high-end device, capable of receiving tiled streams and rendering of multiple high-resolution video streams, as well as channel-based spatial audio. Additionally offers full interaction by end user through remote, tablet or gestures.	High-end device, capable of receiving very high bandwidth streams and rendering of multiple high-resolution video streams, as well as 3D audio. Additionally offers full interaction by end user through remote, tablet or gestures.

We foresee two ways in which this format-specific approach could evolve. First, the technology could be applied purely at the production side, to help create conventional programmes more cheaply, or to create multiple versions of programmes for specific platforms or audience groups. Second, with advances in delivery and consumption technology, the production process could evolve to produce a hierarchically layered audio-visual scene representation, including scripts to indicate 'curated' views of the content, for onward delivery. In the latter case the production process would have to be designed to support user interaction, such as the example application scenarios outlined in the previous section. The following two sections consider these different scenarios and the impact they would have on the way in which programmes are produced.

9.3.2   Use of Format-Agnostic Technology to Aid Conventional Production

An initial way of introducing the format-agnostic approach is to help with the production of conventional TV programming, to make the production of programmes cheaper or easier by using the technology at the production side alone. While this will not directly support the end user application scenarios proposed before, it could make it easier to provide multiple versions for viewers to select themselves. For example, a different version might be available to mobile devices compared to that available via conventional broadcast. This approach does not involve delivering the scripting metadata to the end user.

The ways in which the roles of the production team might change, compared to those of today, (see Chapter 2, section 2.4) are discussed below using the example of an outside sports broadcast.

Director

In sports production the team will usually sit in an outside broadcast truck (referred to as a 'scanner') watching the input from the cameras, without a direct view of the venue. These cameras do not necessarily present an easy-to-interpret picture of what is going on at an event. It is only through skill and experience that a director can translate this view into a coherent understanding of what is happening on the pitch. Having access to a high-resolution panoramic camera view could help improve situational awareness. An indication on the panoramic camera view of the location of cameras and other key parts of the production is also likely to be useful.

Camera Operator

Camera operators at a live event would continue working as they do today, choosing shots based upon a combination of their own experience and the director's instructions. However, they could be complemented or partially replaced by remote 'virtual' camera operators, choosing areas of the output from one or more panoramic cameras. Such 'virtual' cameras could be controlled using a 'pan bar' remote pan/tilt head controller (Figure 9.1), which offers the camera operator the controls and 'feel' of a real camera on a pan/tilt head.

Semi-automated methods of tracking and framing regions-of-interest in a panoramic image (Chapter 5) could also be used to offer additional 'virtual' camera shots, overseen by an

Figure 9.1 Example of a remote control for a pan/tilt head that could be used to control a 'virtual' camera. Reproduced by permission of Shotoku Ltd.

operator. Furthermore, it would be possible to automatically extract a corresponding shot from the point-of-view of each panoramic camera, using knowledge of where the manned broadcast cameras are looking and the approximate depth of the region-of-interest, as discussed in Chapter 3, Section 3.3.

Vision Mixer

The role of a Vision Mixer (VM) would remain largely unchanged from today, although they may have access to a larger number of possible views, thanks to the multiple 'virtual' views available from panoramic cameras. If these views are being automatically generated using the content analysis and Virtual Director tools described in Chapters 5 and 6, it may be useful to provide some indication as to how 'interesting' these views were thought to be, to assist the VM in making a choice.

Audio Engineer

As with the other roles, the audio engineer's job changes very little in this scenario. The linear programme output is the audio engineer's main focus. As long as the production gallery is just making one TV programme, then the audio engineer will mix audio appropriate to that programme. As with video, some of the automated or semi-automated tools described in this book might provide the audio engineer with more tools that they can use to create content. It might also alleviate some more straightforward tasks, for example, an assisted mixing tool based on the output from an audio object extractor (providing content and position of the sound sources) could help inform level and panning decisions to dynamically match a given visual cut or could automatically add a given microphone into the mix if it were to contain salient content. Mixing the 3D audio content would not present any significant changes to the audio engineer's task as it already involves mixing in 2D. However, for a standard linear programme the task of mixing the audio objects with a sound field in 3D could be aided with visual tools to allow the audio engineer to visualise where the sound sources are in the sound scene and to produce the desired spatial mix by manipulating their position.

Replay Operator

We can imagine two different ways that a Replay Operator's (RO) job could change. Either they see the usual set of cameras, and have fixed captured video to work with, or they have complete access to the video from the panoramic camera.

In the first case, the job of the RO is more or less the same as it is today. They will be watching their feeds, and when incidents occur, they will be ready to play back some of their shots should the director wish it. The view from the panoramic camera would probably help with the RO's comprehension of what was going on at a sports event, so they should be positioned where they can see the panoramic camera feed(s).

In the second (and more interesting) case, the panoramic camera data is available in full to the RO. This presents a significant departure from the existing situation. It allows the RO to compose their own camera shot based on the data coming from the panoramic camera. For example, if one of the camera operators has missed a part of a developing incident, the RO can fill in the necessary footage, as they would have control over both time and space for selecting incidents to show.

One way this could be made to work would be for the RO to see the broadcast camera views overlaid on the panoramic camera image, and to be able to adjust the motion of the view so that the interesting incident is included. Alternatively, the RO could have free access to the panoramic camera image to compose shots independently. This would be useful where the conventional cameras have missed an incident altogether, or where there are not enough cameras to cover a large area at once. For instance, in motor racing, there is often only one camera at a corner, covering both the entrance and exit to that corner. A panoramic camera would be able to cover a wider field of view, prolonging the captured action between the entrance and the exit, which would provide more material for the replay operator. Using soccer as another example, the replay operator could provide details of action that took place outside the action immediately around the ball, and where the only option would otherwise be to use the wide overview camera for replays. Some commercial systems such as All22 and Endzone, developed by Camargus (2013) have already begun to offer this kind of functionality for high-definition video. Similarly for still images, automated panoramic heads such as the Clauss Rodeon VR (Clauss, 2013) have enabled ultra-high resolution images of scenes, although not for action footage, since the image is stitched from a matrix of images taken in sequence.

While the view of the scene from the panoramic camera is typically provided at a lower angular resolution than the view from the camera it is replacing, the increased level of interaction and control can still provide a service that is compelling to the user.

9.3.3   Production of Format-Agnostic Content to Support End User Interaction

The full benefits of the format-agnostic approach would be realised when multiple components of the layered scene representation are sent to the end user, along with scripting information to provide one or more 'curated' views. The creation of the raw audio and video streams would proceed much as in Section 9.3.2, where additional capture devices such as panoramic cameras are used to acquire a fuller view of the scene, without being limited at capture time by a particular framing decision. However, these streams could all be delivered to the end user, supporting free navigation within the panoramic video, and free selection of camera views, with scripting information to allow views telling a particular 'story' to be selected automatically. As before, the example of an outside sports broadcast studio is used.

Director

The director might use the outputs of the various scripting systems to create a preferred view. This might provide a similar experience to watching the directed version of the show. However, their role would also include overseeing the production of other scripts, specifying for example, which regions of the action they thought should be offered as alternative scripts (such as identifying the key people that viewers should have the option of following). They might also provide the final decision about which views are transmitted for selection by the viewer, thereby limiting the viewer's ability to freely explore all captured content to those parts that the director considers relevant.

Camera Operator

With the potential of offering multiple options to the viewer, there may be more need for 'virtual' cameras, generated by extracting portions from panoramic camera views. This would require more use of automated content analysis and scripting technology (described in Chapters 5 and 6). The relationship and ratio between manned and virtual camera operators depends to some extent on how they work collaboratively to produce close-up footage of events, on the technical limitations of doing this within the panoramic image alone, and on the technology for juxtaposing manned and virtual cameras in real time to produce detailed shots. Discussions with producers indicated that this may depend on the scale of production (Zoric, Engström and Hannerfors, 2012). Producers saw great potential in using this kind of technology to enable low-end productions with primarily remote camerawork, but they were hesitant to replace close-up camerawork in larger productions, for example, up to 30 manned cameras, with footage they feared would be flatter due to bigger depth-of-field and less detailed compared to images from a manned camera with a big zoom lens.

Vision Mixer

The role most resembling that of the VM would, in the format-agnostic approach with end user interaction, be that of a dedicated *Virtual View Operator* (VVO). This role would resemble that of a VM, in that the VVO would oversee all available visual resources in the production gallery: virtual camera views as well as any manned broadcast cameras. However, as pointed out by professional producers when presented with this scenario (Zoric, Engström and Hannerfors, 2012), the VVO's role would also, in practice, resemble that of an RO, in that the operator would work instrumentally with multiple visual sources. In this case, the operator would be using annotation of events in the scene and adjusting properties of the virtual views to be delivered. For small- or medium-scale productions the VVO role may also include actual replay production work, based on the same visual resources and an additional set of replay production tools, in effect merging the roles of VM and RO into one VVO. As an alternative, several operators could work in parallel sharing the tasks. In a more task-separated scenario, where multiple operators are available, the VVO could attend to the live views while a dedicated RO would produce replays. The VVO would be generating scripts by following certain regions-of-interest, or creating packages of replay content to illustrate certain interesting events. In addition to having an edited broadcast feed from the production gallery, their work would be to select different views of the game from different cameras (including 'virtual' cameras extracted from the panoramic camera), and curate the presentation of the multiple views available to the end user, through annotation and manual adjustments of the views. For example, they may be able to set up preconfigured user settings (e.g., 'prefer the red team', 'select views suited to a small screen') that would use particular scripts, or viewers could be explicitly offered particular 'cuts' of the programme (e.g., 'follow player X').

Audio Engineer

With the requirement to pass scripts to the user, the individual components of the audio mix need to be passed down the transmission chain, along with metadata to allow a suitable mix to be created at the device. This would allow a separate audio mix to be generated to match each view, in a way designed to match the reproduction capabilities of the end user device, such as mono, stereo, 5.1, binaural audio for headphones and so on. From a technical point of view, this implies automated audio mixing at the user end based on control parameters from the mixing desk and other metadata about sound source location. Using object-based audio, rather than using the geometric placement of microphones, increases the creative options for rendering audio. For instance, it is possible to render audio based on perceptual models that may be more suitable than complex geometric transformations between 2D and 3D space for calculating distances between microphones, for example, for zooming in an acoustic scene (see Chapter 4, Section 4.7). Methods of capturing metadata about sound source location were discussed in Chapter 4, Section 4.5. Extracting relevant parameters from the mixing desk means recording all its operational parameters and outputting them, as well as responding to control from outside. The need to produce metadata in addition to the audio itself may require additional people. There are two candidate approaches: using staff to create semantic metadata about the subject of a microphone signal, or using staff to create audio mixes suitable to the video streams being transmitted. In discussions with production staff it was pointed out that mixing correctly is very difficult to do right, and it is easy to spot if it is wrong. This implies that fully-automated intelligent mixing based on semantic metadata may not be practical. Instead, operators can create mixing metadata. These metadata can provide a basic mix, which then might be tweaked in the renderer.

Replay Operator

The role of the Replay Operator (RO) will, like those of the VM and audio engineer, focus on the production of metadata describing relevant parts of the scene. However, the role includes a temporal element, as they can identify key incidents after they have happened, rather than having to do this in real time. They may still provide replays on demand to the director, for inclusion in a 'preferred' view, but would also generate metadata describing interesting areas in the full range of audio-visual feeds available. Indeed, the RO (or someone with a similar set of skills) is likely to be the main source of manually-generated information about footage being captured. The RO would work at a terminal that is used for capturing metadata about what is going on in a given camera shot. This is likely to be semi-automated, using the approaches described in Chapter 5. In order to present the end user with a wide selection of interesting choices, the metadata could include information about which people and events are in which shots, which cameras have the best view of a particular person or event, and what the main focus of a particular shot is.

9.4 Deployment in the Network Domain

For the delivery of format-agnostic media towards end users, as highlighted by the application scenarios, the rationale to push more processing functions into the delivery network is based on both business aspects as well as technical limitations. Hence, service providers are a potential class of users of format-agnostic media technology. Here the term service provider is to be understood in a broad sense, as it encompasses not just network and video service providers, but also local broadcasters or any other third-party that can benefit from the flexibility offered by format-agnostic audio-visual content to create new services, such as linear TV programmes, personalised, interactive services and so on.

9.4.1 Network Requirements

The impact of deploying format-agnostic media on delivery networks is mainly influenced by their technical limitations. The most important limitations are the available bandwidth, the latency and the network scalability with regards to storage, processing and functionality. First and foremost, format-agnostic interactive and immersive media services require a significant increase in bandwidth compared to existing TV or video services. Bandwidth requirements are largely related to the distribution of the video signals, in either production format, intermediate format or a specific pre-rendered view. Other data flows that are transmitted over the delivery network are audio signals, scripting metadata and the user input and commands, but the respective bandwidths of these signals are much lower compared to those of video. The main factor for video bandwidth is the compression or codec scheme that is used. Chapter 7 discussed the trend towards improved coding technologies, which are required for coping with increasing video data rate at the production side. Second, latency requirements for fully interactive scenarios are higher compared to existing TV services. Latency covers the delays that are introduced in the delivery network and can be an important constraint for live video services, for example, a soccer match and services involving (user) interactivity. Four types of latency can be distinguished, with the notion that these are not entirely independent of each other.

The *end-to-end service delay* relates to the time difference between the capturing of audio and video at the capture location and the presentation of these signals to the user by the device. The allowed end-to-end service delay depends on the application scenario. For example, for home viewers watching live events the allowed delay to the end user may be in the order of seconds. In contrast, the allowed delay for on-site devices may require real-time behaviour, for example, when offering a mobile view to visitors during a music concert. The *session setup delay* is related to the time between a user requesting a media stream and the presentation of the content on a screen or through speakers. For certain delivery modes (i.e., unicast and multicast delivery), a device must request the audio-visual media before transmission of the media starts, introducing session setup delay. For broadcast delivery modes, the media streams will already be transmitted to the device when a user requests a media stream, so the time between user input and content presentation is minimal. The *delivery network latency* relates to the amount of delay the delivery network introduces, for example, the transport, routing, conversion and rendering of media streams that are provided by the production domain and delivered to a device. In other words, it is the time difference between the ingestion of audio-visual data into the delivery network and the reception of this data by the device. The *responsiveness* relates to the time between a command input and the generation of the result of that command. For interactive scenarios, that is, where an end user controls what is being displayed via user commands, responsiveness will be a dominant factor.

An additional limitation to consider is the network scalability. That is, the underlying delivery mechanisms should scale to a large number of users, allowing for interactivity and adaptation of content to a variety of mobile devices. Such delivery mechanisms should be implemented in managed or overlay delivery networks, such as a content delivery network (CDN). Furthermore, storage and caching mechanisms are required to provide on-demand and replay functionalities, as the delivery network must cache audio-visual media segments that are delivered during content consumption, or store them for offline access after a live event. The delivery network should further support the types of transport that relate to different delivery modes, for example, unicast, broadcast and multicast or hybrid combinations of these, and delivery types, for example, unidirectional, unidirectional with a separate feedback channel and bidirectional. The more demanding application scenarios place high processing requirements on the delivery network. Some processing functions are natural evolutions of the ones that can be found today, such as packetisation, filtering and routing mechanisms, while others can be seen as more disruptive, for instance functions that perform audio-visual media processing. These would include audio and video segmentation, (re-)coding and rendering. Given the variety in end user devices, some low-profile devices will allow for some limited forms of processing, for example, combining audio-visual segments at the terminal, whereas network-based components are required to handle high processing demands.

9.4.2   Impact of Application Scenarios

Impact of Digital Cinema Application

This scenario considers a current state-of-the-art delivery situation and has its focus on innovations in the production domain. The distribution of high resolution video and 3D audio content is tailored to a specific delivery format and one or more rendered views in the form of TV channels and/or media streams that are presented to the user. Dedicated channels and streams exist for widescreen views, zoom and ROI views. As the typical delivery chain for digital cinema applications mostly consists of satellite broadcast infrastructure (see Figure 9.2), the primary impact is expected to be higher bandwidth requirements, as well as improved display and loudspeaker facilities in cinemas.

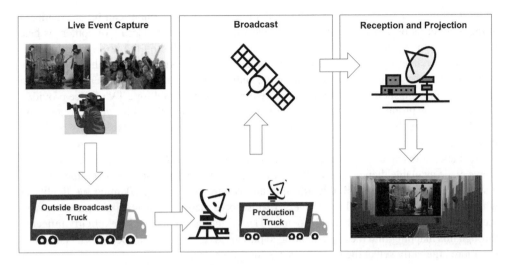

Figure 9.2   Example of a typical delivery chain for live event broadcasting to digital cinema.

Impact of Home Theatre Application

This scenario considers an idealistic delivery network that allows for distribution of all the captured audio-visual streams to an end user device, before rendering and presenting the final view to the user. It is assumed that production metadata is sent towards the device, containing production-side knowledge that specifies the required processing steps. Here, the delivery network provides the necessary functionality to adapt a hierarchical representation coming from the production domain to one or more views suitable for the end user device. Initially, coarse levels of interactivity are supported, where the end user is allowed to select predefined portions of the captured content. The impact will differ depending on the granularity of the interactive offer. Two distinct examples are:

1. Predefined channels, either directly mapped to predefined production cameras and audio mix, or mapped to a view that can be generated from post-processing the captured audio-visual content.
2. Interactive *playmaps*, where the user is offered more complex navigation choices with many switching points defined over time and space. The number of possible combinations does not make it practical to offer every possible combination with its own channel identifier.

The creation of additional views requires some rendering functions after the capture stage. These are best located either as part of the production process or at the service provider's video ingest facilities (e.g., to repurpose content for a regional audience). Rendering such views closer to the end user side (e.g., in a network proxy or in the terminal itself) is also an option, but would lead globally to a waste of processing resources, as many rendering functions in the overall system would turn out to perform exactly the same operations. When evolving to a very high number of navigation scenarios it becomes impossible to render every scenario in advance as the ingest processing and the bandwidth capabilities of TV broadcast networks cannot scale indefinitely. Although some parts of the content can be pre-rendered at the production or at a service-provider's ingest point, the final rendering of content is best located close to the end user. This has the advantage of distributing the processing load for personalised rendering and also minimises the end-to-end service delay of the interactivity loop. The way these views are created from the audio-visual content must either be controlled manually, although such a post-production role is rather unexpected for a TV service provider, or controlled by production metadata that are ingested along with the audio-visual content and that describe how the relevant views can be created for the various contexts (user's preferred type of views, team, etc.).

The bandwidth capacity of the network must be dimensioned for the number of concurrent views made available, hence bandwidth constraints can be an element when deciding whether a service is to be deployed. Bandwidth limitations may also require adapting the fidelity at which the content is packaged for delivery, that is, some mapping must be performed between the available views and the channel streams. The switching points can be explicitly reflected through this mapping into the aforementioned playmaps. In this case, the delivery mechanisms can more efficiently select the relevant views for each group of users.

When supporting the finer-grain level of interactivity foreseen, where an end user can freely navigate around the content, we distinguish three situations that differ in terms of the access bandwidth and device capabilities that a service provider might assume to be available at the end user's side:

1. High-profile delivery: this assumes that all captured audio-visual content can be delivered and processed by end user devices. For the delivery, this essentially requires the development of very high bandwidth transmission technologies. Note that similar requirements apply for the transmission of captured audio-visual content to the digital cinema setup, where the full set of audio-visual media is rendered without user interaction.
2. Main-profile delivery: this assumes traditional residential access bandwidth. In this case, the network is required to transmit only the required portions of the captured audio-visual content to the end user device, so as to respect the bandwidth limit and fulfil the interactivity requests. Therefore an ad hoc segmentation of the captured audio-visual content must be defined, which determines the granularity at which the delivery mechanisms can be optimised. No predefined sequence of the segments is available for steering the delivery of segments. However, in order to optimise the bandwidth and delay performances, metadata can signal the relevance of each segment over time and space. In addition, a dedicated interface is required with the device so as to ensure on-time delivery to the rendering functions of the parts of the captured audio-visual content required to respond to the continuous interactivity requests.
3. Low-profile delivery: this corresponds to the situation with lowest assumptions on audio-visual processing capabilities and access bandwidth, as encountered in set-top boxes and connected TV sets. This requires performing all rendering operations in a network proxy, so that only fully rendered audio-visual data are sent to the end device.

Impact of Mobile Application

In the case of delivery to mobile devices the network capabilities may have to be more advanced in terms of processing than for the previously described applications. The mobile scenario considers a delivery network with limited bandwidth, allowing for distribution of processed and adapted audio-visual streams before rendering and presenting the final view to the user. Here, the delivery network provides the necessary functionality to adapt a hierarchical or layered representation provided by the production domain to mobile devices of consumers at a live event. An important consideration here is the network availability at the venue of a live event. An ad hoc Wi-Fi network or 4th generation mobile broadband network should be available in order to make high bandwidth connection possible. Furthermore, given the variety in mobile devices, some high-profile mobile devices may still allow for some limited forms of processing, for example, combining audio-visual segments on the device. On the production side, this requires functionality to relate the picture taken by the mobile device, to the content captured by the cameras on the scene. It also requires feature and object tracking to create personalised views.

From the above analyses, it can be seen that, in the end, future networks have to be able to cater for all application scenarios. In practice, a hybrid mixture of heterogeneous networks will be encountered, where each network segment or branch may cater for a different set of end user devices.

9.5   Deployment in the Device Domain

Traditional technologies for audio-visual rendering on end user devices in the digital living room, as presented in Section 8.2.1, strongly depend on available processing power as well as storage capacity of the end user device and properties such as the bandwidth of the connection to the service. Here the devices for deploying rendering technologies can be structured into market sections for low-end or high-end complexity and capability. We also differentiate between primary digital living room devices such as TVs, set-top boxes (STB), and home theatre systems as well as secondary devices like digital media receivers, DVD players, Blu-Ray players, digital video recorders and gaming consoles. Typically the margin for low-end products is too small to allow the addition of costly features unless a highly integrated solution is employed. This has resulted in end user devices that support a host of different rendering formats such as video compression standards, but are very limited in adding new features that require a change in the implemented rendering parameters. Therefore, high-end consumer devices now adopt open source platforms to gain from a wide and robust development baseline and even community driven feature extension. To give an overview about the implications and benefits of supporting the format-agnostic approach at the presentation stage, these topics are now discussed in more detail.

9.5.1   Device Capabilities

In the European market for devices in the digital living room, TVs make up more than 75 per cent of the global digital living room market (Market and Markets, 2012). Factors such as evolving technology and changing consumer behaviour are affecting each type of digital living room devices. For instance, a declining trend is observed in the set-top box and Blu-ray player markets. As long as TVs can provide the proper connection to the service such as via cable, satellite or wireless transmission these shares may remain the same or even increase to more than 85 per cent by the end of 2018 (Market and Markets, 2012). However, this requires a future-proof processing platform with generic support of upcoming rendering formats, services and features.

Devices

End user devices for media consumption nowadays range from small mobile clients such as tablets and smartphones, towards large or huge media centres hosted in an STB or in a highly-integrated TV set, typically referred to as a *Smart* or *Connected* TV. The trend towards highly integrated embedded platforms now competes with flexible open source solutions offering an easier route to integrate future functionalities. Nevertheless, as hardware platforms still require evolution in order to provide enhanced performance a new approach is to support functions such as media playback on top of a generic platform. Here, the main platform needs to be equipped with standardised application interfaces to allow cross platform interoperability. These functions are considered both a risk and an opportunity, as it breaks the traditional monetisation channels and business models of media service operators. On the other hand it is getting more difficult to explain to the customer why a certain function is not available on his device.

Rendering Formats

These audio and video specific properties are basically defined to allow a common media production and define image parameters such as aspect ratio, colour gamut, dynamic range or frame rate to guarantee interoperability and thus wide distribution. Additionally, the audio-visual compression standards employed define the scope of achievable audio-visual quality and affect the overall system delay, which must be minimal for comfortable content interaction. Adding multiple sound channels representing sound positions or capabilities to render three-dimensional images improves the media experience.

Processing Performance

A main property for distinction between different products is the processing performance. More and more, these units move from generic central processing units (CPU) to dedicated graphics processing units (GPU). For example, while simpler set-top boxes still contain single core CPUs cores together with specialised digital signal processing (DSP) units, it is becoming more common to have multicore CPUs supported by GPUs introduced by game consoles or workstations. The performance of these processing units roughly follows the Dhrystone Benchmarks (Weiss, 2002), expressed in Dhrystone million instructions per second (DMIPS), ranging from 1000 DMIPS for the basic consumer-end device to 3000 DMIPS for a high-end device. The trend now is to increase processing power by using multiple cores to ensure multiples of those figures. However, processors in set-top boxes have currently only 20 per cent of the processing power available on netbooks and only 2 per cent of the processing power available on high-power workstation processors, a situation that is not likely to change in the near future. GPU benchmarks are strongly dependent on the application but follow similar rules. A common benchmark for these is the GLBenchMark (2013), which is a popular benchmark in the comparison of smartphone and tablet performance. However, having separate

GPUs is costly and current quad core chips such as NVidia's TEGRA 3 combine a popular ARM core (CortexA9) and a GeForce Ultra-Low Power GPU as a competitive processor for media end user devices.

Storage Capacity

The local and remote storage capacity for audio-visual content and the media playing application, including the main operational memory, remain critical. While most of the cost pressure has been relaxed with current memory technologies it is still expensive to move beyond 4GB (gigabyte) for solid state or 2GB of volatile memory. When devices require a larger amount of local storage they need hard disk storage thus adding complexity to file handling in the system. File system complexity is another good reason to go for open source platforms (see also the part on middleware below). Additional freedom gained by using online storage such as provided by cloud-based services are still challenged by the reluctance of customers to store personal content remotely. Additionally, the limited interoperability still supports suppliers of media serving gateways having access to local storage means.

Service Connectivity

The access to the service remains challenging for interactive services when it comes to end user feedback and service response time. Here the challenge is mostly the expense involved in the service infrastructure that will ensure a high-quality customer experience. End user devices allow interoperation by supporting a host of different interfaces and interface standards. These encompass wired links such as the Universal Serial Bus (USB) or Ethernet, and wireless links for the various standards such as Wi-Fi, supporting data rates from 300Mbps to 28Gbps (Parks Associates, 2013b). Several standardisation efforts to homogenise these interfaces have failed. Among the various options Universal Plug and Play (UPnP) (UPnP, 2013), Digital Living Network Alliance (DLNA) (DLNA, 2013) and Multimedia over Coax (MOCA®) (MOCA®, 2013) are some common ones. However, additional interfaces to connect to local storage and for display connections also need to be integrated

Middleware

The platform for media rendering hosts a software framework enabling all the technical features. Traditionally this middleware used proprietary software, however, open source software such as Linux or Android has become more common. This guarantees future-proof development support but also challenges the platform differentiation due to their license policies. This means, it is not easy to come up with a new feature that differentiates a product from a competing one that is not easily replicated by a competitor. Nevertheless, support of various embedded Linux versions is already common in TV and set-top box platforms and the community is also pushing development for future Android-based versions. This supports the business model of service providers, who no longer have to supply the hardware but can run their service on top of available models. Moreover, support for common interfaces and programming languages along with web browsers running HTML5 provides opportunities for cross-platform applications. Other important aspects, apart from the need for widely used frameworks, are common content authoring languages and integrated development environments. New business opportunities have been cultivated by selling additional features in the form of small applications competing with native and HTML5 based implementations (Parks Associates, 2013a).

Flexibility to Feature Adoption

Recently the functionality reconfiguration of end user devices has been enhanced to allow firmware updates to add further features for the user beyond just fixing bugs in a product. As users become more proficient in applying these updates it allows introduction of innovative functions and applications. These functions, referred to as *Apps, Gadgets* or *Widgets* depend on the way they are added to the existing device for their categorisation and, therefore, name. Deriving and supporting a developer community for these functions becomes an important asset in monetisation for high-performance devices in consumer electronics. Examples of such technical additions on TVs or STBs are over-the-top and on-demand video players or special interest applications for weather, maps, news, social networks or video telephony. Furthermore, content encryption and conditional access are critical aspects in order to allow content owners to guarantee their return on investment. While service subscriptions can manage the access, it is hard to provide copy-protected media transmission without support from platform vendors allowing controlled distribution.

To deploy an audio-visual format on current infrastructure the elements discussed above and their application in the selected service infrastructures have to be considered. While video consumption is widely scattered across cable, satellite, wireless and internet services, most of these services do not offer a satisfying flexibility to watch and listen to all the content that has been produced. This is not just the problem of insufficient data throughput of the media channel but may also be due to missing content identifiers or adaptation to the connected end device. For example, the end user is usually required to select the aspect ratio manually instead of getting support from the standardised content specifications such as appropriate metadata that trigger this change in the device automatically. Finally, service providers tend to squeeze more content into available transmission channels than is beneficial for a perfect media quality in order to balance the cost of the service.

Based on the analysis above, adopting a format-agnostic approach in device rendering basically requires several changes. First, having access to a wider view onto a produced scene needs the device to understand the content that is shown in order to frame the requested perspective in a proper fashion. This is basically the default perspective a device wants to present as the most natural, attractive or common view onto the captured scenario. Large-screen TVs offer a wider view of soccer matches than smart phones do. Close-ups are required more often on small screens while on home theatres this can easily get uncomfortable. Imagine, for example, the talking heads of anchor-men filling the screen. Here, the content producers have the chance to provide metadata indicating important elements of the scene to allow framing of a certain size or inclusion of certain scene elements.

This requires standard messaging and metadata schemes to exchange settings and identifiers. Interactive services need to maintain low latency for swift content updates and need media servers and rendering farms close to the audience. Hence, service subscription needs to allow information exchange such as this in an easy fashion. Having access to more scene elements requires careful treatment of user attention and attraction. Depending on the event or program being presented, the user might be overwhelmed by the information provided. Here the service and the end device application need to supply non-intrusive help to support attractive story telling or to allow a lively interaction with the content.

Finally, the value of the service needs to compensate the cost of producing the content, as added value is hard to sell if the benefit to the end user is not obvious. This is the reason why several applications offer a free base functionality funded by advertisements whereas for an extended service the user has to subscribe for a fee. For example, HDTV services only became attractive to consumers with the availability of large-screen TV sets. This holds even more for offering video services with resolutions beyond HDTV, such as 4k. Such services would need to emphasise the relevance of a large screen even further. Additionally, they could make use of processing to capture and deliver scenes, as described in this book, combined with its interactive features for pan-tilt-zoom navigation in specific regions of the content.

9.5.2 User and Producer Expectations

The technology for format-agnostic production demands an integrated workflow of multiple camera feeds and a tight link to the delivery network to provide swift updates on the rendered perspectives. With respect to interactive rendering this technology can be deployed either in production environments supporting the content creator or the end user side providing the freedom to act as a director on its own.

The *producer* wants creative tools to provide attractive content supporting the storytelling and reducing the cost of production. Here, the rendering framework presented in Chapter 8 offers the chance to provide not only a single story line to the end user but to provide several to choose from. Changing the focus parameters of the main actor is a common tool to steer the audience's attention. Controlling blur and other image parameters on this virtual view can further support that. Additionally, since satellite uplinks, or fibre cabling, to a centralised studio are already common at large live events, the cost of remote production is compensated by the chance to cover a wider area of the scene for post-production decisions. That is, controlling real cameras remotely will not be needed when access to the whole captured scene is available to render virtual video feeds.

The *end user* demands the option to either follow a predefined view for a lean-back entertainment, or have the freedom to reach out and explore new views or angles. The technology to have this freedom, which requires good rendering performance and tight network connectivity, will be available in consumer products soon. While some user groups following one program may find it entertaining to interact with the content in a group and risk fighting over who is in control, others will use secondary screens to control individual views or to recommend an interesting view in a social setup.

Depending on the availability of the link back to the producer, the device using this format-agnostic content will evolve to have natural ways of interaction. Different interaction approaches, for example, touch gestures or device-less and device-based gestures will complement each other to provide control of these new rendering capabilities.

9.6   Deployment in the User Domain

The application scenarios presented earlier in this chapter give an idea about what format-agnostic media technology aims to deliver to end users. However, in order to achieve *immersive experiences*, while interacting with this new type of media technology, it is important to understand the quality expected by the user, both in terms of user experience from an interaction perspective, as well as the content being offered. We now consider the impact of deployment in the user domain by discussing viewers' preferences, both from the perspective of what users appreciate in TV viewing as it is today, and what future TV services should offer to them to be at least comparable with the next generation interactive TV sets.

Regarding levels of interaction, systems based on format-agnostic media technology should be designed to support both the passive and active use, that is, they should be designed for various levels of interaction ranging from no interactivity to full interactivity. Watching TV is traditionally a lean-back activity. With enhanced interactivity it becomes a lean-forward activity. However, viewers in many cases want to just lie back and relax. The application scenarios show the potential complexity of new services in the sense that the new services will bring many new possibilities for viewers at the risk of overwhelming them. Thus, systems based on format-agnostic media technology should support *relaxed exploration* instead of *information seeking* by starting with *familiar content* and continuing with browsing of *relevant items* (Chorianopoulos, 2008). Format-agnostic interactive media will enable many viewing options, as well as delivery of personalised content. However, managing it during an event might cause interaction overload. To reduce the effort needed during an event, mechanisms for configuration before an event should be offered, for example, through user profiles. If viewers can customise their viewing preferences in advance, less interaction will be needed during the event, resulting in more relaxed viewing of desired content.

Some people like being part of a larger media context – this usually involves content sharing and communication about the content. With format-agnostic media it could be that consumers follow the same event, but not see the same pictures for example, through choosing different framings or angles. However, social conversation for viewers watching different images of the same event should be enabled. This can include mechanisms, for example, for following a friend or the most popular view of the event, sharing preferences or alarms from a system or other users. The social context in people's homes, for example, in living rooms, is complex and varies over time. Family members and friends watch TV together, and a living room is seen as a shared space where negotiation with regards to the interface (remote control) often happens. A typical setting consists of one or multiple users with hierarchy of users, and a possible use of multiple screens (secondary ones like handheld devices or laptops). A secondary screen usually has more interaction possibilities compared to the first screen. If used to enhance TV viewing, it should display the same image at the same time as the main screen.

Viewing in public places can be either directly at a live event (e.g., concert, sports event or festival), or as live broadcast in cinemas, theatres, open spaces, for example. A typical setting consists of multiple users with one common screen and multiple personal screens: secondary screens. A secondary screen supplements content of the live broadcast/event (enhanced TV), and gives more control. Another form of content control is through members of crowds using interaction via their mobile device and by examining the behaviour and locations of crowds viewing the content.

The system should be able to render the scene on different end user devices like high-definition TVs, tablets, mobile phones in different environments. Each environment and/or terminal has its own features and limitations. However, the choice of an interaction technique, and the manner content is shown will depend on a specific setting – device properties and characteristics of the environment. The applications scenarios differentiate three main environments: mobile, home and public. When a mobile phone is used as the device multitasking is required, mostly because of the communication requirements (Cui, Chipcase and Jung, 2007), for example, watching TV and answering phone calls. Communication requirements should also be taken into consideration when considering the use of new media services, in light of power demands. Looking at the screen properties, compared to home or public screens, a mobile screen is limited in size, and this needs to be taken into consideration when designing services.

9.7   Conclusion

We have studied the application of the techniques that underlie the format-agnostic media approach, through a variety of application scenarios in order to understand better their added value as well as their impact on the respective deployment domains. In the production domain, the impact lies mainly in the change of roles for key production personnel in a production team. The main tasks they have to achieve are in the selection of the appropriate sensors (cameras and microphones) to provide the audio-visual elements for the story, to perform changes to the production process and the selection of metadata describing relevant parts of the scene. To enable deployment in the network domain, current network limitations have to be reduced as future media delivery networks have to be able to cater for all application scenarios. In practice, a hybrid mixture of heterogeneous networks will be encountered, where each network segment or branch may cater for a different set of end user devices. In the device domain, having access to a wider view onto a produced audio-visual scene means that the device must understand the content that is shown in order to frame the requested perspective in a proper fashion. Hence, a production team needs to provide metadata indicating important elements of the scene to allow framing the content to a certain size or to include certain scene elements.

In summary, it appears that both the level of technological innovation, as well as the level of user interaction should be introduced gradually. It is possible to imagine situations where, initially, only limited interactivity is permitted or possible. The introduction of new technologies should be gradual, working up to the point where users have full interaction at their disposal.

References

Camargus (2013) 'Premier Stadium Video Technology Infrastructure', accessed 22 August 2013 at:
    http://www.camargus.com/
Chorianopoulos, K. (2008) 'User Interface Design Principles for Interactive Television Applications'. *International Journal of Human-Computer Interaction*, 24(6), 556–573.
Clauss (2013) Home page, accessed 22 August 2013 at: http://www.dr-clauss.de
Cui, Y., Chipchase, J. and Jung, Y. (2007), 'Personal TV: A Qualitative Study of Mobile TV Users'. In P. Cesar, K. Chorianopoulos and J.F. Jensen (Eds) *Interactive TV: A Shared Experience, Proceedings of 5th European Conference, EuroITV 2007*. New York: Springer, pp. 195–204.

DLNA (2013) Home page, accessed 22 August 2013 at: http://www.dlna.org

GLBenchMark (2013) Home page, accessed 1 March 2013 at: http://www.glbenchmark.com

Market and Markets (2012) 'Global In-Room Entertainment Market, By Products (Smart/3D/LED/LCD/OLED Tvs, Set-Top Box, Home Theater Systems (Projectors, Audio Equipment), Blu-Ray, Gaming Console); By Technology (Processor, Memory, Sensors, Connectivity); and By Geography'. Available 22 August 2013 by payment at: http://www.researchandmarkets.com/reports/2365872/global_inroom_entertainment_market_by_products

MOCA® (2013) Home page, accessed 22 August 2013 at: http://www.mocalliance.org

Parks Associates (2013a) 'App Publishing and Distribution Models', available 22 August 2013 by payment at: http://www.parksassociates.com/report/apps-publishing-distribution

Parks Associates (2013b) 'Connected Appliances – Evolution and Uses', available 22 August 2013 by payment at http://www.parksassociates.com/report/connected-appliances

UPnP (2013) Home page, accessed 22 August 2013 at: http://www.upnp.org/

Weiss, A.R. (2002) 'Dhrystone Benchmark - History, Analysis, 'Scores' and Recommendations'. ECL White Paper, Austin, TX, November 1.

Zoric, G., Engström, A. and Hannerfors, P. (2012) 'FascinatE Deliverable D3.5.1 – Design Brief on User Interface for Semi-automatic Production', February, accessed 22 August 2013 at: http://ec.europa.eu/information_society/apps/projects/logos/8/248138/080/deliverables/001_FascinateD351TIIDesignBriefOnUserInterfaceForSemiautomatic Productionv04.pdf

Index

Media Production, Delivery and Interaction for Platform Independent Systems: Format-Agnostic Media, First Edition. Edited by
Oliver Schreer, Jean-François Macq, Omar Aziz Niamut, Javier Ruiz-Hidalgo, Ben Shirley, Georg Thallinger and Graham Thomas.
© 2014 John Wiley & Sons, Ltd. Published 2014 by John Wiley & Sons, Ltd.